METHODS IN MOLECULAR BIOLOGY

Series Editor
John M. Walker
School of Life and Medical Sciences
University of Hertfordshire
Hatfield, Hertfordshire, UK

For further volumes:
http://www.springer.com/series/7651

For over 35 years, biological scientists have come to rely on the research protocols and methodologies in the critically acclaimed *Methods in Molecular Biology* series. The series was the first to introduce the step-by-step protocols approach that has become the standard in all biomedical protocol publishing. Each protocol is provided in readily-reproducible step-by-step fashion, opening with an introductory overview, a list of the materials and reagents needed to complete the experiment, and followed by a detailed procedure that is supported with a helpful notes section offering tips and tricks of the trade as well as troubleshooting advice. These hallmark features were introduced by series editor Dr. John Walker and constitute the key ingredient in each and every volume of the *Methods in Molecular Biology* series. Tested and trusted, comprehensive and reliable, all protocols from the series are indexed in PubMed.

Mammary Stem Cells

Methods and Protocols

Second Edition

Edited by

Maria dM. Vivanco

Cancer Heterogeneity Laboratory, Centre for Cooperative Research in Biosciences (CIC bioGUNE), Basque Research and Technology Alliance (BRTA), Derio, Spain

Humana Press

Editor
Maria dM. Vivanco
Cancer Heterogeneity Laboratory,
Centre for Cooperative Research in
Biosciences (CIC bioGUNE)
Basque Research and Technology Alliance (BRTA)
Derio, Spain

ISSN 1064-3745 ISSN 1940-6029 (electronic)
Methods in Molecular Biology
ISBN 978-1-0716-2195-0 ISBN 978-1-0716-2193-6 (eBook)
https://doi.org/10.1007/978-1-0716-2193-6

This Humana imprint is published by the registered company Springer Science+Business Media, LLC part of Springer Nature.
The registered company address is: 1 New York Plaza, New York, NY 10004, U.S.A.

Preface

Breast cancer remains the most commonly diagnosed cancer and the leading cause of cancer death in women. In 2015, the first edition of Methods and Protocols for mammary stem cell research was published and since then the field has continued expanding and deepening our understanding of mammary gland biology and tumorigenesis. The ever-growing mammary stem cell research data illustrate their huge potential to impact our knowledge about breast cancer initiation, progression, development of resistance, and metastasis to other organs. This new edition has revisited some of the main techniques used to study the mammary gland. While some methods from the first edition, such as FACS sorting for isolation and characterization of stem cells, as well as the visualization of the side population phenotype, have not significantly changed and thus they have not been repeated here, others have been updated and quite a few new methods have been incorporated. Together, this collection of chapters aims to lead breast researchers through some of the essential and latest state-of-the-art methods to study the full spectrum of mammary gland biology and pathological states.

Breast cancer is a complex disease, often viewed as "caricature" of normal development gone awry. Gaining insights into these early events provides a unique opportunity to develop innovative strategies to define the molecular interactions and genetic foundations of cancers and identify novel biomarkers and therapeutic targets. Collectively, the approaches described in the first three chapters provide a window into mammary developmental dynamics, and the perturbations underlying tissue dysfunction and disease. Thus, Mikkola and her group start from the beginning, focusing on mammary gland development during embryogenesis. They contribute to understand the earliest molecular and cellular events by describing an updated organ culture protocol of the murine embryonic mammary gland that allows monitoring of growth and branching morphogenesis of mammary gland ex vivo by live imaging. They further introduce a novel method that facilitates the identification of the direct stromal cues for branching morphogenesis. The group of Lloyd-Lewis presents a powerful approach that uses the latest multidimensional fluorescence imaging for visualizing mammary branching morphogenesis and epithelial cell fate dynamics during postnatal and embryonic mammary gland development. The Wahl lab continues with the theme of mouse embryo studies and describes single-cell RNA-seq and single-nucleus ATAC-seq methods and discusses the bioinformatic techniques they use to interpret the vast amount of data generated. Del Sol and his group continue to pursue this important challenge by reviewing various computational approaches that rely on single-cell and/or bulk RNA sequencing data for elucidating the molecular foundations of cell subpopulation identities, lineage specification, and cell fate. Further omics analysis is discussed in the chapter by Elortza and colleagues, who present methods for relative protein quantification. They describe sample preparation prior to mass spectrometry analysis, as well as a standard acquisition method, and some common bioinformatics analyses that help extracting biologically relevant information out of the large amount of data generated.

Several chapters are dedicated to different functional experimental approaches. Faraldo and colleagues revisit the classic orthotopic transplantation assay, which has provided important insights into mammary development, stem cell function, and tumorigenesis, and they highlight the strengths and limitations of this in vivo grafting assay. Lineage tracing, which in the previous edition was a relatively novel addition, is now considered the gold standard approach to study cellular hierarchies and cell fate in vivo, as discussed by Fre and Rodilla, who also present the downstream analyses and read-out assays most commonly used. In Chapter 8, Iggo provides a detailed and informative protocol for lentiviral transduction of mammary epithelial cells, which can be used to infect cells simultaneously with multiple vectors, greatly facilitating studies on malignant transformation. In contrast, Jechlinger and colleagues also used lentiviral transduction, but just to mark and track a few cells within organoids derived from genetically modified mice, in order to study events occurring during healthy tissue development, as well as disease progression.

The need for faithful models of breast cancer has been addressed using different systems, all of them helpful to evaluate the response of tumors to potential therapeutic agents and to contribute to identify the mechanisms of resistance to existing therapies. Dobrolecki and Lewis describe the well-established patient-derived xenograft (PDX) model, which involves the orthotopic transplantation of breast cancer tissue (as well as normal or preneoplastic tissue) into the mammary fat pad of immunocompromised mice. A very practical alternative in vivo PDX model is the chicken egg chorioallantoic membrane (CAM), which is presented by Villanueva and Sikora, highlighting its potential as an efficient, scalable precision medicine platform. Another relevant preclinical model is the mammary intraductal xenografting technique, which has been established to inject cells or other substances directly into the mammary ducts of female mice, presented by Howard and colleagues. This refined method provides the possibility of mimicking the normal microenvironment of preinvasive breast lesions to study breast cancer progression in a more physiologically relevant manner than with other xenografting methods. Iggo takes this methodology a step further by describing the creation of breast cancer models by transducing tumor cells and normal mammary epithelial cells with lentiviral vectors in organoid culture and injecting them into the mammary ducts of mice. Sumbal and Koledova continue taking advantage of organoid cultures, which have become vastly popular in developmental and cancer biology, and exploiting their potential, by presenting protocols for visualization of their architecture and composition in high resolution in three dimensions (3D).

The complexity of the human breast tissue is dissected by Arendt, who presents methods to isolate epithelial cells, stromal cells, adipocytes, and macrophages to investigate how risk factors impact distinct cell populations within breast tissue. Koledova and colleagues focus on the epithelial-stromal interactions, by providing a protocol for an organotypic human breast assay to facilitate research in the roles of human breast fibroblasts in epithelial morphogenesis and early tumorigenesis. All models have limitations, such as low take rate, long duration, lack of immune system or an appropriate stroma, among others, and some of those can be addressed using patient-derived explants, which preserve the native tissue architecture and microenvironment, as described by Vivanco and her group. The possibility of monitoring disease progression and responses to therapy through a relatively non-invasive blood-based biopsy is explored by Maheswaran and colleagues. They describe a versatile negative-selection protocol for enrichment of rare circulating tumor cells (CTCs) through removal of blood components including red blood cells, platelets, and white blood cells, which also includes the possibility of ex vivo culture of patient-derived CTCs. The biochemical signaling pathways related to cell structure, proliferation, differentiation, motility, and cell fate are functionally linked to the mechanical properties of cells, and in Chapter 19, Toca-Herrera and his colleagues describe how to perform mechanical experiments on breast cancer cells using the atomic force microscope. Finally, Schneider, in the last chapter, pulls all the knowledge gained through these effective methodologies to revise the view of the clinician about the promise held by mammary stem cell research and the future impact in the clinic.

Revising the classic methods and identifying the excellent new developments has been an exciting journey. I hope that the reader will find these protocols inspiring and helpful to navigate the intricate behavior of mammary stem cells and to gain further knowledge to take us closer to the design of innovative strategies to prevent and treat breast cancer.

Derio, Spain *Maria dM. Vivanco*

Contents

Contributors

Muhammad Ali • *Luxembourg Centre for Systems Biomedicine (LCSB), University of Luxembourg, Belvaux, Luxembourg*

Lisa M. Arendt • *Department of Comparative Biosciences, School of Veterinary Medicine, University of Wisconsin-Madison, Madison, WI, USA*

Mikel Azkargorta • *Proteomics Platform, CIC bioGUNE, Basque Research and Technology Alliance (BRTA), ProteoRed-ISCIII, CIBERehd, Science and Technology Park of Bizkaia, Derio, Spain*

Risa Burr • *Massachusetts General Hospital Center for Cancer Research, Harvard Medical School, Charlestown, MA, USA*

Claudia Carabaña • *Institut Curie, PSL Research University, INSERM U934, CNRS UMR3215, Paris, France; Sorbonne University, UPMC University of Paris VI, Paris, France*

Brian Chirn • *Massachusetts General Hospital Center for Cancer Research, Harvard Medical School, Charlestown, MA, USA*

Antonio del Sol • *Luxembourg Centre for Systems Biomedicine (LCSB), University of Luxembourg, Belvaux, Luxembourg; CIC bioGUNE, Bizkaia Technology Park, Derio, Spain; IKERBASQUE, Basque Foundation for Science, Bilbao, Spain*

Lucia Garcia del Valle • *Cell Biology and Biophysics Unit, EMBL, Heidelberg, Germany*

Marie-Ange Deugnier • *Institut Curie, PSL Research University, CNRS, UMR144, Paris, France*

Lacey E. Dobrolecki • *Departments of Molecular and Cellular Biology and Radiology, Lester and Sue Smith Breast Center, Dan L Duncan Comprehensive Cancer Center, Advanced Technology Cores, Baylor College of Medicine, Houston, TX, USA*

Jon F. Edd • *Center for Engineering in Medicine, Massachusetts General Hospital, Harvard Medical School, Boston, MA, USA*

Felix Elortza • *Proteomics Platform, CIC bioGUNE, Basque Research and Technology Alliance (BRTA), ProteoRed-ISCIII, CIBERehd, Science and Technology Park of Bizkaia, Derio, Spain*

Johanna Englund • *Centre of Excellence in Stem Cell Metabolism, Institute of Biotechnology, Helsinki Institute of Life Science (HiLIFE), University of Helsinki, Helsinki, Finland*

Iraide Escobés • *Proteomics Platform, CIC bioGUNE, Basque Research and Technology Alliance (BRTA), ProteoRed-ISCIII, CIBERehd, Science and Technology Park of Bizkaia, Derio, Spain*

Marisa M. Faraldo • *Institut Curie, PSL Research University, CNRS UMR3215/ INSERM U934, Paris, France*

Silvia Fre • *Department of Genetics and Developmental Biology, Institut Curie, PSL Research University, INSERM U934, CNRS UMR3215, Paris, France*

Marina A. Glukhova • *Institut Curie, PSL Research University, CNRS, UMR144, Paris, France*

Isabel Gris-Cárdenas • *Cancer Heterogeneity Laboratory, Centre for Cooperative Research in Biosciences (CIC bioGUNE), Basque Research and Technology Alliance (BRTA), Derio, Spain*

THORARINN GUDJONSSON • *Stem Cell Research Unit, Biomedical Center, Department of Anatomy, Faculty of Medicine, School of Health Sciences, University of Iceland, Reykjavik, Iceland; Department of Laboratory Hematology, Landspitali - University Hospital, Reykjavik, Iceland*

DANIEL A. HABER • *Massachusetts General Hospital Center for Cancer Research, Harvard Medical School, Charlestown, MA, USA; Howard Hughes Medical Institute, Bethesda, MD, USA; Department of Medicine, Massachusetts General Hospital, Harvard Medical School, Boston, MA, USA*

BEATRICE A. HOWARD • *The Breast Cancer Now Toby Robins Research Centre, The Institute of Cancer Research, London, UK*

RICHARD IGGO • *INSERM U1218, Institut Bergonié, University of Bordeaux, Bordeaux, France; DRMCRL Lab, University of Adelaide Medical School, Adelaide, SA, Australia*

IBON ILORO • *Proteomics Platform, CIC bioGUNE, Basque Research and Technology Alliance (BRTA), ProteoRed-ISCIII, CIBERehd, Science and Technology Park of Bizkaia, Derio, Spain*

MARTIN JECHLINGER • *Cell Biology and Biophysics Unit, EMBL, Heidelberg, Germany*

BEATA KACZYŃSKA • *Cell and Tissue Dynamics Research Program, Institute of Biotechnology, Helsinki Institute of Life Science (HiLIFE), University of Helsinki, Helsinki, Finland*

ZUZANA KOLEDOVA • *Department of Histology and Embryology, Faculty of Medicine, Masaryk University, Brno, Czech Republic*

QIANG LAN • *Cell and Tissue Dynamics Research Program, Institute of Biotechnology, Helsinki Institute of Life Science (HiLIFE), University of Helsinki, Helsinki, Finland*

MICHAEL T. LEWIS • *Departments of Molecular and Cellular Biology and Radiology, Lester and Sue Smith Breast Center, Dan L Duncan Comprehensive Cancer Center, Advanced Technology Cores, Baylor College of Medicine, Houston, TX, USA*

RIITTA LINDSTRÖM • *Cell and Tissue Dynamics Research Program, Institute of Biotechnology, Helsinki Institute of Life Science (HiLIFE), University of Helsinki, Helsinki, Finland*

BETHAN LLOYD-LEWIS • *School of Cellular and Molecular Medicine, University of Bristol, Bristol, UK*

NIKKI K. LYTLE • *Gene Expression Laboratory, Salk Institute for Biological Studies, La Jolla, CA, USA*

ZHIBO MA • *Gene Expression Laboratory, Salk Institute for Biological Studies, La Jolla, CA, USA*

SHYAMALA MAHESWARAN • *Massachusetts General Hospital Center for Cancer Research, Harvard Medical School, Charlestown, MA, USA; Department of Surgery, Massachusetts General Hospital, Harvard Medical School, Boston, MA, USA*

MARJA L. MIKKOLA • *Cell and Tissue Dynamics Research Program, Institute of Biotechnology, Helsinki Institute of Life Science (HiLIFE), University of Helsinki, Helsinki, Finland*

AVANISH MISHRA • *Center for Engineering in Medicine, Massachusetts General Hospital, Harvard Medical School, Boston, MA, USA*

MARTA G. MONTERO • *Cell Biology and Biophysics Unit, EMBL, Heidelberg, Germany*

SATU-MARJA MYLLYMÄKI • *Cell and Tissue Dynamics Research Program, Institute of Biotechnology, Helsinki Institute of Life Science (HiLIFE), University of Helsinki, Helsinki, Finland*

RAZIA F. NAEEM • *Gene Expression Laboratory, Salk Institute for Biological Studies, La Jolla, CA, USA*

RICHARD NEWMAN • *The Breast Cancer Now Toby Robins Research Centre, The Institute of Cancer Research, London, UK*

ERIK OLIEMULLER • *The Breast Cancer Now Toby Robins Research Centre, The Institute of Cancer Research, London, UK*
MIRIAM RÁBANO • *Cancer Heterogeneity Laboratory, Centre for Cooperative Research in Biosciences (CIC bioGUNE), Basque Research and Technology Alliance (BRTA), Derio, Spain*
CYNTHIA RAMOS • *Gene Expression Laboratory, Salk Institute for Biological Studies, La Jolla, CA, USA*
MARIANA MESSIAS RIBEIRO • *Luxembourg Centre for Systems Biomedicine (LCSB), University of Luxembourg, Belvaux, Luxembourg*
VERÓNICA RODILLA • *Cancer Heterogeneity and Hierarchies Group, Josep Carreras Leukaemia Research Institute (IJC), Badalona, Spain*
JYOTI SATTA • *Cell and Tissue Dynamics Research Program, Institute of Biotechnology, Helsinki Institute of Life Science (HiLIFE), University of Helsinki, Helsinki, Finland*
JOSE SCHNEIDER • *Department of Obstetrics and Gynecology, University of Valladolid, Valladolid, Spain*
ANDREW G. SIKORA • *Otolaryngology - Head and Neck Surgery, Advanced Technology Cores, Baylor College of Medicine, Houston, TX, USA; Department of Head and Neck Surgery, MD Anderson Cancer Center, Houston, TX, USA*
JAKUB SUMBAL • *Department of Histology and Embryology, Faculty of Medicine, Masaryk University, Brno, Czech Republic*
JOSÉ L. TOCA-HERRERA • *Department of Nanobiotechnology (DNBT), Institute of Biophysics, University of Natural Resources and Life Sciences Vienna (BOKU), Vienna, Austria*
MEHMET TONER • *Center for Engineering in Medicine, Massachusetts General Hospital, Harvard Medical School, Boston, MA, USA; Shriners Hospital for Children, Boston, MA, USA; Department of Surgery, Massachusetts General Hospital, Harvard Medical School, Boston, MA, USA*
GUNNHILDUR ASTA TRAUSTADOTTIR • *Stem Cell Research Unit, Biomedical Center, Department of Anatomy, Faculty of Medicine, School of Health Sciences, University of Iceland, Reykjavik, Iceland*
EWELINA TRELA • *Cell and Tissue Dynamics Research Program, Institute of Biotechnology, Helsinki Institute of Life Science (HiLIFE), University of Helsinki, Helsinki, Finland*
HUGO VILLANUEVA • *Otolaryngology - Head and Neck Surgery, Advanced Technology Cores, Baylor College of Medicine, Houston, TX, USA; Department of Head and Neck Surgery, MD Anderson Cancer Center, Houston, TX, USA*
MARIA DM. VIVANCO • *Cancer Heterogeneity Laboratory, Centre for Cooperative Research in Biosciences (CIC bioGUNE), Basque Research and Technology Alliance (BRTA), Derio, Spain*
FRANCISCO J. VIZOSO • *Research Unit and Department of Surgery, Fundación Hospital de Jove, Gijón, Spain*
GEOFFREY M. WAHL • *Gene Expression Laboratory, Salk Institute for Biological Studies, La Jolla, CA, USA*
ANDREAS WEBER • *Department of Nanobiotechnology (DNBT), Institute of Biophysics, University of Natural Resources and Life Sciences Vienna (BOKU), Vienna, Austria*
BARBARA ZBIRAL • *Department of Nanobiotechnology (DNBT), Institute of Biophysics, University of Natural Resources and Life Sciences Vienna (BOKU), Vienna, Austria*

Chapter 1

Protocol for Studying Embryonic Mammary Gland Branching Morphogenesis Ex Vivo

Qiang Lan, Jyoti Satta, Satu-Marja Myllymäki, Ewelina Trela, Riitta Lindström, Beata Kaczyńska, Johanna Englund, and Marja L. Mikkola

Abstract

Mammary gland development starts during embryogenesis, and the process continues after birth. During development, the mammary gland undergoes massive morphological and physiological alterations including growth, invasion, and branching morphogenesis providing an ideal model for stem cell and cancer biology studies. Great efforts have been made in understanding mammary gland development during puberty and adulthood; however, the process during embryogenesis is still elusive. One reason is that the tools to study tissue dynamics during development are limited, which is partially due to the lack of an ex vivo culture method. Here we describe an updated organ culture protocol of the murine embryonic mammary gland. This powerful tool allows monitoring of growth and branching morphogenesis of mammary gland ex vivo by live imaging. In addition, we introduce a novel method for culturing intact, stroma-free mammary rudiments from late gestation mouse embryos in 3D in Matrigel. This approach can be used to identify the direct stromal cues for branching morphogenesis.

Key words Organ culture, ex vivo, Explant culture, 3D culture, Organoid, Mammary gland, Mammary rudiment

1 Introduction

Mammary gland is a unique organ, which exists only in mammals and functions as a secretory organ to provide nutrition as well as immunological benefits to the newborn [1]. In order to achieve its function, mammary gland undergoes finely regulated massive dynamic alterations during fetal stage, puberty, estrus cycle, pregnancy, lactation, and involution. Thus, mammary gland provides a perfect model for studying organ development, stem cell function, and cell fate specification and differentiation [2, 3]. Much attention has been given to the molecular and cellular regulation of mammary gland development during puberty and pregnancy, whereas

Maria dM. Vivanco (ed.), *Mammary Stem Cells: Methods and Protocols*,
Methods in Molecular Biology, vol. 2471, https://doi.org/10.1007/978-1-0716-2193-6_1,

morphogenesis during embryogenesis is still poorly described. Increasing evidence supports the similarity between embryonic development and cancer progression [4–6]. Thus, understanding embryonic mammary gland development could provide new insights into cancer biology and result in novel opportunities for breast cancer therapy.

Mammary gland is a skin appendage whose development in mice starts at embryonic day (E) 10.5 in the ventro-lateral region extending from groin to axilla [7–9]. By E12.5, five pairs of mammary placodes have formed symmetrically but asynchronously [10]. A day later, placodes invaginate into the underlying mesenchyme to form mammary buds, which are surrounded by condensed and specified mammary mesenchyme [1, 11]. In male mice, androgen-mediated destruction of the mammary buds begins at E13.5 leading to disconnection of the bud from the overlying epidermis and ultimately disappearance of the mammary rudiment [9]. In females, the bud gives rise to a sprout that invades into the secondary mammary mesenchyme, the fat pad precursor tissue at E15.5. Branching morphogenesis begins at E16.5, and by birth, a small arborized gland has formed.

Due to the technical challenges in monitoring developing embryos inside the uterus, information of embryonic mammary morphogenesis is mostly gathered by analyzing fixed samples from wild-type or genetically modified mice. However, the dynamic nature of morphogenesis and epithelial-mesenchymal tissue interactions, a key determinant of organ development [1], is difficult to capture from fixed samples. To overcome these limitations, ex vivo organ culture methods have been developed and adopted for embryonic mammary gland studies. The first application was reported already in 1960s [12]; however, this method has not been widely used, likely due to technical challenges. We have previously established whole organ culture technique for embryonic mammary glands and routinely culture E12.5 to E13.5 embryonic mammary glands up to 7 days [13, 14]. We now report how this method can be combined with time-lapse imaging allowing monitoring epithelial morphogenesis in its native mesenchyme ex vivo.

Three-dimensional (3D) organoid culture methods have also been widely used in the studies of pubertal mammary gland branching morphogenesis, as well as stem/progenitor cell maintenance and differentiation [15, 16]. In a typical organoid setup, small pieces of epithelial tissue are cultured in a 3D matrix such as Matrigel or collagen, supplemented with a cocktail of growth factors. However, the application of organoid culture techniques in embryonic mammary gland research has not been reported previously. One challenge in this setting is the dependency of embryonic mammary progenitor cell maintenance on mesenchymal cues that are missing in the culture conditions. Spike et al. have reported that dissociated mammary epithelial cells from E13.5 to E15.5 mouse embryos contain very few cells that are able to reconstitute the mammary gland in a cleared fat pad, while cells isolated from E16.5 mammary rudiments were able to reconstitute the gland comparable to adult cells [5]. Here, we describe a Matrigel-based 3D culture protocol for the intact, mesenchyme-free mammary rudiments isolated from E16.5 female embryos. The epithelial rudiment undergoes branching morphogenesis for several days, although the resulting ductal tree is more compact than in vivo, similar to organoids generated from the postnatal glands [17]. Combining the organ culture and stroma-free mammary rudiments 3D culture provides an opportunity to gain a comprehensive understanding of the roles of epithelial and mesenchymal tissue compartments during fetal mammary gland branching morphogenesis.

2 Materials

2.1 Dissection of E13.5 Embryo Flank with Mammary Gland

1. Steri 250 glass bead sterilizer (Simon Keller Ltd., Switzerland).
2. Sterile Dulbecco's phosphate-buffered saline (DPBS; D5652, Sigma-Aldrich).
3. Sterile phosphate-buffered saline (PBS; P4417, Sigma-Aldrich).
4. DMEM/F12 full medium: F-12-GlutaMAX (31765, Thermo Fisher) with DMEM-GlutaMAX (61965, ThermoFisher) at a 1:1 ratio, 10% (vol/vol) heat-inactivated fetal bovine serum (FBS; SV30160.03, HyClone/Thermo Fisher), and 20 U/mL penicillin-streptomycin (PS; 15140-122, Gibco/Thermo Fisher). Store at 4 °C. Penicillin and streptomycin are stable in culture medium for 2 weeks at 4 °C.
5. 35 mm and 10 cm plastic Petri dishes.
6. 35 mm and 10 cm diameter glass Petri dishes. Autoclaved before use.
7. Glass Petri dishes with silicon coating: Coat the 10 cm glass Petri dishes with approximately 5 mm layer of silicone (SYLGARD 184 Silicone Elastomer Kit, 04019862, Dow Silicones Corporation, USA). Autoclave the dishes before use. Silicon Petri dishes can be reused multiple times by washing and autoclaving them.
8. Vannas Spring Scissors with 4 mm cutting edge (15018-10, Fine Science Tools).
9. Scissors (14090-09, Fine Science Tools).
10. Forceps (Dumont #3, 11293-00, Fine Science Tools).
11. Perforated spoon (10370-18, Fine Science Tools).

12. Watchmaker's forceps number 3 (07.60.03, Medicon Instruments, Germany).
13. Two disposable 27- or 25-gauge needles attached to 1 mL plunger-removed disposable plastic syringes.
14. Glass Pasteur pipettes. Autoclaved before use.
15. Horizontal laminar flow hood.
16. Stereomicroscope.

2.2 Separation of E13.5 Skin Epithelium from the Mammary Gland Explant

1. Dispase II solution: Dispase II powder (4942078001, Roche) to a final concentration of 1.25 or 2.5 U/mL in PBS and filter-sterilize with a 0.2 μm filter. Use only freshly made working solution.
2. Thyrode's solution (pH 7.4): 8.0 g NaCl, 0.2 g KCl, 0.05 g $NaH_2PO_4 \cdot H_2O$, 1.0 g D-(+)-Glucose, 1.0 g $NaHCO_3$ in 900 mL MilliQ water. Adjust the pH to 7.4 and add MilliQ water to final volume of 1000 mL. Filter-sterilize the solution using a 0.22 μm filter and store at 4 °C.
3. 10× Pancreatin stock solution: 0.85 g NaCl and 2.5 g Pancreatin in 100 mL of MilliQ water on ice for 3–4 h (or at 4 °C o/n) using magnetic stirrer. Centrifuge the solution at 5000 rpm (3000 × *g*) for 10 min and filter-sterilize the supernatant with a 0.22 μm filter (due to the viscosity of the solution at least two 0.22 μm filters will be needed to pass a volume of 100 mL). Divide the solution into 1 mL aliquots and store in −20 °C (it can be used for up to 1 year).
4. Pancreatin-trypsin working solution (pH 7.4): 0.225 g of trypsin in 6 mL of Thyrode's solution on ice using a magnetic stirrer. Add 1 mL of 10× pancreatin stock solution and 20 μL of PS (10,000 U/mL in stock) and adjust pH to 7.4 with NaOH. Obtain the final volume of 10 mL with Thyrode's solution and filter-sterilize the solution with a 0.22 μm filter. Divide the solution into 1 mL aliquots and store at −20 °C. The working solution can be kept in −20 °C for 2–3 weeks. Discard the aliquot after thawing as it cannot be reused.

2.3 Ex Vivo Culture of E13.5 Embryonic Mammary Gland Explants

1. Ascorbic acid (100 mg/mL stock solution): Reconstitute powder in water and filter-sterilize the solution with a 0.2 μm filter. Divide the solution into single-use aliquots and store at −20 °C. Ascorbic acid is light sensitive and needs to be protected from light.
2. Organ culture medium: DMEM/F12 full medium (*see* Subheading 2.1) supplemented with 100 ng/mL ascorbic acid (1: 1000 diluted from stock) right before use.
3. Nuclepore polycarbonate filters (WHA110605, Whatman). Sterilize and store the filters in 70% ethanol at room

temperature. Before use, wash the filter with sterilized PBS inside a laminar hood 3 times and cut the filter into approximately 5 mm squares in PBS. The filter squares can be stored in PBS at 4 °C for 1–2 weeks.

4. Metal grids: Cut corrosion-resistant stainless-steel mesh (0.7 mm mesh size) into approximately 30 mm diameter disks. Measure approximately 3 mm from the edge of the disk and bend the mesh to obtain self-supporting grid. To facilitate the passage of light and the observation of the explant in the microscope, make small holes (e.g., 2 rows of 3) into the grid. Autoclave the grids before use. Grids can be washed and autoclaved multiple times. Alternatively, commercial cell culture inserts could be used to replace the grid and Nuclepore filters (*see* Subheading 2.4).
5. Humidified incubator, set at 5% CO_2, 37 °C.
6. Stereomicroscope equipped with camera.

2.4 Live Imaging of Cultured E13.5 Mammary Gland Explants

1. Trans-well inserts of 24 mm, 0.4 μm Polyester membrane.
2. 6-well transparent flat bottom plate.
3. Organ culture imaging medium: DMEM/F-12, Hepes medium without phenol red, 10% heat-inactivated FBS, 20 U/mL penicillin-streptomycin, and 100 ng/μL of ascorbic acid (*see* Subheading 2.3).
4. 3I Marianas wide-field microscope with Slidebook 6 Reader software (3I intelligent Imaging Innovations), or an equivalent inverted microscope equipped with a LED light source (CoolLED pE2 with 490 nm/550 nm set transmitted LED in our setting) and air objective with long working distance.
5. Objective: 10×/0.30 N.A. EC Plan-Neofluar Ph1 WD = 5.2 M27.
6. Sterile water.
7. CO_2 Incubator for tissue culture use, equilibrated to 37 °C, 5% CO_2 atmosphere.
8. Fiji imaging analysis software [18].

2.5 3D Stroma-Free Organoid Culture of E16 Mammary Rudiments

1. Growth factor reduced Matrigel (Matrigel; 354230, Corning).
2. Serum-free organoid culture medium: F12-GlutaMAX and DMEM-GlutaMAX at a 1:1 ratio, 20 U/mL PS, 1:100 ITS (I3146, Sigma-Aldrich), and 2.5 nM hFGF2 (F0291, Sigma-Aldrich). Store at 4 °C.

3 Methods

Tissue dissection and all the other operations should be performed in the laminar flow hood if possible. All materials and equipment should be sterilized before use. The dried forceps and scissors should be sterilized with Steri 250 glass bead sterilizer after cleaning with 70% ethanol.

3.1 Dissection of E13.5 Embryo Flank with Mammary Gland

In this section, we will describe the dissection method optimized for E13–E13.5 embryos. At this stage, mammary buds isolated from both male and female embryos can be used. The developmental stage of the embryo is defined by the presence of the vaginal plug (noon of the plug day corresponding to Day E0.5), limb morphology, and other external criteria [19]. Dissection procedures are carried out in a laminar flow hood with a dissection stereomicroscope.

1. After euthanizing the pregnant mouse, remove the mouse uterus and place it in a 10 cm diameter plastic Petri dish containing DPBS. Dissect the embryos out from the uterus by removing parietal yolk sac, visceral yolk sac, and amnion under stereomicroscope. Once the embryos are liberated, transfer them with a spoon to a new 10 cm plastic Petri dish with DPBS. Use two needles attached to syringes to decapitate the embryo and move them to a silicon-coated glass Petri dish (Fig. 1a) filled with DPBS. Attach the embryos to silicon bottom with two needles around neck area and above the tail, respectively, so that the ventral side of the embryo faces upwards (Fig. 1b, c). Do not stretch the embryo while attaching it to the silicon as this might damage it.
2. Remove the fore limb with small scissors (Fig. 1d).
3. At the site where the fore limb has been removed, slide 4 mm cutting edge scissors below the skin and incise it from the fore limb continuing along the flank of the embryo towards the hind limb (Fig. 1e). Be cautious not to penetrate too deep into the underlying tissues while incising, as this will lead to difficulties in subsequent dissection.
4. Use the small scissors to cut the skin connected to the hind limb and then remove the hind limb.
5. Cautiously slide in one tip of forceps between skin and underneath tissues starting from the middle side of the embryo flank and moving towards the midline. Repetitively move the forceps in parallel to the embryo between the fore limbs and the hind limbs to detach the skin (Fig. 1g). Continue detaching the skin towards the midline until reaching the top of the liver. Once the skin is separated from the underlying tissue, liberate it using small scissors (Fig. 1h, i).

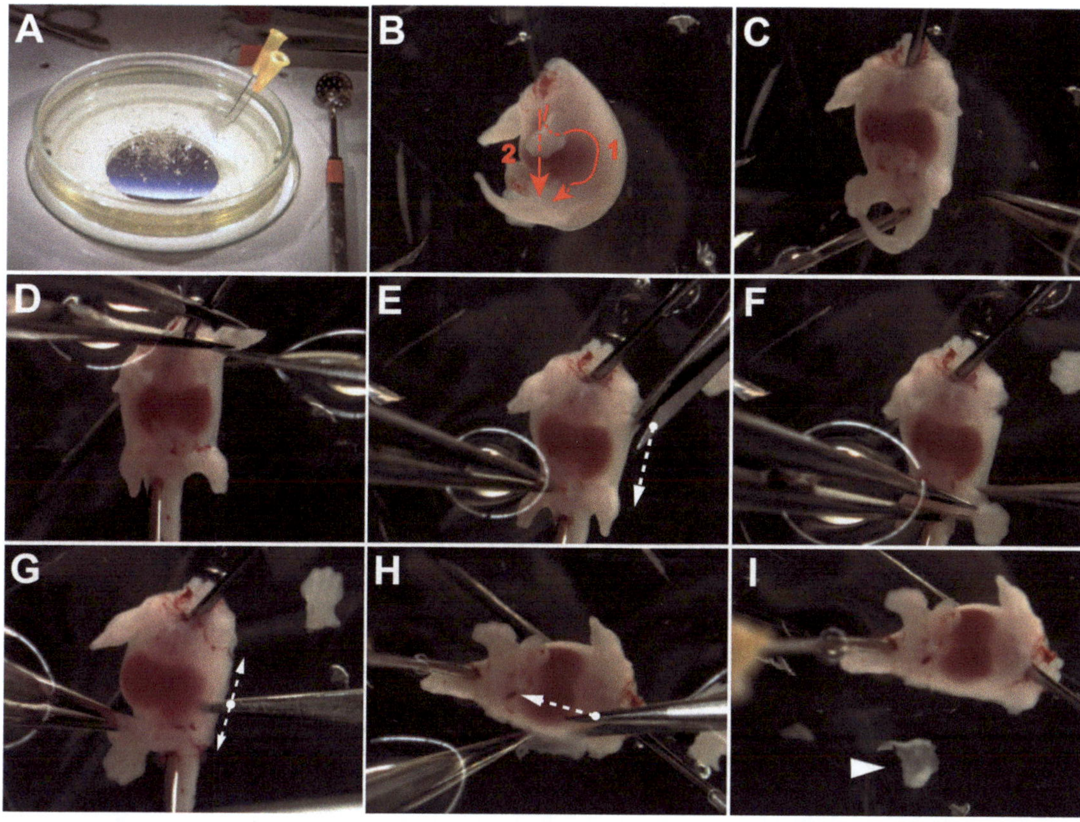

Fig. 1 Dissection of mammary bud-bearing flanks from E13–E13.5 embryos. (**a**) Silicon-coated glass Petri dish and two needles are used for dissection. (**b**) One decapitated embryo is transferred to the Petri dish. The dissection area is marked with red lines. (**c**) The embryo is attached to the silicon with the needles. (**d**) Fore limbs are removed with small scissors. (**e**) The skin is incised from the fore limb continuing along the white arrow towards the hind limb (red line 1 in **b**). (**f**) Skin is cut at the bottom of hind limb before it was removed. (**g**) The flank skin is separated from the body by gently moving the forceps along the white arrows repeatedly and gradually towards the midline. (**h**, **i**) When the skin overlaying the liver has been separated, cut it off with small scissors along the white dashed arrow (**h**, red line 2 in **b**), and collect separated explants (white arrowhead) for further treatment (**i**)

6. Repeat the **steps 2–5** to dissect the other side of the skin.
7. After processing all embryos, transfer the explants to a 35 mm plastic Petri dish with a modest volume of DPBS using a glass Pasteur pipette. To avoid attaching of the explants to the glass Pasteur pipette, pre-coat with FBS prior to use. Proceed to epithelial-mesenchymal separation step.

3.2 Separation of E13.5 Skin Epithelium from the Mammary Bud Explant

To avoid the interference of developing hair follicles on mammary gland outgrowth, the skin epidermis must be removed leaving only the mammary buds with ample amounts of mesenchyme. This is achieved by performing an enzyme treatment followed with mechanical separation. For the enzyme treatment, a mixture of pancreatin and trypsin (pancreatin-trypsin), Dispase II, or a combination of both have been successful. In this protocol, we combine Dispase II-treatment with a short incubation with pancreatin-trypsin (*see* **Note 1**).

1. Remove the DPBS from the Petri dish with explants and add freshly prepared Dispase II solution (1.25 U/mL in PBS) and keep the Petri dish on a linear shaker mixer with speed around 40 rpm in 4 °C for 25–35 min. Proceed to the next step when the epithelium and mesenchyme compartment at the edge of the explant appear separated (Fig. 2a).
2. Replace the Dispase II solution with pancreatin-trypsin working solution and incubate for 2–7 min at room temperature (*see* **Note 2**). Gently rotate the dish horizontally to facilitate the treatment.
3. Carefully monitor the separation process under stereomicroscope. When a big piece of epithelium appears to be separated from mesenchyme (*see* **Note 2**) (Fig. 2b), remove the enzymes with a glass Pasteur pipette and inactivate the enzyme activity by adding 2–3 mL DMEM/F12 full medium (with 10% FBS).
4. Incubate the Petri dish with explants on ice for 30–40 min for recovery. At this point, the detachment of the skin epidermis should be more obvious (Fig. 2c). Protect from light if necessary.
5. Transfer the explants to a glass Petri dish and use forceps or needles to separate the skin epithelium so that only buds remain with the mesenchyme (Fig. 2d) (*see* **Note 3**).
6. Use forceps and small scissors to remove extra tissue and expose the mammary gland if necessary (*see* **Note 4**).

3.3 Explant Culture of E13.5 Embryonic Mammary Gland Explants

Here we describe how to culture the mammary glands in their natural positions in the mesenchyme in a Trowell-type system [20].

1. Prepare a 35 mm plastic culture dish with a metal grid at the bottom and add 1–2 mL of organ culture medium. To prevent explants from floating, the culture media should contact the grid, but not flood over it. Try not to trap air bubbles underneath the grid and filters as this will interfere with the growth of the explants.
2. Transfer the filter squares to the Petri dish containing the explants with forceps.
3. Either with Watchmaker's forceps or with needles, gently push one explant onto the Nuclepore filter squares in such a way that mesenchyme should face the filter.
4. Lift the filter with the explant carefully from the edges (do not touch the explant) with the help of the Watchmaker's forceps

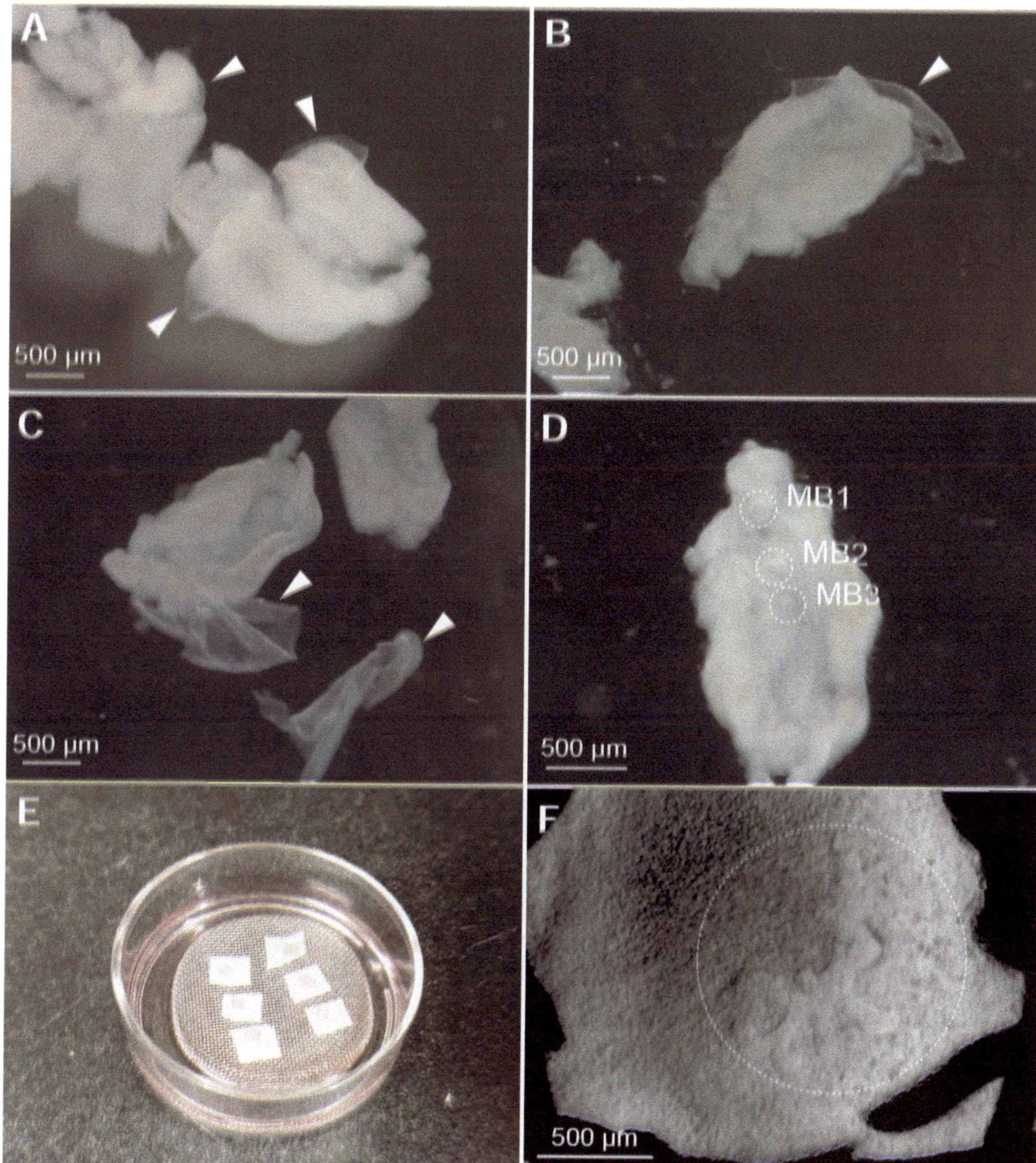

Fig. 2 Epithelial-mesenchymal separation and mammary explant culture from E13–E13.5 embryos. (**a**) Dissected flanks of skin are incubated with Dispase II at 4 °C until the more translucent epidermis appears separated at the edges (white arrowhead). (**b**) Dispase II-treated explants are further incubated with pancreatin-trypsin at room temperature until bigger pieces of epidermis have become loose (white arrowhead). (**c**) Enzyme-treated explants are recovered on ice for 30 min. During this time, the separation usually proceeds further (white arrowhead). (**d**) The epidermis is mechanically separated, and the mammary buds remain attached to the mesenchyme. (**e**) The explants are placed on top of small squares of filter paper and cultured on top of a grid on a 35 mm Petri dish filled with culture media. The grid contains small holes, on top of which the explants can be better visualized and imaged. (**f**) Example of phase contrast image of a branched mammary gland (inside the white circle) after 6 days of culture

and transfer it above one hole of the metal grid (up to six explants can be cultured on one grid with six holes) (Fig. 2e).

5. Culture the explants in the incubator with 5% CO_2 and 37 °C for 5–7 days depending on the purpose of the experiment (Fig. 2f) (*see* **Note 5**).
6. Change the culture medium every other day and image every day to monitor the mammary gland growth and branching.

3.4 Live Imaging of Cultured E13.5 Mammary Gland Explants

Here we describe a tissue-scale time-lapse imaging setup to follow branching morphogenesis of mammary gland explants for up to 5 days, starting on Day 3 of culture. For imaging, we use an inverted fluorescence microscope with a humidified environmental chamber, equipped with a motorized stage for multipoint imaging, and fitted with a 6-well plate sample holder. Explants expressing an epithelial fluorescent reporter are prepared as described in Subheading 3.2 and 3.3 and cultured in Transwell® Permeable Supports, sitting on a 6-well plate, which allows imaging from below with a 10× objective.

1. Put one or two inserts in the central wells of a 6-well plate. Transfer 6–10 explants prepared from Subheading 3.2 with the help of Pasteur glass pipette to the apical side of Trans-well inserts and place them gently on the membrane and avoid scratching the membrane during the transfer (Fig. 3a).
2. Add 1–1.5 mL organ culture medium to the wells, taking care not to allow the liquid to rise above the membrane surface of the insert (Fig. 3b).
3. Culture the explants for 3 days in standard conditions (Subheading 3.3, **steps 5** and **6**) allowing the explants to settle on the inserts.
4. On Day 3, when branching morphogenesis has initiated, glue the inserts into the wells with double-sided tape to avoid the shaking of the insert during stage movement and replace the organ culture media to phenol red free imaging media and fill the empty wells with sterile water to prevent dehydration during imaging and culture (Fig. 3c) and proceed for live imaging.
5. Heat up the environmental chamber of the microscope to +37 °C 2–3 h before imaging and turn on the CO_2 supply.
6. Set up the gas/humidity control for an atmosphere of 6% CO_2 with a humidity of 90%, using a minimal gas flow of 6 L/min.
7. Position the 10× objective on the light path and clean the lens with lens paper.
8. Clean the bottom of the sample plate with tissue and 70% ethanol.

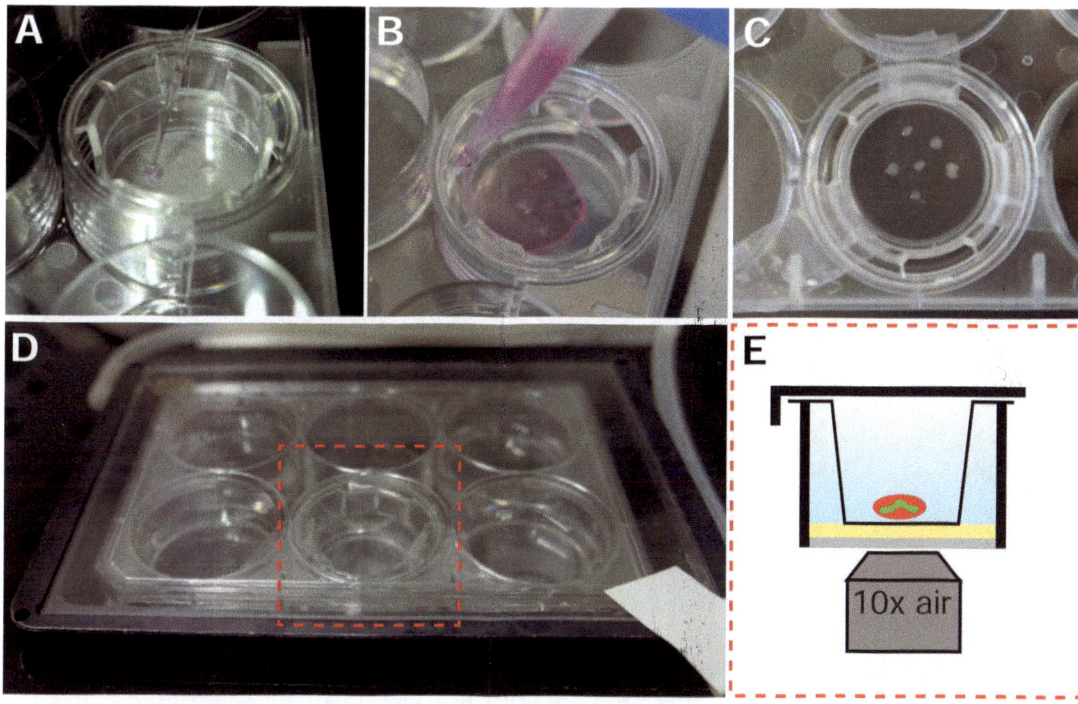

Fig. 3 Time-lapse imaging of mammary glands in explant culture using an inverted microscope. (**a**) Mammary explants are transferred to the inner membrane of a Trans-well insert, sitting on a 6-well plate, using a Pasteur pipette. (**b**) Medium is added to the 6-well underneath the insert until the liquid surface reaches the level of the insert membrane. (**c**) The insert is secured to the 6-well using double-sided tape from three sides. (**d**) The 6-well plate with the explants is mounted to a sample holder of an inverted fluorescence microscope with atmospheric and temperature controls preset to +37 °C and 5% CO_2. (**e**) Schematic of the imaging setup shown in (**b**) (red dashed line): imaging is performed using an inverted microscope through the medium (yellow) using a 10× air objective. (**f**) Live imaging of a mammary gland explant with 3i Marianas wide-field microscope. Expression of membrane-localized GFP (mGFP) of mT/mG Cre reporter mice was activated in the epithelium with K14-Cre transgene allowing visualization of mammary epithelial cells in green. Snapshots are at 8 h intervals from a 64-h video

9. Lower the objective to the lowest position and secure the sample plate on the sample holder (make sure sample holder is equilibrated at least 30 min before imaging) and apply the metal frame with the gas input and cover it with a transparent lid (Fig. 3d).
10. Open the microscope software to set up time-lapse imaging and activate the appropriate light sources (*see* **Note 6**).
11. Locate the sample on the oculars using bright field illumination. Systematically find each gland and record their positions using the multiple positions function of the software, minimizing stage movement between positions.
12. Inspect the glands under fluorescent light and set up the light path and illumination settings for each fluorophore.

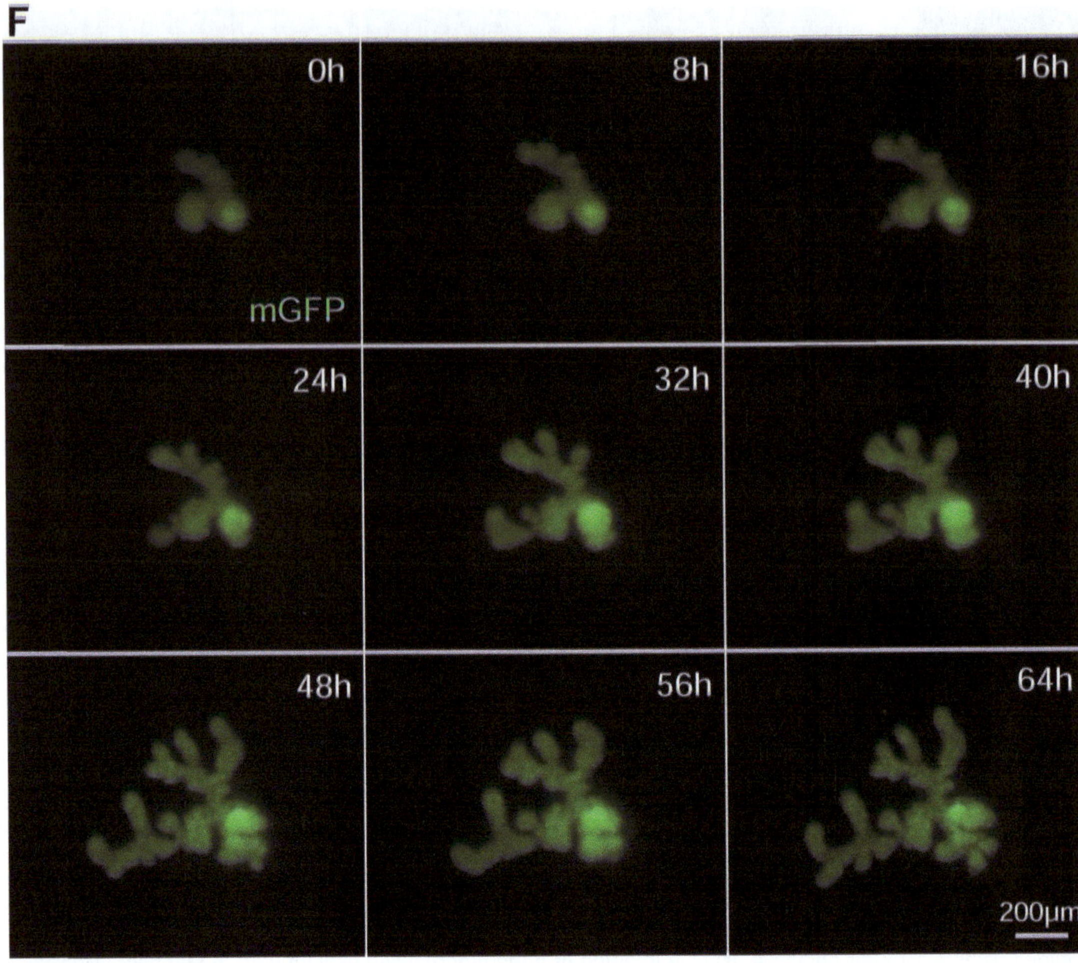

Fig. 3 (continued)

13. Go through all positions, refocus and reposition if necessary.
14. Set time-lapse imaging with 4 h intervals for 5 days (*see* **Note 7**).
15. Before starting the capture, check that the microscope arm is lowered, CO_2 flow is constant, and the openings to the environmental chamber are properly closed. Turn off all external lights.
16. Before starting capture, let the system rest for 1–2 h and check for the quality of images for each position once more.
17. The imaging positions need to be adjusted for drifting each day. If glands have moved outside of the field of view, pause the time lapse and re-configure the positions in XYZ. Resume imaging.
18. Medium has to be changed on every other day to replenish the explants. To change medium, remove the sample plate from

the sample holder, take out media with a 1 mL disposable plastic pipette, and add 1–1.5 mL fresh imaging medium. Tilt the plate to get rid of air bubbles and insert the plate back to the microscope as described above.

19. Reposition and refocus all the positions and resume imaging.
20. On the last day, stop the live imaging and save the data for further analysis of the well-grown mammary glands. Cultured glands may also be processed, e.g., for immunostaining.

3.5 3D Stroma-Free Organoid Culture of E16 Mammary Rudiments

Here, we describe a method to culture intact mammary rudiments dissected at E16–E16.5 stage when the outgrowth has just initiated (*see* **Note 8**). Due to the developmental failure of male mammary gland, it is important to determine the sex of the embryos before culture.

3.5.1 Dissection of the Mammary Tissue

1. Dissect the embryo from the uterus and fix it in the glass Petri dish coated with silicon as described in Subheading 3.1 **step 1**.
2. Use the forceps to make a small opening along one side of the inguinal fossa to expose the gonad. Determine the gender of the embryo based on the morphology of the gonad (*see* **Note 9**) (Fig. 4a–d). Discard the male embryos.
3. Remove the limbs with scissors.
4. Slide the cut edge of the scissors under the skin along the ventral midline to cut the skin (Fig. 4e). Be cautious not to cut the muscle layer.
5. Use Watchmaker's forceps to lift the edge of ventral skin containing mammary gland and gently peel off the skin with the help of the scissors (Fig. 4f). Make sure the fat pad precursor tissue stays together with the skin. Repeat the step to detach the ventral skin in the other side of the embryo.
6. Liberate the skins containing mammary gland and fat pad precursors with small scissors (Fig. 4g).
7. Repeat the **steps 2–6** to process all the embryos.
8. Collect all the skins containing mammary glands into a new 35 mm Petri dish, and proceed for enzyme treatment.

3.5.2 Epithelial-Mesenchymal Separation

In order to dissect the mammary rudiments out from their surrounding mesenchyme, Dispase II and pancreatin-trypsin enzyme treatments described in Subheading 3.2 will be performed, but with higher concentration of the enzyme and prolonged time of treatment as the samples are much bigger and thicker. In addition, further mechanical dissection is needed to separate mammary rudiments and surrounding mesenchyme.

1. Follow the steps described in Subheading 3.2 to separate the skin epithelium and mesenchyme with the following

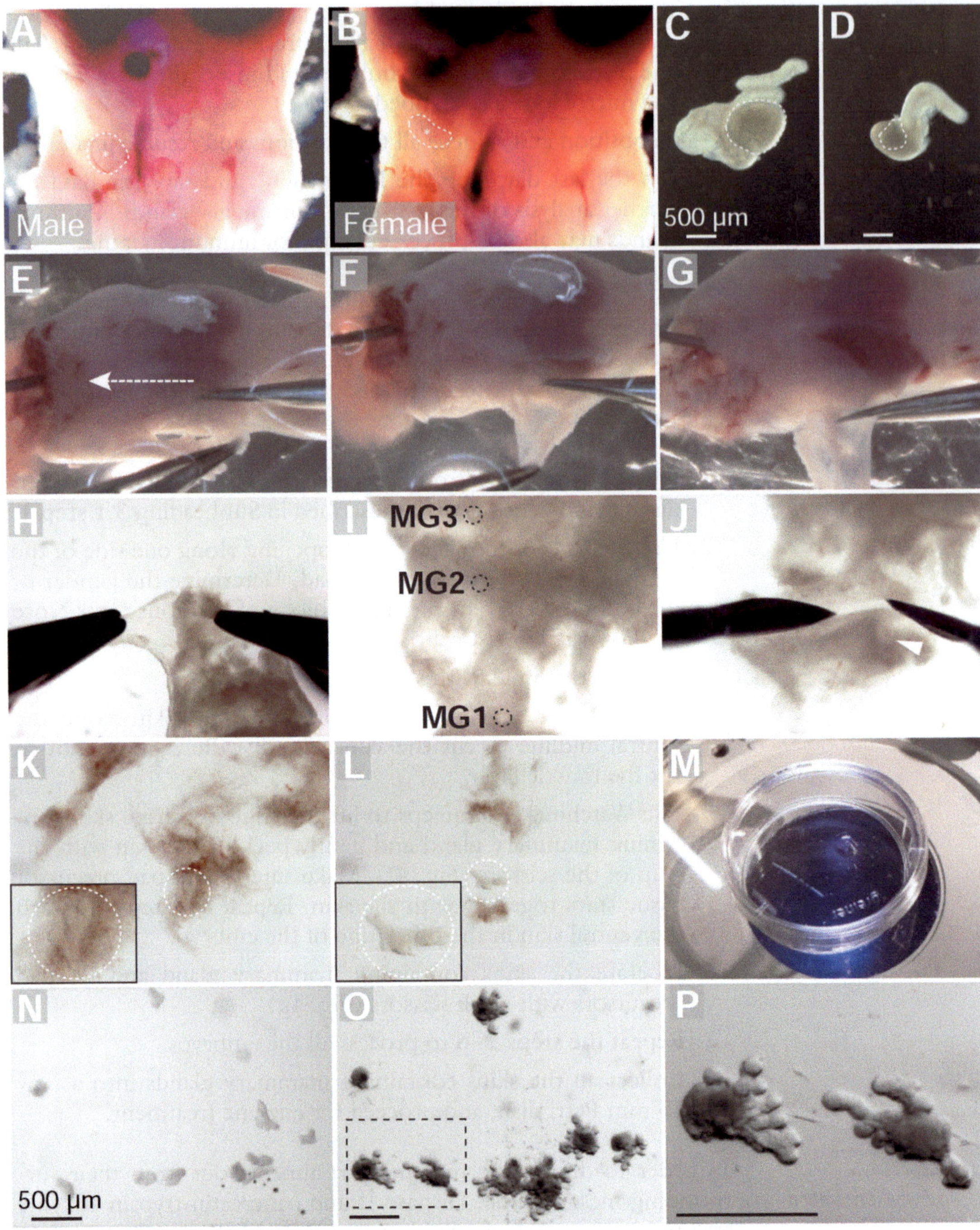

Fig. 4 Dissection and 3D culture of E16.5 mammary rudiments in Matrigel without mesenchyme. (**a**, **b**) A decapitated embryo fixed to the silicon plate with the help of needles showing the presence of ball-shaped gonads indicating a male embryo (**a**) or spatula-shaped ovary indicating a female embryo (**b**). (**c**) Closer view of a dissected out male gonad (**c**) or female ovary (**d**). (**e**) Dissection of the ventral skin starting from the midline. (**f**) The ventral skin is gently peeled off using forceps with the help of small scissors. (**g**) The separated skin is cut off with small scissor. (**h**) The skin epidermis and mesenchyme after enzymatic treatment with Dispase II and pancreatin-trypsin are further separated using forceps. (**i**) The position of mammary gland can

modifications. Treat the sample with 2.5 U/mL Dispase II for 40–60 min in 4 °C depending on the age of the embryos and the activity of the enzyme. Thereafter, incubate samples in pancreatin-trypsin working solution for 5–7 min at room temperature. The mesenchyme will be softer and start detaching from the skin. Typically, no further separation of the skin epidermis can be observed at this point, but this treatment is important for separating the mammary rudiments and their surrounding mesenchyme in **step 6**.

2. Incubate the Petri dish with explants on ice for 30–40 min for recovery. Protect from light if necessary.
3. Collect the samples into a new 35 mm glass Petri dish and use forceps to separate the skin epithelium from the rest of the tissue (Fig. 4h) (*see* **Note 10**).
4. Use forceps or scissors to remove extra tissue and expose the mammary gland.
5. Localize the mammary gland (Fig. 4i). The mammary epithelium is more condensed than the mesenchyme. With proper setting of the transmitted light, the mammary rudiments can be observed under a stereomicroscope. In case it is challenging, mammary rudiments may also be located by the position of the nipple (Fig. 4i) by placing the epidermal side of the mesenchyme upwards. Dissect the mammary gland out together with surrounding mesenchyme with the small scissors.
6. To isolate the mammary rudiments from surrounding mesenchyme, use two needles to tear apart the surrounding mesenchyme gently. Do not touch the mammary epithelium with needles as the epithelium is very fragile and damaged mammary gland will not grow. Continue the process to get rid of all the surrounding mesenchyme (Fig. 4j–l). An experienced researcher can process 30 mammary rudiments (**steps 1–6** for MG1-3 from five female embryos) in 3.5 h with proper enzyme treatments.
7. Collect the isolated intact mammary rudiments using small pipette with 10 μL pipette tip to a new 35 mm Petri dish and wash with DMEM/F12 full medium to remove the residual mesenchyme cells.

Fig. 4 (continued) be identified by the nipple which is evident as a small hole in the epidermal side of the mesenchyme. (**j**) Tear apart the mesenchymal tissue surrounding the mammary gland (white arrowhead) using needles. (**k**) Dissected mammary glands with little surrounding mesenchyme. A close-up is shown in the corner. (**l**) Mammary gland has been cleared off from all remaining; a close-up is shown in the corner. (**m**) Isolated intact mammary rudiments suspended in Matrigel. (**n–p**) E16.5 mesenchyme-free mammary rudiments on the first (**n**) and fourth day (**o**) of culture, with a close-up of the area in the dashed box (**p**)

8. Transfer the mammary rudiments to the center of a clean 35 mm Petri dish using small pipette with 10 μL pipette tip. Remove extra medium.
9. Gently resuspend the mammary rudiments with 25 μL growth factor reduced Matrigel by using cold pipette tip (Fig. 4m).
10. Gently place the dish upside down in the 37 °C humidified incubator with atmosphere of 5% CO_2 for 30 min.
11. Add 2–3 mL serum-free organoid culture medium and return the plate to the 37 °C humidified incubator with atmosphere of 5% CO_2.
12. Change the medium every 2–3 days. Branching is typically resumed in 2 days and can be followed up to 5 days culture (Fig. 4n–p).

4 Notes

1. The enzyme activity varies from batch to batch, especially for pancreatin. Insufficient treatment will lead to low efficiency in the mechanical separation step. The combination of the pancreatin-trypsin and Dispase II treatments greatly improves the separation efficiency.
2. To obtain the optimal results of enzyme treatment, the treatment procedure should be carefully monitored. Incubation time varies between 2 and 7 min, due to batch-to-batch differences in enzyme activity and should be optimized for every batch. Prolonged treatment may impair future growth of the mammary gland. Inactivate the enzyme when the solution became viscous and large piece of mesenchyme starts to separate from the explant.
3. The explants can be pipetted back and forth 2–3 times with the Pasteur glass pipette to facilitate the separation. The mammary buds should stay with the mesenchyme. If not, the enzyme treatment time could be shortened.
4. In some cases, it is beneficial to remove some of the buds, leaving only one to grow without interference from the others (we routinely culture only gland number 3 for live imaging purposes). This can be done by dissecting out other mammary buds with the help of needles.
5. It is possible to add growth factors or chemical components into the medium to examine their effect on morphogenesis.
6. Usually, the microscope software will calibrate the motorized stage upon startup. It is important to make sure that the CO_2 tube is properly positioned, and the sample will not hit the objective during calibration, so that the movement of the stage

is unobstructed. If this is a concern, turn on the software before fixing the sample to the stage.

7. Four hours intervals are frequently enough to resolve individual branching events in mammary explants.
8. In theory, it is possible to culture the mammary rudiments from older embryos. However, it is technically very difficult to dissect the branched mammary epithelium intact out from its surrounding mesenchyme. In the other hand, due to the asynchronous development of five pairs of mammary glands, we normally only culture first three pairs of mammary rudiments.
9. The male reproductive organ can be distinguished easily from the female organs. The male gonad is ball-shaped and located superficially in the lower abdomen and pops up easily after opening of the inguinal fossa.
10. Optimal enzyme treatment facilitates the dissection of the mammary rudiments. Some mammary rudiments may detach from their surrounding mesenchyme with the skin during separation of the epidermis. In such situation, use scissors to cut the mammary gland off from the main duct close to the nipple area and proceed to **step 7** directly.

Acknowledgment

We thank Raija Savolainen and Maria Voutilainen for technical assistance and helpful discussions, and all the Mikkola lab members for the great support.

This work was supported by the Academy of Finland, the Cancer Society of Finland, the Jane and Aatos Erkko Foundation, and the Sigrid Jusélius Foundation.

References

1. Watson CJ, Khaled WT (2008) Mammary development in the embryo and adult: a journey of morphogenesis and commitment. Development 135(6):995–1003. https://doi.org/10.1242/dev.005439
2. Inman JL, Robertson C, Mott JD, Bissell MJ (2015) Mammary gland development: cell fate specification, stem cells and the microenvironment. Development 142(6):1028–1042. https://doi.org/10.1242/dev.087643
3. Fu NY, Nolan E, Lindeman GJ, Visvader JE (2020) Stem cells and the differentiation hierarchy in mammary gland development. Physiol Rev 100(2):489–523. https://doi.org/10.1152/physrev.00040.2018
4. Manzo G (2019) Similarities between embryo development and cancer process suggest new strategies for research and therapy of tumors: a new point of view. Front Cell Dev Biol 7:20. https://doi.org/10.3389/fcell.2019.00020
5. Spike BT, Engle DD, Lin JC, Cheung SK, La J, Wahl GM (2012) A mammary stem cell population identified and characterized in late embryogenesis reveals similarities to human breast cancer. Cell Stem Cell 10(2):183–197. https://doi.org/10.1016/j.stem.2011.12.018

6. Ingthorsson S, Briem E, Bergthorsson JT, Gudjonsson T (2016) Epithelial plasticity during human breast morphogenesis and cancer progression. J Mammary Gland Biol Neoplasia 21(3–4):139–148. https://doi.org/10.1007/s10911-016-9366-3
7. Biggs LC, Mikkola ML (2014) Early inductive events in ectodermal appendage morphogenesis. Semin Cell Dev Biol 25-26:11–21. https://doi.org/10.1016/j.semcdb.2014.01.007
8. Mikkola ML, Millar SE (2006) The mammary bud as a skin appendage: unique and shared aspects of development. J Mammary Gland Biol Neoplasia 11(3–4):187–203. https://doi.org/10.1007/s10911-006-9029-x
9. Veltmaat JM, Mailleux AA, Thiery JP, Bellusci S (2003) Mouse embryonic mammogenesis as a model for the molecular regulation of pattern formation. Differentiation 71(1):1–17. https://doi.org/10.1046/j.1432-0436.2003.700601.x
10. Propper AY, Howard BA, Veltmaat JM (2013) Prenatal morphogenesis of mammary glands in mouse and rabbit. J Mammary Gland Biol Neoplasia 18(2):93–104. https://doi.org/10.1007/s10911-013-9298-0
11. Cowin P, Wysolmerski J (2010) Molecular mechanisms guiding embryonic mammary gland development. Cold Spring Harb Perspect Biol 2(6):a003251. https://doi.org/10.1101/cshperspect.a003251
12. Kratochwil K (1969) Organ specificity in mesenchymal induction demonstrated in the embryonic development of the mammary gland of the mouse. Dev Biol 20(1):46–71. https://doi.org/10.1016/0012-1606(69)90004-9
13. Voutilainen M, Lindfors PH, Lefebvre S, Ahtiainen L, Fliniaux I, Rysti E, Murtoniemi M, Schneider P, Schmidt-Ullrich-R, Mikkola ML (2012) Ectodysplasin regulates hormone-independent mammary ductal morphogenesis via NF-kappaB. Proc Natl Acad Sci U S A 109(15):5744–5749. https://doi.org/10.1073/pnas.1110627109
14. Voutilainen M, Lindfors PH, Mikkola ML (2013) Protocol: ex vivo culture of mouse embryonic mammary buds. J Mammary Gland Biol Neoplasia 18(2):239–245. https://doi.org/10.1007/s10911-013-9288-2
15. Padmanaban V, Grasset EM, Neumann NM, Fraser AK, Henriet E, Matsui W, Tran PT, Cheung KJ, Georgess D, Ewald AJ (2020) Organotypic culture assays for murine and human primary and metastatic-site tumors. Nat Protoc 15(8):2413–2442. https://doi.org/10.1038/s41596-020-0335-3
16. Srivastava V, Huycke TR, Phong KT, Gartner ZJ (2020) Organoid models for mammary gland dynamics and breast cancer. Curr Opin Cell Biol 66:51–58. https://doi.org/10.1016/j.ceb.2020.05.003
17. Jamieson PR, Dekkers JF, Rios AC, Fu NY, Lindeman GJ, Visvader JE (2017) Derivation of a robust mouse mammary organoid system for studying tissue dynamics. Development 144(6):1065–1071. https://doi.org/10.1242/dev.145045
18. Schindelin J, Arganda-Carreras I, Frise E, Kaynig V, Longair M, Pietzsch T, Preibisch S, Rueden C, Saalfeld S, Schmid B, Tinevez JY, White DJ, Hartenstein V, Eliceiri K, Tomancak P, Cardona A (2012) Fiji: an open-source platform for biological-image analysis. Nat Methods 9(7):676–682. https://doi.org/10.1038/nmeth.2019
19. Martin P (1990) Tissue patterning in the developing mouse limb. Int J Dev Biol 34(3):323–336
20. Munne PM, Narhi K, Michon F (2013) Analysis of tissue interactions in ectodermal organ culture. Methods Mol Biol 945:401–416. https://doi.org/10.1007/978-1-62703-125-7_24

Chapter 2

Multidimensional Fluorescence Imaging of Embryonic and Postnatal Mammary Gland Development

Claudia Carabaña and Bethan Lloyd-Lewis

Abstract

Multidimensional fluorescence imaging represents a powerful approach for studying the dynamic cellular processes underpinning the development, function, and maintenance of the mammary gland. Here, we describe key multidimensional imaging strategies that enable visualization of mammary branching morphogenesis and epithelial cell fate dynamics during postnatal and embryonic mammary gland development. These include 4-dimensional intravital microscopy and ex vivo imaging of embryonic mammary cultures, in addition to methods that facilitate 3-dimensional imaging of the ductal epithelium at single-cell resolution within its native stroma. Collectively, these approaches provide a window into mammary developmental dynamics, and the perturbations underlying tissue dysfunction and disease.

Key words Intravital microscopy, Multidimensional imaging, 3D imaging, 4D imaging, Mammary gland morphogenesis, Mammary embryonic development, Ex vivo culture, Optical tissue clearing, In vivo imaging

1 Introduction

Biological imaging is a fundamental and universal tool in the life sciences. For over 80 years, researchers have harnessed microscopy to reveal the inner workings of the mammary gland, a secretory organ essential for the survival of over 5000 mammalian species [1]. Light and electron microscopy of fixed tissue sections and cells are mainstay tools in a mammary gland biologist's armory. While immensely useful, two-dimensional (2D) static analyses are limited in their ability to provide detailed topological and/or dynamic functional information. Thus, to fully comprehend the intricate organization of the branched mammary epithelium, including the dazzling architecture of the lactating mammary gland, three-dimensional (3D; *x*-, *y*-, *z*-) imaging is essential. Moreover, interrogating the inherently dynamic cellular processes underpinning this complex tissue demands the ability to visualize mammary

Maria dM. Vivanco (ed.), *Mammary Stem Cells: Methods and Protocols*,
Methods in Molecular Biology, vol. 2471, https://doi.org/10.1007/978-1-0716-2193-6_2,

epithelial and stromal cells by live, four-dimensional (4D; *x*-,*y*-, *z*-, *t*-) imaging [2].

Due to its widespread accessibility and flexible multicolor acquisition capabilities, confocal laser scanning microscopy (CLSM) is a commonly used optical sectioning technique for volumetric fluorescence imaging [3]. However, conventional confocal modalities rely on single-photon excitation wavelengths in the visible range for imaging, which suffer from tissue light scattering and absorption. This limits confocal microscopy to depths of ~100 μm in most tissues [3, 4]. As such, microscopes equipped with pulsed, infrared multiphoton lasers are favored for deep tissue imaging. Alongside decreased photo-toxicity, multiphoton infrared excitation is less prone to light scattering and absorption, enabling deeper penetration and tissue imaging depths of up to 1 mm [5, 6]. These features make multiphoton systems particularly useful for in vivo fluorescence imaging by intravital microscopy (IVM).

In this chapter, we describe strategies for high-resolution, multidimensional imaging of live and fixed mammary tissues using CLSM and multiphoton microscopy. These include step-by-step protocols for 4D imaging of the adult and embryonic mammary gland using IVM and ex vivo imaging of embryonic cultures, respectively. We also describe approaches that facilitate high-resolution volumetric imaging of the mammary epithelium within its native stroma. These methodologies are particularly useful in contexts that demand the ability to visualize large regions of the mammary epithelium at high spatiotemporal resolutions, such as in genetic-fate mapping studies focused on delineating the differentiation potential and capacity of distinct populations of mammary progenitor cells [2, 7, 8]. Importantly, the approaches outlined enable in situ visualization of the dynamic interplay between mammary epithelial cells and their stromal microenvironment. While not discussed herein, many of the described protocols are also compatible with mammary tumor imaging, providing important insights into breast cancer cell biology [9–12].

IVM is a powerful technique that facilitates high-resolution, real-time fluorescence imaging of cells deep inside live animals [2, 13]. For short-term, acute IVM under non-recovery anesthesia, mammary gland tissues and tumors can be made accessible to the microscope's objective via a surgical "skin-flap" incision [14, 15]. Longitudinal IVM studies, however, require the surgical implantation of an imaging window to protect and provide optical access to tissues during repeated imaging sessions [9, 16–20]. Here, we describe the surgical procedure for implanting conventional metal mammary imaging windows (MIW), based on protocols developed for similarly designed models [17, 20–22]. These imaging windows consist of a glass cover-slipped titanium frame that can be fixed above tissues using a purse-string suture. Alongside, we suggest methods for retracing regions of interest in consecutive imaging sessions, facilitating in situ visualization of specific mammary epithelial cells and tissue structures for extended periods of time. For studies focused on the embryonic mammary gland, we describe methods for high-resolution, time-resolved CLSM and multiphoton imaging of embryonic mammary buds established in ex vivo culture [23–25]. This represents a highly tractable system for interrogating epithelial cell fate dynamics and behaviors during embryonic mammogenesis, a fundamental phase in mammary gland development that remains inaccessible to IVM.

Finally, we outline strategies for high-resolution, wholemount immunostaining and 3D visualization of fixed mammary tissues, aimed at circumventing issues associated with antibody penetration and depth of imaging in this optically opaque organ. This includes an enzymatic digestion-based procedure that facilitates improved deep tissue immunostaining and whole-gland imaging of slide-mounted tissues [26–29]. This approach, however, risks proteolytic-mediated damage to mammary epithelial and stromal cells, limiting its utility in some contexts. Consequently, we also provide two optical tissue clearing methods that enable high-resolution, deep tissue mammary gland imaging in the absence of enzymatic digestion or mechanical dissection, adapted from "see deep brain" (SeeDB) [30] and "clear unobstructed brain imaging cocktails" (CUBIC) [31] protocols. By mitigating tissue light scattering caused by cellular and extracellular components with different refractive indices (RIs), these methods allow for the visualization of expansive regions of the mammary epithelial tree at single-cell resolution within its native stroma [11, 32–37]. These fixed tissue methods are particularly valuable when applied downstream of in vivo/ex vivo live-imaging, allowing mammary cells visualized in 4D to be further characterized by immunostaining for biomarkers of interest. The refinement of the imaging approaches described herein—and their application to novel experimental models and methods for visualizing distinct cells and cellular processes—will continue to provide important insights into mammary gland and breast cancer biology.

2 Materials

2.1 Imaging Window (MIW) Preparation

1. Custom-made titanium MIW (*see* **Note 1**).
2. Glass coverslips (12 mm).
3. 100% Ethanol—prepare 70% (v/v) for use.
4. 100% acetone.
5. Cyanoacrylate-based glass glue.
6. Cotton swabs.
7. Sterile 0.9% saline.

8. *Optional step*: Sterile 1 ng/mL Poly(L-amino acid)-polyethylene glycol/poly(L-lysine)-polyethylene glycol (PLL-g-PEG) solution prepared in 10 mM HEPES buffer, pH 7.4.

2.2 Imaging Window Implantation

1. Extra fine Graefe forceps.
2. Straight Bonn surgical scissors.
3. *Optional*: Needle holder.
4. Autoclave pouches.
5. Bead sterilizer.
6. Sterile surgical gloves, gown, hair net, and face mask.
7. Sterile surgical drapes.
8. Sterile cotton gauze swabs.
9. Isoflurane inhalation anesthetic.
10. Isoflurane vaporizer with oxygen supply.
11. Anesthetic scavenger unit.
12. Anesthesia induction box/cage.
13. Heat pad/heated operating stage.
14. Analgesics, e.g., 0.3 mg/mL buprenorphine hydrochloride.
15. Physiological saline solution.
16. 1 mL syringe and 25-guage needle.
17. Ophthalmic ointment.
18. Tape.
19. Pet hair clipper.
20. *Optional*: depilatory cream.
21. Antiseptic surgical scrub—e.g., Betadine.
22. Sterile non-absorbable polypropylene surgical suture (4-0) (*see* **Note 2**).

2.3 Intravital Microscopy (IVM)

1. Multiphoton confocal microscope surrounded by a heated dark box (e.g., Nikon A1R MP, equipped with a Spectra-Physics Insight Deepsee laser, conventional and resonant scanners, and GaAsP non-descanned detectors).
2. Long-working distance objectives (e.g., Nikon, 25×/NA 1.1 2.0 mm WD water immersion objective).
3. Custom-made imaging window stabilizers, e.g., microscope stage insert or imaging box.
4. Physiological saline solution.
5. Syringe and butterfly-wing needle (for long-term (>3 h) imaging).

6. Mouse vital sign monitor (e.g., MouseOx Plus, Starr Life Sciences).
7. *Optional*: Injectable fluorescent probes to label mammary cells or structures of interest.
8. Image analysis software and processing tools: e.g., ImageJ (https://imagej.nih.gov/ij/), Imaris (http://www.bitplane.com/).

2.4 Embryonic Mammary Gland Dissection

1. Timed-pregnant mice, e.g., fluorescent reporter mouse models (*Lgr5-GFP* [38] or *Rosa26-mTmG* reporter strains [39]).
2. 1 mL syringe and 25-gauge needles.
3. Dissection tools: micro-dissecting scissors, spring scissors, Dumont #5 forceps, and curved micro-dissecting forceps.
4. 100 mm diameter Petri dish.
5. 24-well plates.
6. Stereoscopic dissecting microscope with transmitted illumination.
7. Sterile phosphate-buffered saline solution (PBS).
8. Embryonic culture medium: DMEM/F-12, 2 mM GlutaMAX™, 10% Fetal Bovine Serum (FBS) (v/v), and 20 U/mL Penicillin-Streptomycin (PS).
9. *Optional (for Cre-inducible reporter models)*: Tamoxifen-free base (MP Biomedicals), dissolved in sunflower oil.

2.5 Ex Vivo Embryonic Mammary Gland Culture and 4D Live-Cell Imaging

1. Tissue Culture Dish with Cover Glass Bottom (35 mm).
2. Cell culture inserts (0.4 μm, 30 mm diameter; Millicell).
3. 10× Pancreatin solution pH 7.0: 2.5 g of pancreatin from porcine pancreas and 0.85 g NaCl dissolved in 100 mL of cold MilliQ water and filter-sterilized (for preparation guidelines, *see* **Note 3**).
4. Porcine Trypsin, 1:250 (*see* **Note 4**).
5. Thyrode's solution (pH 7.4): 8 g/L NaCl, 0.2 g/L KCl, 0.05 g/L $NaH_2PO_4 + H_2O$, 1 g/L glucose, 1 g/L $NaHCO_3$ dissolved in 1 L of distilled water and filter-sterilized.
6. Pancreatin-trypsin working solution pH 7.4: 0.225 g porcine trypsin made up in a final volume of 10 mL Thyrode's solution containing 1 mL 10× pancreatin stock solution and 20 μL of PS. Filter-sterilize before use (for preparation guidelines, *see* **Note 4**).
7. Ascorbic acid: 10 mg/mL in distilled water and filter-sterilized.
8. Horizontal laminar flow hood.
9. Inverted CLSM microscope (e.g., Zeiss LSM780/880) or two-photon confocal microscope (e.g., Leica SP8 microscope

with a Chameleon Vision II laser), equipped with a heated environmental chamber.

10. Long-working distance objective (e.g., Zeiss 40×/1.0 NA water immersion objective, Leica 25×/0.95 NA water immersion objective).
11. Image analysis software and processing tools: e.g., ImageJ (https://imagej.nih.gov/ij/), Imaris (http://www.bitplane.com/).

2.6 Enzymatic Digestion and Wholemount Immunostaining of Mammary Tissues

1. Enzyme digestion (ED) mix: HBSS containing 300 U/mL collagenase Catologue numbers and suppliers for digestive enzymes are important information for researchers to replicate protocols as there are many types available (Cat. No. C0130, Sigma) and 300 μg/mL hyaluronidase Same comment as per collagenase - cat numbers/supplier details are helpful for protocols requiring digestive enzymes (Cat. No. 4272, Sigma).
2. 4% Paraformaldehyde in PBS.
3. 0.5 M NH_4Cl.
4. ED Blocking buffer: 1% (w/v) bovine serum albumin (BSA), 5% (v/v) horse serum, and 0.8% (w/v) Triton X-100 in PBS.
5. PBST: PBS + 0.2% (w/v) Tween20.
6. 4′,6-diamidino-2-phenylindole (DAPI) dilactate.
7. Primary and fluorescent secondary antibodies (as required).
8. Microscope slides and coverslips (No. 1.5).
9. Aqua Poly/Mount.

2.7 CUBIC-Based Optical Tissue Clearing

1. 10% Neutral Buffered Formalin (NBF).
2. Card or glass microscope slides.
3. CUBIC Reagent 1A: 10% (w/w) urea, 5% (w/w) *N*,*N*,*N*′,*N*′-tetrakis(2-hydroxypropyl)ethylenediamine, 10% (w/w) Triton X-100 and NaCl (25 mM) in distilled water (*see* **Notes 5** and **6**).
4. CUBIC Reagent 2: 44% (w/w) sucrose, 22% (w/w) urea, 9% (w/w) 2,2′,2″-nitrilotriethanol, and 0.1% (w/w) Triton X-100 in distilled water (*see* **Note 5**).
5. CUBIC Blocking buffer: 10% (v/v) goat serum, 0.5% (w/v) Triton X-100 in PBS.
6. 4′,6-diamidino-2-phenylindole (DAPI) dilactate.
7. *Optional*: Primary and fluorescent secondary antibodies as necessary.
8. Glass-bottomed imaging dishes.

2.8 SeeDB-Based Optical Tissue Clearing

1. 10% Neutral Buffered formalin (NBF).
2. Card or glass microscope slides.
3. α-thioglycerol.
4. Serial fructose solutions made up in distilled water—20%, 40%, 60%, 80%, 100% (w/v), SeeDB (115% (w/v)) (*see* **Note 7**).
5. SeeDB Blocking buffer: 10% (w/v) bovine serum albumin (BSA), 1% (w/v) Triton X-100 in PBS.
6. 4′,6-diamidino-2-phenylindole (DAPI) dilactate.
7. *Optional*: Primary and fluorescent secondary antibodies as necessary.
8. Glass-bottomed imaging dishes.

3 Methods

3.1 Mammary Imaging Window Implantation for 4D-IVM

All animal studies must be approved by institutional ethical committees and national authorities as required. Perform surgeries according to the LASA Guiding Principles of Preparing and Undertaking Aseptic Surgery to minimize infection risk (*see* **Note 8**). Due to its fast induction and recovery, alongside the ability to precisely control anesthesia depth and length, inhalation anesthetics should be used for window implantation and intravital imaging [14, 17]. While acute intravital imaging of the mammary gland under non-recovery anesthesia (i.e., where the animal is euthanized immediately at the end of imaging) can be performed via a skin-flap incision [14, 40, 41], this approach poses challenges for maintaining optimal mouse hydration and physiology. Thus, implanting a MIW is also beneficial when performing non-recovery imaging studies spanning several hours. When applied to fluorescent reporter mouse models, IVM is a powerful approach for tracing the fate, migration, proliferation, geometry, and re-arrangements of individual cells in real time during mammary gland postnatal development and tissue homeostasis.

3.1.1 Mammary Imaging Window (MIW) Preparation

1. Apply a thin layer of cyanoacrylate-based glass glue on the etched inset of the MIW that will contact the coverslip.
2. Using forceps, place the coverslip on the glue-covered inset and apply gentle pressure using a cotton swab for 1 min. If necessary, apply glue at the interface of the coverslip and the window.
3. Allow the glue to dry completely by placing it on its side in a flow cabinet for a minimum of 2 h.
4. If required, use a cotton swab soaked in 100% acetone to remove excessive and condensed glue from the coverslip. Use a cotton swab soaked in 70% (v/v) ethanol to remove the acetone.

5. Place the window glass-side down on a dry tissue and fill with physiological saline to assess for leaks. If the window is watertight, the tissue will remain dry (*see* **Note 9**).
6. Sterilize the MIW by steam sterilization, or by placing it in 70% (v/v) ethanol for a minimum of 30 min. Do not autoclave cover-slipped windows as it will degrade the glue.
7. *Optional*: In a sterile flow cabinet, coat the interior side of the coverslip with PLL-g-PEG solution. Incubate for 1 h at room temperature. Subsequently, wash the window in sterile PBS (*see* **Note 10**).

3.1.2 Surgery Preparation

1. Sterilize surgical tools by autoclaving or dry sterilization (*see* **Note 11**). Thoroughly disinfect the entire operating station (including heat pads, anesthesia nose cones, and adjustment controls) using 70% ethanol.
2. Anaesthetize the mouse in an induction chamber using 3–4% (v/v) gaseous isoflurane (*see* **Note 12**).
3. Once anaesthetized, transfer the mouse onto a heat pad situated away from the sterile operating area, and continue to supply anesthesia via a nose cone. Reduce the isoflurane concentration to 1.5–2% (v/v) (*see* **Note 13**).
4. Apply ophthalmic ointment to both eyes to prevent corneal drying.
5. Administer Buprenorphine at a dose of 0.1 mg/kg mouse body weight by sub-cutaneous injection (*see* **Note 14**).
6. Shave the skin over the fourth mammary gland using electronic pet clippers. As stray hairs under the window pose an infection risk and can obstruct imaging, it is important to remove as much residual hairs as possible from the surgical field. Brief exposures to depilatory cream can be used if necessary.
7. Disinfect the shaved skin using Betadine solution and a cotton gauze. Begin at the center of the surgical site and move out towards the periphery using an ever-widening circular motion. Repeat 3 times using a clean gauze each time.
8. Gently transfer the mouse to the disinfected heated operating stage. Continue to supply the mouse with 1.5–2% (v/v) isoflurane via a nose cone to maintain anesthesia.
9. Loosely immobilize the hind legs using tape. Cover the mouse with a sterile surgical drape equipped with an opening to allow access to the prepared surgical field.

3.1.3 Surgical Implantation of the MIW

1. Verify that the mouse is sufficiently anaesthetized by pinching a hind paw through the overlying sterile drape. If the mouse is unresponsive, proceed to the next step. Otherwise, adjust the isoflurane dose and wait until reflex behaviors are absent.

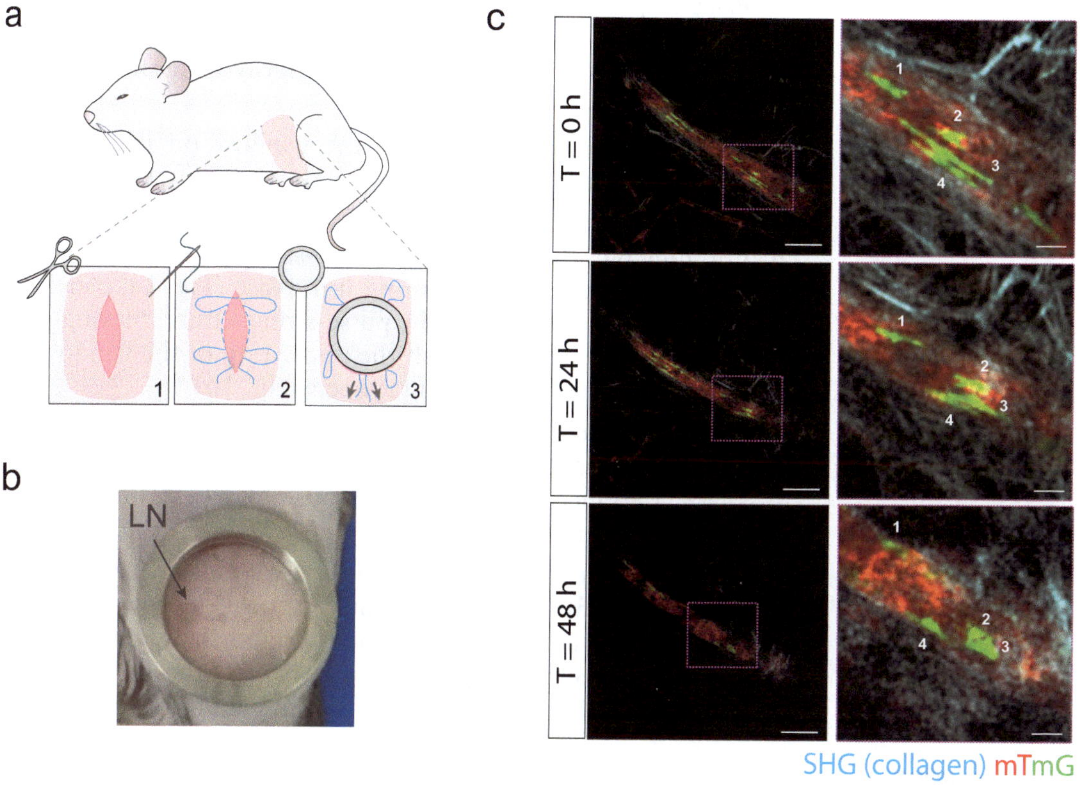

Fig. 1 Intravital imaging through the mammary imaging window (MIW). (**a**) Schematic representation of the surgical procedure. (1) Make an incision above the fourth mammary gland. (2) Place a purse-string suture around the incision. Dotted line denotes internal suture position. (3) Insert the MIW above the mammary gland and place the sutured skin within the groove. Secure the skin around the MIW by tightening the purse-string suture (protocol based on [17, 19]). (**b**) Photograph of the MIW implanted over the fourth abdominal mammary gland. *LN;* lymph node. (**c**) Longitudinal imaging of mammary epithelial cell dynamics by multiphoton IVM. Images show a mammary epithelial duct in a 5-week-old *K5-Cre*^{ERT2}*;Rosa26-mTmG* female mouse. In this model, all cells are labeled with a membrane-bound Tomato fluorescent protein (mT, red). Tamoxifen administration induces Cre-mediated recombination of the reporter allele in sporadic Keratin (K)5-expressing mammary myoepithelial cells, resulting in membrane-bound EGFP fluorescent protein expression (mG, green). Images show the migratory behavior of four EGFP^{+ve} mammary myoepithelial cells over time. Second harmonic generation (SHG) microscopy reveals the collagen organization surrounding the ductal structure. Scale bars: 100 μm (left panels), 25 μm (right panels)

2. Use sterile forceps to gently lift the skin away from the abdominal wall and make a ~12 mm long incision in the flank area above the fourth abdominal mammary gland using sterile scissors (Fig. 1a, panel 1, *see* **Notes 15** and **16**).
3. Use sterile forceps to blunt dissect the skin away from the underlying mammary gland, taking care not to damage the tissue.

4. Place a circular purse-string suture around the incision, approximately 4 mm from skin edges. Keep the external sections of the suture loose at this stage, leaving 4 butterfly-wing shaped outer loops (Fig. 1a, panel 2).
5. Using forceps, carefully insert the window on the exposed mammary gland and gently place the sutured skin into the window groove (Fig. 1a, panel 3).
6. Carefully pull the loops of the purse-string to tighten the skin in the window groove, securing it in place (Fig. 1a, panel 3). Place a double knot to fix the purse-string suture, making sure to hide the knot underneath the upper ring of the window to prevent the mice from biting it open (Fig. 1b).
7. Proceed immediately to intravital imaging, or allow the mouse to recover from the anesthesia in a cage placed on a heating pad at 37 °C. Once recovered, the mouse may be group-housed with other window-bearing mice.
8. Post-operative care: Closely monitor mice for signs of pain and discomfort after surgery. If required, provide post-operative analgesics in accordance with local veterinary guidance (*see* **Note 17**). Inspect the surrounding skin and the tissue underneath the MIW daily for signs of inflammation and necrosis.

3.2 Longitudinal 4D-IVM by Multiphoton Microscopy

1. If not proceeding directly from MIW surgery, anaesthetize the mouse in an induction chamber using 3–4% (v/v) gaseous isoflurane.
2. *Optional*: Administer fluorescent probes by tail vein injection to label mammary cells and structures as necessary (*see* **Note 18**).
3. To maintain mouse hydration during short-term experiments (<3 h), administer a maximum of 500 μL saline by sub-cutaneous injection prior to imaging. In experiments exceeding 3 h, saline should be provided periodically (~50–100 μL/h) using an indwelling intraperitoneal line [14, 17] (*see* **Note 19**).
4. Transfer the mouse onto the stage insert of a multiphoton microscope for imaging, continuing to supply 1.5–2% (v/v) isoflurane via a nose cone to maintain anesthesia.
5. Position and immobilize the MIW for imaging. On an inverted setup, a microscope insert or box customized with a hole that precisely fits the window can be used for stabilization [19]. To image in the upright configuration, the window can be fixed using custom-made holders or microstage devices [40] (*see* **Note 20**). Maintain the mouse's body temperature at 37 °C during imaging, ideally using a dark heated chamber that surrounds the microscope stage.

6. Once the mouse is secure and stabilized, reduce the isoflurane concentration to between 0.8% and 1.2% (*see* **Note 21**).
7. Closely monitor the mouse's vital parameters during imaging:
   (a) Regularly check the breathing rate and adjust the isoflurane supply accordingly if irregular or abnormal breathing is observed.
   (b) Measure the mouse's temperature during imaging using a rectal probe. Adjust the temperature of the heated imaging chamber as necessary.
   (c) Monitor capillary blood flow to ensure optimal conditions for imaging.
   (d) Continuous assessment of vital parameters can be performed using a non-invasive pulse oximetry monitoring system (e.g., MouseOx system) that measures the mouse's temperature, arterial oxygen saturation, breathing distension, and heart and respiratory rates during imaging.
8. When anesthesia levels are tightly controlled and animal vitals well-maintained, 4D-IVM can be performed continuously for up to ~40 h under non-recovery anesthesia [41]. Alternatively, longitudinal IVM can be performed whereby mice undergo repeated, shorter imaging sessions over extended periods of time (*see* **Note 22**).
9. For longitudinal imaging of specific cells and tissue structures, a number of approaches can be used to help retrace regions of interest in consecutive imaging sessions, including:
   (a) Using a motorized stage: In configurations where the relative position of the imaging window is fixed in repeated imaging sessions, the coordinates of individual regions can be stored and used to relocate imaging areas.
   (b) Morphological landmarks: The branched mammary epithelial network is structurally heterogeneous, providing unique patterns of ductal structures and adjacent blood vessels that can be readily recognized in repeat imaging sessions though the eyepiece or by manual scanning (*see* **Note 23**). Intravenous injection of fluorescent agents to label vasculature (e.g., fluorescently labeled dextrans), in addition to second harmonic generation (SHG) imaging of tissue collagen organization, can also assist with retracing regions of interest in consecutive imaging sessions (Fig. 1c).
   (c) Fluorescent reporter mouse models: Heterogeneous labeling of mammary epithelial cells gives rise to unique and identifiable color patterns for serial imaging, e.g., using fluorescent reporter mouse models such as *R26R*-

Confetti [9, 42] or *Rosa26-mTmG* [39] mouse strains (Fig. 1c).

10. *Optional*: To aid the re-identification of intravitally imaged mammary tissue regions in downstream 3D or 2D histological analyses, at the end of the final imaging session use focused high laser power to generate distinct autofluorescent "photo-tattoos" in nearby tissue areas, taking care not to damage cells and tissue structures of interest [43].
11. At experimental endpoints, euthanize the mouse according to institutional/national guidelines.
12. Taking note of the orientation of the tissue relative to the coverslip, harvest and fix the intravitally imaged mammary gland for downstream histological analysis (e.g., by 3D whole-mount immunostaining as described in Subheading 3.5). Harvest the contra-lateral, non-window mammary gland as a control.
13. To reuse the MIW, clean the titanium ring using soap and hot water. Incubate overnight in 100% acetone to remove the glue and release the used coverslip. Prepare the MIW for future implantation following the instructions detailed in Subheading 3.1.1.
14. Process and analyze intravital images using ImageJ (https://imagej.nih.gov/ij/) and/or commercial software such as Imaris (http://www.bitplane.com/), depending on availability.

3.3 Establishing Mammary Embryonic Buds in Culture for 4D Ex Vivo Imaging

The mammary epithelium is first specified as placodes at approximately embryonic (E) Day 11, which invaginate into the underlying mesenchyme to form mammary buds by E12. After E15.5, buds undergo sprouting and branching morphogenesis to give rise to a rudimentary epithelial tree by E18.5 [44]. To visualize this process in real time, mammary embryonic buds must be established in ex vivo culture [23–25]. In this section, we describe methods for high-resolution longitudinal or time-lapse fluorescence imaging of ex vivo cultured embryonic mammary buds by CLSM and multiphoton microscopy. Similarly to IVM, when applied to fluorescent reporter mouse models this approach allows for the dynamic behavior and fate of individual embryonic mammary cells to be traced in real time during embryonic mammary morphogenesis.

3.3.1 Dissection of the Embryonic Mammary Gland

1. Set up timed mouse matings to obtain pregnant females bearing embryos of the desired genotype and age. Embryo stage is determined by the detection of a vaginal plug the following day (detection at mid-day defined as 0.5 days-post-coitus, i.e., E0.5).

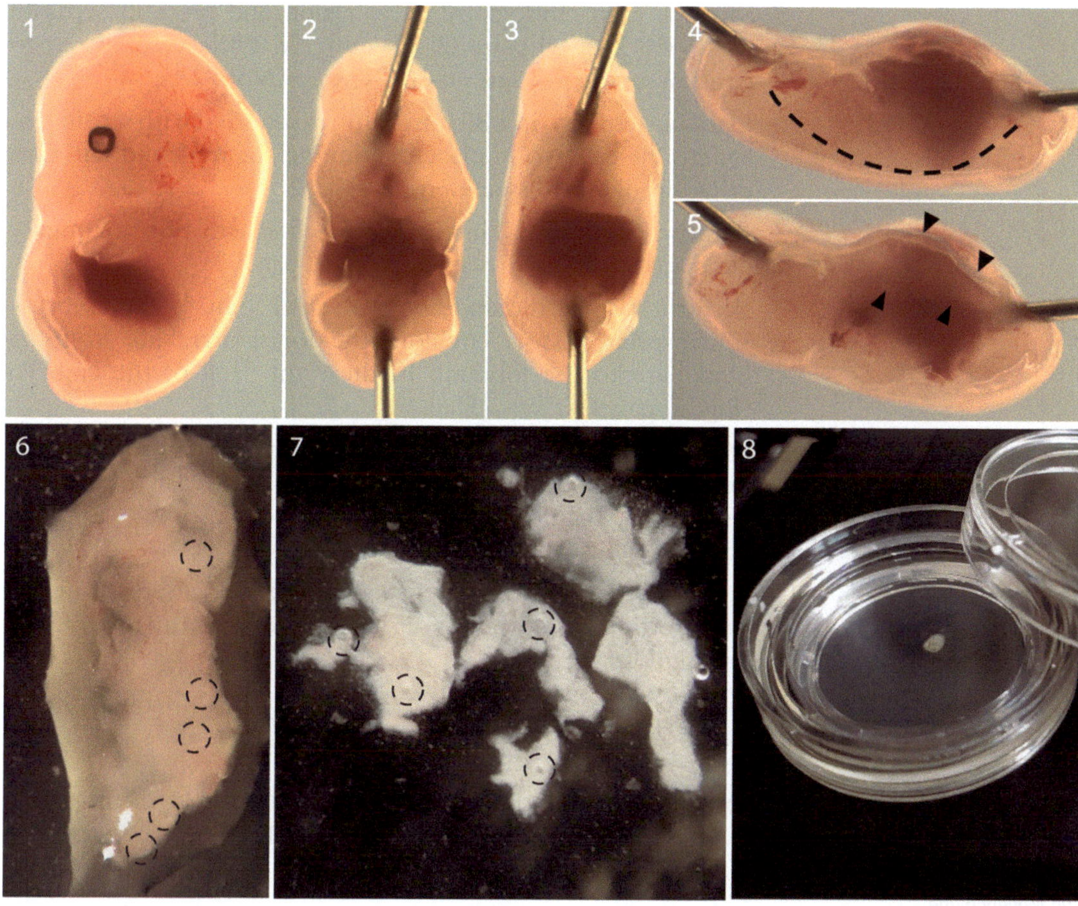

Fig. 2 Embryonic mammary gland dissection protocol. (1) Collect all embryos from the uterus of a pregnant female mouse. (2) Attach the decapitated embryo to a dish filled with set silicon. (3) Remove the limbs. (4) Cut along the dorsal-lateral line of the embryo (dashed line). (5) Detach the flank (see arrowheads) to the midline. (6) Collect both flanks of the embryo containing the embryonic mammary buds (marked with dashed circles). (7) Remove skin epithelium after enzymatic digestion with pancreatic-trypsin working solution. Dashed circles denote the location of embryonic mammary buds within the mesenchyme. (8) Culture the isolated tissues on an air-liquid interface. (Protocol based on [25])

2. *Optional*: If using Cre-inducible fluorescent reporter mice, administer a low dose of tamoxifen to pregnant females 24 h before tissue dissection to induce fluorescent labeling in embryos. While a tamoxifen dose of 0.1 mg/g of mouse body weight is commonly used, precise doses will vary depending on specific study requirements [28].
3. At the desired embryonic stage, sacrifice the pregnant female and harvest the embryos from the uterus in a 100 mm diameter Petri dish filled with cold PBS.
4. Remove the yolk sac and separate each embryo from its placenta, taking care not to damage the tissue (Fig. 2, panel 1).

5. Sacrifice each embryo by decapitation and place in separate wells of a 24-well plate filled with cold PBS. Keep tissues to confirm genotypes and sex by PCR (*see* **Note 24**).
6. Place one embryo in a 35 mm culture dish filled with set silicon. Secure the embryo in place by pinning the neck and tail joint using needles or dissection pins (Fig. 2, panel 2).
7. Remove the limbs. This makes the first and fifth mammary buds accessible for dissection (Fig. 2, panel 3).
8. Perform a small incision above the tail joint of the embryo. Using spring scissors, cut along the dorsal-lateral line from the hind limb to the forelimb in the right flank of the embryo (Fig. 2, panel 4).
9. Detach the flank of the embryo from the incision along the dorsal-lateral line to the midline (Fig. 2, panel 5). Hold the tissue using Dumont #5 forceps, and use the spring scissors to trim the right flank (*see* **Note 25**).
10. Repeat **steps 8** and **9** with the left flank of the embryo, but this time cutting along the dorsal-lateral line from the forelimb to the hind limb.
11. Transfer both flanks to a new 24-well plate with PBS (Fig. 2, panel 6).

3.3.2 Separating the Embryonic Skin Epithelium and Mesenchyme

This procedure entails proteolytic digestion of dissected embryonic flanks, based on a protocol developed by the laboratory of M. Mikkola [25].

1. Replace the PBS with pancreatic-trypsin working solution and incubate for 4–5 min. Optimal incubation times are heavily dependent on embryo stage, in addition to the particular enzyme batches used [25]. Thus, closely monitor the tissue under the stereomicroscope during enzyme treatment.
2. When the skin epithelium starts to detach from the edges of the mammary mesenchyme, inactivate the enzymes by replacing the pancreatic-trypsin solution with DMEM/F-12 embryonic culture medium (*see* **Note 26**).
3. Incubate the mammary tissue on ice for 30–45 min.
4. Place the mammary tissue in a 35 mm culture dish. Using two needles, gently peel the skin epidermis away from the mesenchyme containing the embryonic mammary buds. The mesenchyme is required for embryonic mammary epithelial development [23] and should be maintained as intact as possible around mammary buds (Fig. 2, panel 7).
5. Once isolated, use a plastic pipette to transfer the intact mesenchyme containing the embryonic mammary buds to fresh

DMEM/F-12 embryonic culture medium in a 24-well plate. Repeat this process for all harvested tissues.

3.3.3 Establishing Mammary Buds in Ex Vivo Culture

1. Per embryo, prepare a 35 mm cover glass-bottomed tissue culture dish containing 1 mL of DMEM/F-12 embryonic culture medium freshly supplemented with 75 μg/mL ascorbic acid.
2. Place the embryonic mammary tissue on a cell culture insert (containing 0.4 μm pores) using a 1000 mL pipette.
3. Using curved micro-dissecting forceps, carefully and slowly place the cell culture insert into the prepared glass-bottomed tissue culture dish to avoid bubble formation. This ensures mammary embryonic buds are cultured on an air-liquid interface, whereby the tissue remains exposed to air while maintaining contact with the embryonic cell culture medium through the pores of the cell culture insert (Fig. 2, panel 8).
4. Maintain mammary cultures in a tissue culture incubator at 37 °C and 5% CO_2 atmosphere.
5. Exchange the culture media with fresh media every second day for the duration of the experiment. Cultures can be maintained ex vivo for up to 2 weeks.

3.4 4D Time-Lapse and Longitudinal Imaging of Ex Vivo Embryonic Mammary Cultures

Once established in ex vivo culture, embryonic mammary branching development can be recorded daily by 3D fluorescence imaging for up to 2 weeks. Use an inverted CLSM or multiphoton microscope (e.g., Zeiss LSM780/880 or Leica SP8) equipped with long-working distance objectives to acquire high-resolution 3D image stacks at each time-point (Fig. 3a). Alternatively, time-lapse imaging can be performed for up to 48 h (Fig. 3b). Below, we provide an overview of the experimental conditions that facilitate high-resolution, ex vivo imaging of fluorescently labeled embryonic mammary epithelial cells over time.

1. Isolate embryonic mammary buds from a fluorescent reporter mouse model and establish in ex vivo culture as described in Subheading 3.3. Figure 3a, b show a cultured embryonic mammary bud established from a *Lgr5-GFP* [38] or *Rosa26-mTmG* [39] reporter mouse embryo, respectively. In the absence of Cre-mediated recombination in the *Rosa26-mTmG* mammary bud (Fig. 3b), all embryonic mammary epithelial and stromal cells are labeled with a membrane-bound Tomato fluorescent protein.
2. Due to superior depth of imaging and low photo-toxicity, a multiphoton confocal microscope is preferred for time-lapse imaging of ex vivo embryonic mammary cultures, e.g., an inverted Leica SP8 microscope equipped with a femtosecond Chameleon Vision II multiphoton laser (680–1350 nm;

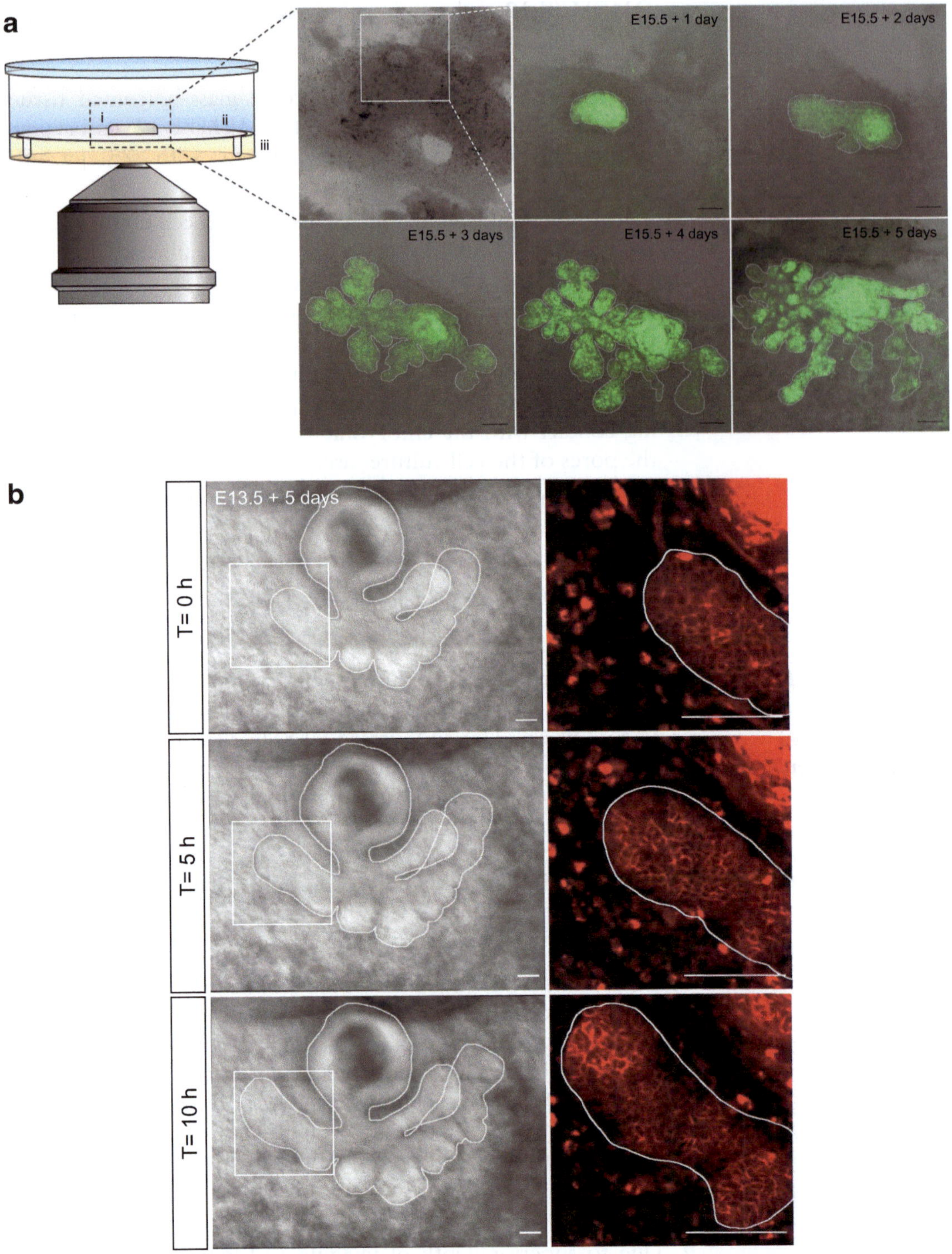

Fig. 3 4D ex vivo imaging of mammary embryonic bud cultures (**a**) Longitudinal fluorescence imaging of embryonic mammary branching morphogenesis. Schematic representation of the microscope configuration for live-cell imaging of embryonic mammary bud cultures. The tissue (i) is deposited on a cell culture insert (ii) placed in contact with the culture medium (iii) in a glass bottom dish. Use an inverted confocal or multiphoton microscope equipped with a long-working distance objective for imaging. Images show the growth of an E15.5 embryonic mammary bud isolated from a *Lgr5-GFP* [38] mouse embryo over 5 days.

Coherent, Inc.) and long-working distance objectives (e.g., Leica 25×/0.95 NA water immersion objective).

3. The microscope should be equipped with an incubation chamber that maintains tissues at 37 °C, 5% CO_2 atmosphere and 95% humidity (*see* **Note 27**). To avoid sample drift in the *Z*-axis, allow environmental conditions of the microscope chamber to stabilize for 30–60 min before commencing image acquisition.
4. Acquire image sequences as *Z*-stacks of the desired volume. Image acquisition time may vary between 5 and 60 min, depending on tissue thickness, required z resolution and area imaged, as well as the number of fluorophores to detect.
5. To image the whole embryonic mammary epithelium, overview tile scans can be acquired. Ensure at least a 10% overlap between tiled images, which can be stitched into larger mosaics using the microscope's acquisition software, or ImageJ Plugins (e.g., the Grid Collection/Stitching plugin) (https://imagej.nih.gov/ij/).
6. Time-lapse processing and analysis can be performed using ImageJ and/or commercial software such as Imaris (http://www.bitplane.com/), depending on availability.

3.5 3D Fluorescence Imaging of Fixed Mammary Gland Tissues

In this section, we provide three protocols for wholemount immunostaining and 3D visualization of fixed mammary gland tissues, encompassing proteolytic digestion and two optical tissue clearing methods based on published CUBIC [31] and SeeDB [30] protocols. These can be applied to freshly harvested tissues, or after completing 4D in vivo/ex vivo imaging to further characterize live-imaged tissue regions by immunostaining. While proteolytic digestion methods are particularly useful for whole-gland imaging of immunostained mammary tissues [26–29], CUBIC and SeeDB-basedoptical tissue clearing methods facilitate visualization of the mammary epithelial tree at single-cell resolutions within its intact stroma [11, 32–36] (*see* **Note 28**). For an overview of the optical

Fig. 3 (continued) Maximum intensity projections of image sequences are displayed, and the rendered surface of the mammary epithelium is outlined in white. Scale bars: 100 μm. (**b**) Time-lapse imaging of embryonic mammary cultures. Close-up of one embryonic mammary bud dissected from a *Rosa26-mTmG* [39] mouse embryo at Day E13.5, and cultured ex vivo for 5 days prior to time-lapse imaging. Images show the growth of an epithelial branch over 10 h (white box). In this reporter mouse strain, all cells are labeled with a membrane-bound Tomato fluorescent protein (red). Inset images show a thin optical slice. Individual mammary epithelial cells can be clearly visualized in ex vivo cultures by high-resolution multiphoton microscopy. Branching morphogenesis can be monitored for up to 48 h by time-lapse confocal or multiphoton imaging. The rendered surface of the mammary epithelium is outlined in white. Scale bars: 100 μm

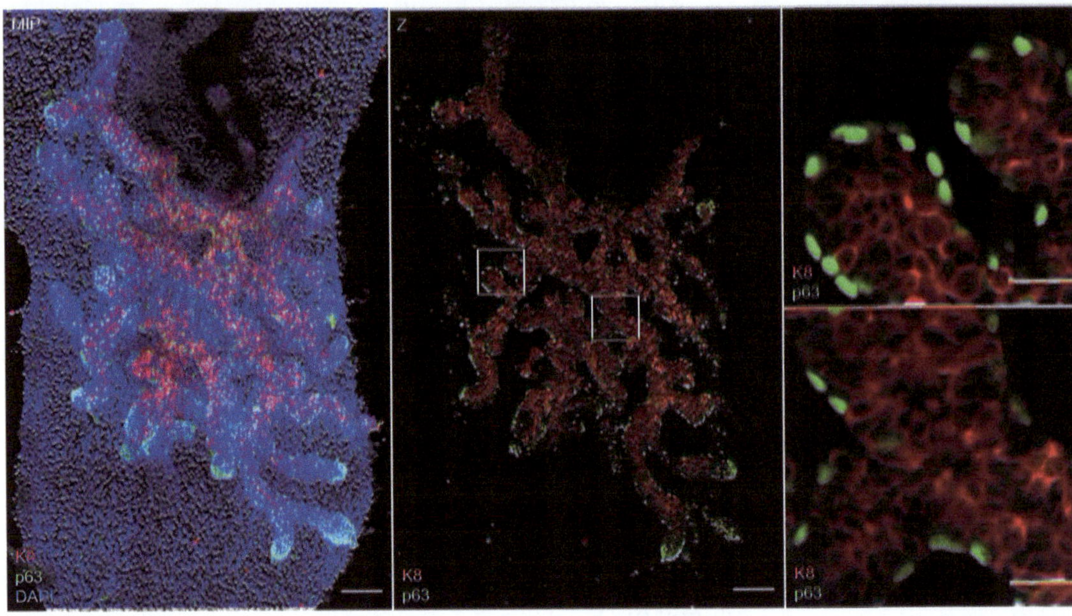

Fig. 4 Wholemount immunostaining of embryonic mammary cultures. Immunofluorescence staining of an embryonic mammary gland dissected at Day E13.5 and grown in ex vivo culture for 8 days. Left panel: Maximum intensity projection (MIP) showing the expression of the luminal epithelial marker protein Keratin 8 (K8, red), and the basal epithelial marker protein p63 (green) in embryonic mammary cells. Nuclei are stained with DAPI (blue). Center panel: A single optical section (*z*). White boxes mark regions displayed in the right panels. Scale bar: 100 μm. Right panel: Close-up images of K8 (red) and p63 (green) expressing cells in a tip (upper panel) and branch (lower panel) region of the embryonic mammary epithelium. Scale bar: 25 μm

tissue clearing methods previously applied to the mammary gland, in addition to their advantages and disadvantages, *see* [2, 11].

3.5.1 Proteolytic Digestion-Based Immunostaining of Mammary Gland Tissues

1. Euthanize mice according to institutional/national guidelines.
2. Excise mammary glands and immerse in Enzyme Digestion solution for 30–60 min (depending on tissue size) at 37 °C with gentle agitation. Due to its small size, high quality wholemount immunostaining can be performed in embryonic mammary gland tissues and explants without the need for enzymatic digestion [45] (Fig. 4).
3. Wash tissues 3 times in HBSS (5 min each time).
4. Fix tissues for 2 h at room temperature in 4% PFA.
5. Wash samples twice in 0.5 M NH_4Cl (10 min each time).
6. Wash tissues 3 times in PBS (10 min each time).
7. Incubate tissues in ED blocking buffer for 3 h at room temperature, or overnight at 4 °C with gentle agitation.
8. Incubate samples in primary antibodies diluted in blocking buffer overnight at room temperature.
9. Wash samples 3 times in PBST (10 min each time).

10. Incubate samples with secondary antibodies diluted in blocking buffer for 5 h at room temperature, or overnight at 4 °C with gentle agitation.
11. Wash samples 3 times in PBST (10 min each time).
12. Incubate tissues with DAPI (10 μM) for 30 min to 1 h at room temperature with gentle agitation to stain nuclei.
13. Mount tissues on microscope slides using Aqua Poly/Mount, taking care to avoid bubbles.
14. Acquire images using CLSM (e.g., Zeiss LSM780/880 or Leica SP8) with a long-working distance objective to facilitate deep tissue imaging (e.g., Zeiss 25×/0.8 oil immersion objective or Leica 25×/0.95 NA water immersion objective).
15. Imaging considerations: Adjust laser power and gain manually to give optimal fluorescence intensity for each fluorophore with minimal photobleaching. Acquire *Z*-stacks using step-sizes and line averaging appropriate for the desired resolution. If performing tile scans, at least a 10% overlap is recommended for optimal stitching of tiled images into larger mosaics (Fig. 4). Stitching can be performed using the microscope's acquisition software, or using ImageJ Plugins (e.g., the Grid Collection/Stitching plugin) (https://imagej.nih.gov/ij/).
16. Process and analyze image stacks using ImageJ (https://imagej.nih.gov/ij/) and/or commercial software such as Imaris (http://www.bitplane.com/), depending on availability.

3.5.2 Modified CUBIC Tissue Clearing and Immunostaining of Mammary Gland Tissues

1. Euthanize mice according to institutional/national guidelines.
2. Excise mammary glands and spread immediately on card (Tetra Pak) or glass microscope slides (*see* **Note 29**).
3. Fix tissues by immersing in 10% NBF for 6–9 h (according to tissue size/thickness) at room temperature (*see* **Note 30**).
4. Wash tissues briefly in PBS with gentle agitation to remove residual NBF. Cut tissues into large (~15 × 15 × 2 mm) pieces if necessary (*see* **Note 31**).
5. Immerse tissues in CUBIC Reagent 1A at 37 °C for 2–3 days, depending on the size of the tissue, exchanging the solution with fresh R1A each day. For timeline, *see* Fig. 5a.
6. Wash samples 3 times in PBS (10 min each time) to remove excess R1A solution. If tissues are to be imaged for genetically encoded fluorescent proteins and do not require immunostaining, proceed to **step 12**.
7. *Optional immunostaining*: Immerse tissues in CUBIC blocking buffer and incubate overnight at 4 °C with gentle agitation.

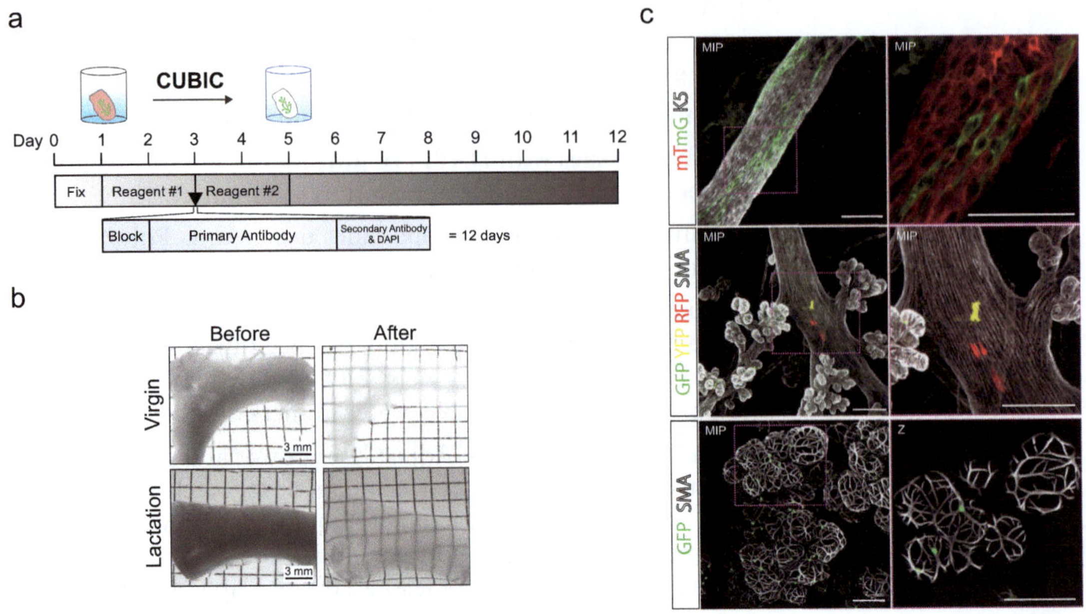

Fig. 5 Clear unobstructed brain imaging cocktails (CUBIC) optical clearing and 3D imaging of mouse mammary tissues. (**a**) CUBIC optical tissue clearing and immunostaining protocol and timeline. Black arrow shows the stage at which (optional) immunostaining may be performed. The experimental timeline may be adapted depending on the desired degree of transparency, and the size and nature of the tissue. (**b**) Transmission images of whole abdominal (fourth) virgin and lactating mammary glands, before and after CUBIC clearing (adapted from [11] with permission from Springer under http:/creativecommons.org/licenses/by/4.0/). (**c**) Example three-dimensional confocal images of cleared mammary tissues immunostained with basal mammary epithelial cell markers (K5 or smooth muscle actin (SMA)) showing compatibility of CUBIC clearing with high-resolution imaging of genetically encoded fluorescent proteins. Top panels show a close-up image of a mammary duct in a virgin *SMA-Cre^{ERT2}; Rosa26-mTmG* reporter mouse. Membrane labeling in recombined (mG, green) mammary basal cells can be observed at high resolutions by CUBIC clearing. Non-recombined cells express membrane-bound Tomato fluorescent protein (mT, red). Middle and bottom panels show 3D images of cleared mammary tissues from involuting and lactating *R26R-Confetti* fluorescent reporter mice, respectively. Reporter expression (nuclear GFP, cytosolic YFP, and cytosolic RFP) is induced at low, sporadic levels. Scale bars: 50 μm (top two panels), 100 μm (middle and bottom panels). MIP, maximum intensity projection; *z*, single optical section

8. Incubate tissues with primary antibodies diluted in CUBIC blocking buffer at 4 °C for 4 days with gentle agitation. Incubation times may be optimized for specific antibodies.
9. Wash tissues 3 times in PBS (1 h each time) with gentle agitation at room temperature.
10. Incubate samples with fluorescent-dye conjugated secondary antibodies made up in PBS for 2 days at 4 °C with gentle agitation (*see* **Note 32**).
11. Wash tissues 3 times in PBS (1 h each time) with gentle agitation at room temperature.

12. Incubate with DAPI (10 μM) for 2–3 h at room temperature with gentle agitation.
13. Transfer samples to CUBIC Reagent 2 and incubate at 37 °C for at least 24 h for refractive index matching before imaging (Fig. 5b).
14. Image samples in CUBIC Reagent 2 by CLSM or multiphoton microscopy within 1 week (e.g., Zeiss LSM780/880 or Leica SP8 equipped with long-working distance objectives to facilitate deep tissue imaging). Samples may be imaged on inverted microscopes using glass-bottomed iBidi dishes, or mounted in CUBIC Reagent 2 for imaging (Fig. 5c) (*see* **Note 33**).
15. Imaging considerations: Adjust laser power and gain manually to give optimal fluorescence intensity for each fluorophore with minimal photobleaching. Acquire *Z*-stacks using step-sizes and line averaging appropriate for the desired resolution. If performing tile scans, at least a 10% overlap is recommended for optimal stitching of tiled images into larger mosaics. Stitching can be performed using the microscope's acquisition software, or using ImageJ Plugins (e.g., the Grid Collection/Stitching plugin) (https://imagej.nih.gov/ij/).
16. If tissues were previously imaging by 4D intravital imaging, place cleared samples in the same orientation for imaging. To retrace regions of interest, some of the strategies suggested for longitudinal IVM may be used (*see* Subheading 3.2, **step 9**), e.g., using morphological landmarks or laser-induced autofluorescent photo-tattoos.
17. Process and analyze image stacks using ImageJ (https://imagej.nih.gov/ij/) and/or commercial software such as Imaris (http://www.bitplane.com/), depending on availability.

3.5.3 SeeDB Tissue Clearing and Immunostaining of Mammary Gland Tissues

1. Euthanize mice according to institutional/national guidelines.
2. Excise mammary glands and spread immediately on card (Tetra Pak) or glass slides (*see* **Note 29**).
3. Fix tissues by immersing in 10% neutral buffered formalin (NBF) for 6–9 h (according to tissue size/thickness) at room temperature (*see* **Note 30**).
4. Subsequently, wash tissues briefly in PBS with gentle agitation to remove residual NBF. Cut tissues into large (~15 × 15 × 2 mm) pieces if necessary (*see* **Note 31**).
5. If tissues are to be imaged for genetically encoded fluorescent proteins and do not require immunostaining, proceed to **step 11**. For timeline, *see* Fig. 6a.
6. *Optional immunostaining*: Immerse tissues in SeeDB blocking buffer overnight at 4 °C with gentle agitation.

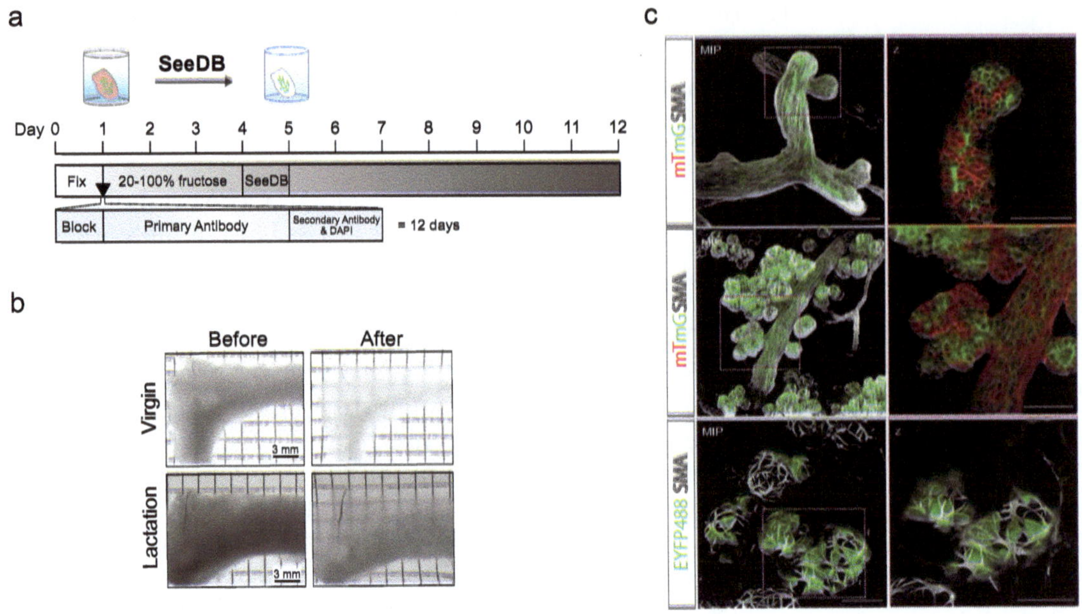

Fig. 6 See deep brain (SeeDB)-clearing and 3D imaging of mouse mammary tissues. (**a**) SeeDB tissue clearing and immunostaining protocol and timeline. Black arrow shows the stage at which (optional) immunostaining may be performed. The experimental timeline may be adapted depending on the desired degree of transparency, and the size and nature of the tissue. (**b**) Transmission images of whole abdominal (fourth) virgin and lactating mammary glands, before and after SeeDB clearing (adapted from [11] with permission from Springer under http:/creativecommons.org/licenses/by/4.0). (**c**) Example three-dimensional confocal images of SeeDB-cleared mammary tissues immunostained with a basal cell marker (SMA) showing compatibility of SeeDB clearing with high-resolution imaging of genetically-encoded fluorescent proteins. Top and middle panels show images of the mammary epithelium in virgin and pregnant *N1-Cre*^{ERT2}*; Rosa26-mTmG* reporter mice, respectively. Membrane labeling in recombined (mG, green) and unrecombined (mT, red) mammary luminal cells can be observed at high resolutions. Bottom panels show images of SeeDB-cleared mammary tissues from a lactating *Rosa26-YFP* mouse model immunostained with a GFP antibody. MIP, maximum intensity projection; *z*, single optical section. Scale bars: 50 μm

7. Incubate tissues with primary antibodies diluted in SeeDB blocking buffer at 4 °C for 4 days with gentle agitation. Incubation times may be optimized for specific antibodies.
8. Wash tissues 3 times in PBS (1 h each time) with gentle agitation at room temperature.
9. Incubate samples with fluorescent-dye conjugated secondary antibodies made up in PBS for 2 days at 4 °C with gentle agitation.
10. Wash tissues 3 times in PBS (1 h each time) with gentle agitation at room temperature.
11. Incubate with DAPI (10 μM) for 2–3 h at room temperature with gentle agitation.

12. Subsequently, serially incubate samples for 8–16 h (twice daily changes, i.e., at beginning and end of the day) at room temperature with gentle agitation in increasing fructose solutions: 20%, 40%, 60%, and 80% (w/v) fructose solutions containing freshly added 0.5% (v/v) α-thioglycerol to inhibit the Maillard reaction [30].
13. Incubate samples in 100% (w/v) fructose solution containing freshly added 0.5% (v/v) α-thioglycerol for 24 h at room temperature.
14. Incubate samples in SeeDB (115% (w/v)) fructose solution containing freshly added 0.5% (v/v) α-thioglycerol for 24 h at room temperature before imaging (Fig. 6b).
15. Image samples in SeeDB reagent by CLSM or multiphoton microscopy within 2 weeks (e.g., Zeiss LSM780/880 or Leica SP8 equipped with long-working distance objectives to ensure deep tissue imaging). Samples may be imaged on inverted microscopes using glass-bottomed iBidi dishes, or mounted in SeeDB reagent for imaging (Fig. 6c) (*see* **Note 33**).
16. Imaging considerations: Adjust laser power and gain manually to give optimal fluorescence intensity for each fluorophore with minimal photobleaching. Acquire *Z*-stacks using step-sizes and line averaging appropriate for the desired resolution. If performing tile scans, at least a 10% overlap is recommended for optimal stitching of tiled images into larger mosaics. Stitching can be performed using the microscope's acquisition software, or using ImageJ Plugins (e.g., the Grid Collection/Stitching plugin) (https://imagej.nih.gov/ij/).
17. If tissues were previously imaging by 4D intravital imaging, place cleared samples in the same orientation for imaging. To retrace regions of interest, some of the strategies suggested for serial IVM may be used, e.g., using morphological landmarks or laser-induced autofluorescent photo-tattoos.
18. Process and analyze *Z*-stack images using ImageJ (https://imagej.nih.gov/ij/) and/or commercial software such as Imaris (http://www.bitplane.com/), depending on availability.

4 Notes

1. This protocol describes a re-usable titanium MIW that is based on the design of the abdominal imaging window [17, 21, 22]. This consists of an upper and a lower titanium ring (outer diameter 14 mm), separated by a 0.9 mm width groove. Titanium possesses superior biocompatibility compared to other stainless alloys. More recent advances in imaging window designs include a titanium version with a replaceable lid [22],

and a silicone-based model consisting of a fully flexible, sutureless design that is ideally suited for long-term IVM of growing tissues and tumors, including the mammary gland [21].

2. Suture material may depend on user preference. Non-absorbable 5-0 nylon and 6-0 silk sutures are also used for imaging window implantation.
3. Dissolve pancreatin from porcine pancreas (P3292-25G, Sigma) and NaCl in MilliQ water using a magnetic stirrer on ice for 3–4 h (or at 4 °C overnight). Centrifuge at approximately 3200 × *g* for 10 min and filter using suction prior to use.
4. Dissolve 0.225 g porcine trypsin (cat. no: 85450C-25G, Sigma) in 6 mL of ice-cold Thyrode's solution using a magnetic stirrer on ice. Once in solution add 1 mL of 10× pancreatin stock solution, in addition to 20 μL of PS. Using NaOH, adjust the pH of the solution to 7.4. Make up the solution to a final volume of 10 mL using additional ice-cold Thyrode's solution. Filter and prepare single-use aliquots.
5. CUBIC Reagent 1A is a modified, unpublished version of CUBIC Reagent 1 [31] available at http://cubic.riken.jp/. CUBIC Reagent 2 has been slightly modified from the original formulation 31 as described in Ref. [11] CUBIC Reagent 1 and 2 require intermittent cycles of heating (60–65 °C, 30 minutes) and agitation (using a magnetic stirring bar and stirrer) to dissolve (typically over 2–3 h).
6. *N*,*N*,*N*′,*N*′-tetrakis(2-hydroxypropyl)ethylenediamine is highly viscous. Dispense this reagent first and adjust other reagent amounts accordingly. In the original protocol [31], the authors state that a specific brand of Triton X-100 (Nacalai Tesque Inc., 25987-85, Japan) is required to avoid quenching endogenous EGFP fluorescence. However, in our hands, Triton X-100 purchased from VWR International is compatible with EGFP fluorescent protein in 10% NBF-fixed mammary gland tissues. If quenching is suspected, consider immunostaining tissues with a GFP antibody, or use an alternative optical tissue clearing method [2, 46].
7. Fructose solutions greater than 60% are difficult to get into solution. Use heat (60 °C) and agitation to dissolve.
8. To avoid contaminating the surgical station, prepare the mouse for surgery in a separate area before transferring to the sterilized operating station. Autoclaved tools should only come into contact with sterile gloves, surfaces, and the disinfected skin area undergoing surgery. To safeguard sterility throughout the procedure, the surgeon should only contact sterile tools and the prepared surgical field. Mouse preparation and

handling, adjustment of anesthesia flow rates, and similar tasks should be performed by an assistant.

9. For optimal results, it is essential that the seal between the titanium ring and coverslip is air and watertight.
10. PEG-based coating of the glass coverslip can be performed to prevent cell attachment and to improve its biocompatibility, limiting the risk of an inflammatory or immune reaction in response to implantation.
11. Ideally, a new set of sterilized tools should be used per mouse. If this is unfeasible, a bead sterilizer may be used to sterilize tools between limited numbers of mice.
12. Isoflurane—which requires oxygen as a carrier gas—is a well-tolerated inhalation anesthetic with minimal side-effects and a short recovery time. Isoflurane induction and maintenance doses may require optimization depending on animal parameters (including strain, age, and condition) and the available setup. It is advisable to use relatively high doses of isoflurane for rapid induction (up to 4%), reducing doses to maintenance levels as quickly as possible to minimize the time exposed to high concentrations [14]. Isoflurane should be handled in well-ventilated areas using systems equipped with a gas scavenger to minimize user exposure to exiting gases.
13. Maintaining the physiological body temperature of the mouse is critical for long-term survival under anesthesia. This can be achieved using adjustable heated anesthesia posts and induction boxes, heat pads, or environmental chambers during intravital imaging. Conversely, hyperthermia may amplify the effect of the inhalation anesthetic, depressing respiration rates. Monitor the mouse's body temperature throughout surgery and intravital imaging using a rectal probe.
14. Buprenorphine may cause mild respiratory depression. Multimodal regimes consisting of Buprenorphine and a NSAID (e.g., Carprofen at a 5 mg/kg) may be recommended by your local veterinarian. However, depending on the nature of the study, suppression of the immune system by NSAIDs may impact important experimental parameters.
15. Gently handle the skin at incision edges when using forceps to avoid compression-associated damage. Consider using non-serrated forceps to minimize this risk. For optimal implantation, the incision size should closely match the size of the window.
16. While the MIW can be implanted over inguinal (third) mammary glands, the fourth is less impacted by respiratory movements during imaging.
17. Post-operative buprenorphine (0.1 mg/kg for up to 3 days) may be administered to provide additional pain relief. NSAIDs

(e.g., Carprofen at a 5 mg/kg dose and/or ibuprofen administered in drinking water) may also be used if appropriate (*see* **Note 14**). Local application of a topical analgesic (e.g. 1% Xylocaine) at the surgical site can be used to provide additional pain relief [17].

18. Endogenous and injectable fluorescent probes may be used to study aspects of mammary gland/tumor cell biology and tissue morphology [10, 47]. For example, blood vessels can be labeled by injecting fluorescently conjugated dextrans into the circulation to investigate vascular flow and permeability, in addition to the invasion of mammary tumorigenic cells into nearby vessels. Vessel labeling can also aid identification of imaging regions in repeated IVM sessions, in addition to the registration of serially acquired image stacks.

19. A winged infusion set attached to a syringe can be used to administer saline during imaging. This can be performed manually, or by using a programmable syringe pump for controlled, continuous administration.

20. An inverted microscope is preferable as it provides better stabilization, reducing image distortions arising from respiration-induced tissue movement. It is important not to compress the underlying tissue and impair blood flow when immobilizing the MIW.

21. These anesthesia levels are optimal for long-term maintenance of mice in a nonresponsive state with a constant and non-forced breathing pattern. Irregular and abnormal breathing patterns are associated with persistent anesthesia greater than 1.5%, which can decrease survival times [14, 41].

22. Due to the limited tolerance of mice to repeated anesthesia, the duration and frequency of imaging sessions in longitudinal experiments should be adapted according to the study parameters under investigation, and in line with institutional/national ethical rules. For instance, if mice are to be anesthetized daily for imaging, this should be restricted to shorter study time periods, with imaging sessions kept as brief as possible to aid recovery.

23. Generating a tile scan of the imaging field using a low-magnification objective also aids retracing of mammary regions of interest.

24. Before E14, there are no obvious morphological differences between female and male mammary embryonic buds. To distinguish the embryos' sex, perform a PCR using the following primer sequences and cycling conditions:

    5′- TGGATGGTGTGGCCAATG -3′, 3′- CACCTG CACGTTGCCCTT-5′

    94 °C for 2 min; then 35 cycles of: 94 °C for 30 s, 58 °C for 30 s, 72 °C for 30 s; finally 72 °C for 5 min.

Run PCR products on a 2% agarose gel. One band is observed for female embryos and two bands for male embryos.

25. Avoid cutting the tissue into small pieces as this will make the next steps challenging.
26. To exchange the medium hereafter, use pipette tips with narrow orifices (e.g., gel loading tips) to carefully remove the medium without contacting mammary tissues.
27. Humidified conditions are required to minimize evaporation of the culture medium during imaging.
28. Of a number of tissue clearing techniques tested in the mammary gland, SeeDB and CUBIC-based protocols provided optimal results [11]. These protocols have subsequently been further developed [48, 49], although they have yet to be tested in mammary tissues. In the event of issues with penetration and staining performance of some antibodies, other tissue clearing protocols are available [46], including FUnGI, a method recently developed for the mammary gland [50].
29. Well-spread tissues are thinner and easier to render transparent. Fibers from card/paper can transfer to tissues after fixation, hampering imaging from that side. Using Tetra Pack card (e.g., milk carton card) overcomes this issue. Alternatively, foam biopsy pads may be used.
30. Fixation time depends on the thickness and size of tissues, and may require antibody-specific optimization. Thinner or smaller tissue pieces and embryonic tissues may be fixed after only 2–3 h at room temperature. A ~10:1 ratio v/v of NBF to tissue is required for adequate fixation. Alternatively, fresh solutions of 4% PFA (for 2–4 h at room temperature) may be used. In our hands, 10% NBF is compatible with several genetically encoded fluorescent proteins, including TdTomato, EGFP, YFP, RFP, and CFP in *Rosa26-mTmG*, *R26R-Confetti* and *Rosa26-tdTomato* reporter mouse strains [11, 32, 33]. If quenching of fluorescent proteins is suspected, consider using methanol-free formaldehyde solutions.
31. While best to process samples immediately after harvesting, fixed tissues can be stored at 4 °C in PBS containing 0.05% (w/v) sodium azide for up to 8 weeks. Cutting samples into large pieces may improve antibody penetration and immunostaining, in addition to allowing more immunostainings to be performed in tissues harvested from the same mouse.
32. CUBIC clearing is also compatible with wholemount immunohistochemistry using HRP-conjugated secondary antibodies and horseradish peroxidase-3,3-diaminobenzidine detection, in addition to the detection of β-glucosidase expression (a magenta histochemical stain) (*see* [11, 32]).

33. Tissues may be mounted between coverslips using iSpacer chambers or concave glass microscope slides. However, in these contexts, samples can be difficult to adjust or repositioned for optimal illumination, making working distance a limiting factor in image acquisition [11]. To mitigate this, sample thickness must be closely matched to the thickness of the iSpacer, or concave chamber, and/or specialized imaging objectives with long-working distances must be used. Alternative refractive index matching solutions may also be considered.

Acknowledgments

We are grateful to members of the Fre laboratory, particularly Dr Silvia Fre (Institut Curie), for support and constructive discussions. Protocols for optical tissue clearing and 3D imaging in the mammary gland were developed in collaboration with Dr Felicity Davis (Mater Research Institute, University of Queensland) and Professor Christine Watson (University of Cambridge). We thank Marie Irondelle and Lucie Sengmanivong (Institut Curie) for intravital imaging support. We are also grateful to the Institut Curie In Vivo Experimental Facility for help in the maintenance and care of mouse colonies. We acknowledge the Cell and Tissue Imaging Platform (PICT-IBiSA) of the Genetics and Developmental Biology Department (UMR3215/U934) and the Nikon Imaging Centre of Institut Curie, member of the French National Research Infrastructure France-BioImaging (ANR10-INBS-04). B.L-L is supported by the Academy of Medical Sciences and the University of Bristol (Vice-Chancellor's Research Fellowship). C.C is supported by funding from the European Union's Horizon 2020 research and innovation program under the Marie Skłodowska-Curie grant agreement No 666003.

Ethics Statement *All animal experimentation were carried out in accordance with the Animal (Scientific Procedures) Act 1986/609 and with local ethics committee approval.*

References

1. Neville MC (2009) Classic studies of mammary development and milk secretion: 1945-1980. J Mammary Gland Biol Neoplasia 14:193–197
2. Lloyd-Lewis B (2020) Multidimensional imaging of mammary gland development: a window into breast form and function. Front Cell Dev Biol 8:203
3. Follain G, Mercier L, Osmani NL et al (2017) Seeing is believing-multi-scale spatio-temporal imaging towards in vivo cell biology. J Cell Sci 130:23. https://doi.org/10.1242/jcs.189001
4. Conchello JA, Lichtman JW (2005) Optical sectioning microscopy. Nat Methods 2:920–931
5. Helmchen F, Denk W (2005) Deep tissue two-photon microscopy. Nat Methods 2:932–940

6. Dunn KW, Young PA (2006) Principles of multiphoton microscopy. Nephron Exp Nephrol 103:e33–e40. https://doi.org/10.1159/000090614
7. Lloyd-Lewis B, Harris OB, Watson CJ, Davis FM (2017) Mammary stem cells: premise, properties, and perspectives. Trends Cell Biol 27:556
8. Fumagalli A, Bruens L, Scheele CLGJ, van Rheenen J (2019) Capturing stem cell behavior using intravital and live cell microscopy. Cold Spring Harb Perspect Biol 12:a035949. https://doi.org/10.1101/cshperspect.a035949
9. Zomer A, Ellenbroek SIJ, Ritsma L et al (2013) Brief report: intravital imaging of cancer stem cell plasticity in mammary tumors. Stem Cells 31:602–606. https://doi.org/10.1002/stem.1296
10. Ellenbroek SIJ, Van Rheenen J (2014) Imaging hallmarks of cancer in living mice. Nat Rev Cancer 14:406–418
11. Lloyd-Lewis B, Davis FM, Harris OB et al (2016) Imaging the mammary gland and mammary tumours in 3D: optical tissue clearing and immunofluorescence methods. Breast Cancer Res 18:127. https://doi.org/10.1186/s13058-016-0754-9
12. Suijkerbuijk SJE, van Rheenen J (2017) From good to bad: intravital imaging of the hijack of physiological processes by cancer cells. Dev Biol 428:328–337
13. Pittet MJ, Weissleder R (2011) Intravital imaging. Cell 147:983–991. https://doi.org/10.1016/j.cell.2011.11.004
14. Ewald AJ, Werb Z, Egeblad M (2011) Preparation of mice for long-term intravital imaging of the mammary gland. Cold Spring Harb Protoc 6:pdb.prot5562. https://doi.org/10.1101/pdb.prot5562
15. Ewald AJ, Werb Z, Egeblad M (2011) Dynamic, long-term in vivo imaging of tumor-stroma interactions in mouse models of breast cancer using spinning-disk confocal microscopy. Cold Spring Harb Protoc 6:pdb.top97. https://doi.org/10.1101/pdb.top97
16. Kedrin D, Gligorijevic B, Wyckoff J et al (2008) Intravital imaging of metastatic behavior through a mammary imaging window. Nat Methods 5:1019–1021. https://doi.org/10.1038/nmeth.1269
17. Alieva M, Ritsma L, Giedt RJ et al (2014) Imaging windows for long-term intravital imaging: general overview and technical insights. Intravital 3:e29917. https://doi.org/10.4161/intv.29917
18. Gligorijevic B, Kedrin D, Segall JE et al (2009) Dendra2 photoswitching through the mammary imaging window. J Vis Exp. https://doi.org/10.3791/1278
19. Jacquemin G, Benavente-Diaz M, Djaber S, Bore A, Dangles-Marie V, Surdez D, Tajbakhsh S, Fre S, Lloyd-Lewis B (2021) Longitudinal high-resolution imaging through a flexible intravital imaging window. Science Advances 7(25) https://doi.org/10.1126/sciadv.abg7663
20. Messal HA, van Rheenen J, Scheele CLGJ (2021) An Intravital Microscopy Toolbox to Study Mammary Gland Dynamics from Cellular Level to Organ Scale. J Mammary Gland Biol Neoplasia 26, 9–27
21. Ritsma L, Steller EJA, Ellenbroek SIJ et al (2013) Surgical implantation of an abdominal imaging window for intravital microscopy. Nat Protoc 8:583. https://doi.org/10.1038/nprot.2013.026
22. Ritsma L, Steller EJA, Beerling E et al (2012) Intravital microscopy through an abdominal imaging window reveals a pre-micrometastasis stage during liver metastasis. Sci Transl Med 4: 158ra145. https://doi.org/10.1126/scitranslmed.3004394
23. Kratochwil K (1969) Organ specificity in mesenchymal induction demonstrated in the embryonic development of the mammary gland of the mouse. Dev Biol 20:46–71. https://doi.org/10.1016/0012-1606(69)90004-9
24. Hens JR, Dann P, Zhang JP et al (2007) BMP4 and PTHrP interact to stimulate ductal outgrowth during embryonic mammary development and to inhibit hair follicle induction. Development 134:1221–1230. https://doi.org/10.1242/dev.000182
25. Voutilainen M, Lindfors PH, Mikkola ML (2013) Protocol: ex vivo culture of mouse embryonic mammary buds. J Mammary Gland Biol Neoplasia 18:239–245. https://doi.org/10.1007/s10911-013-9288-2
26. Scheele CLGJ, Hannezo E, Muraro MJ et al (2017) Identity and dynamics of mammary stem cells during branching morphogenesis. Nature 542:313–317. https://doi.org/10.1038/nature21046
27. Wuidart A, Ousset M, Rulands S et al (2016) Quantitative lineage tracing strategies to resolve multipotency in tissue-specific stem cells. Genes Dev 30:1261–1277. https://doi.org/10.1101/gad.280057.116
28. Lilja AM, Rodilla V, Huyghe M et al (2018) Clonal analysis of Notch1-expressing cells reveals the existence of unipotent stem cells

that retain long-term plasticity in the embryonic mammary gland. Nat Cell Biol 20: 677–687. https://doi.org/10.1038/s41556-018-0108-1
29. Wuidart A, Sifrim A, Fioramonti M et al (2018) Early lineage segregation of multipotent embryonic mammary gland progenitors. Nat Cell Biol 20:666–676. https://doi.org/10.1038/s41556-018-0095-2
30. Ke M-T, Fujimoto S, Imai T (2013) SeeDB: a simple and morphology-preserving optical clearing agent for neuronal circuit reconstruction. Nat Neurosci 16:1154–1161. https://doi.org/10.1038/nn.3447
31. Susaki EA, Tainaka K, Perrin D et al (2014) Whole-brain imaging with single-cell resolution using chemical cocktails and computational analysis. Cell 157:726–739. https://doi.org/10.1016/j.cell.2014.03.042
32. Davis FM, Lloyd-Lewis B, Harris OB et al (2016) Single-cell lineage tracing in the mammary gland reveals stochastic clonal dispersion of stem/progenitor cell progeny. Nat Commun 7:13053. https://doi.org/10.1038/ncomms13053
33. Lloyd-Lewis B, Davis FM, Harris OB et al (2018) Neutral lineage tracing of proliferative embryonic and adult mammary stem/progenitor cells. Development 145:dev164079. https://doi.org/10.1242/dev.164079
34. Elias S, Morgan MA, Bikoff EK, Robertson EJ (2017) Long-lived unipotent Blimp1-positive luminal stem cells drive mammary gland organogenesis throughout adult life. Nat Commun 8:1714. https://doi.org/10.1038/s41467-017-01971-w
35. Hitchcock JR, Hughes K, Harris OB, Watson CJ (2019) Dynamic architectural interplay between leucocytes and mammary epithelial cells. FEBS J 287:250. https://doi.org/10.1111/febs.15126
36. Stewart TA, Hughes K, Hume DA, Davis FM (2019) Developmental stage-specific distribution of macrophages in mouse mammary gland. Front Cell Dev Biol 7:250. https://doi.org/10.3389/fcell.2019.00250
37. Stevenson AJ, Vanwalleghem G, Stewart TA et al (2019) Multiscale activity imaging in the mammary gland reveals how oxytocin enables lactation. bioRxiv:657510. https://doi.org/10.1101/657510
38. Barker N, Van Es JH, Kuipers J et al (2007) Identification of stem cells in small intestine and colon by marker gene Lgr5. Nature 449: 1003–1007. https://doi.org/10.1038/nature06196
39. Muzumdar MD, Tasic B, Miyamichi K et al (2007) A global double-fluorescent Cre reporter mouse. Genesis 45:593–605. https://doi.org/10.1002/dvg.20335
40. Masedunskas A, Milberg O, Porat-Shliom N et al (2012) Intravital microscopy: a practical guide on imaging intracellular structures in live animals. BioArchitecture 2:143–157. https://doi.org/10.4161/bioa.21758
41. Ewald AJ, Werb Z, Egeblad M (2011) Monitoring of vital signs for long-term survival of mice under anesthesia. Cold Spring Harb Protoc 6:pdb.prot.5563. https://doi.org/10.1101/pdb.prot5563
42. Snippert HJ, van der Flier LG, Sato T et al (2010) Intestinal crypt homeostasis results from neutral competition between symmetrically dividing Lgr5 stem cells. Cell 143: 134–144. https://doi.org/10.1016/j.cell.2010.09.016
43. Ritsma L, Vrisekoop N, van Rheenen J (2013) In vivo imaging and histochemistry are combined in the cryosection labelling and intravital microscopy technique. Nat Commun 4:2366. https://doi.org/10.1038/ncomms3366
44. Watson CJ, Khaled WT (2008) Mammary development in the embryo and adult: a journey of morphogenesis and commitment. Development 135:995–1003. https://doi.org/10.1242/dev.005439
45. Panchal H, Wansbury O, Howard BA (2010) Embryonic mammary anlagen analysis using immunolabelling of whole mounts. Methods Mol Biol 585:261–270. https://doi.org/10.1007/978-1-60761-380-0_18
46. Matryba P, Kaczmarek L, Gołąb J (2019) Advances in ex situ tissue optical clearing. Laser Photonics Rev 13:1800292. https://doi.org/10.1002/lpor.201800292
47. Perrin L, Bayarmagnai B, Gligorijevic B (2019) Frontiers in intravital multiphoton microscopy of cancer. Cancer Rep 3:1192. https://doi.org/10.1002/cnr2.1192
48. Tainaka K, Murakami TC, Susaki EA et al (2018) Chemical landscape for tissue clearing based on hydrophilic reagents. Cell Rep 24: 2196–2210.e9. https://doi.org/10.1016/j.celrep.2018.07.056
49. Ke MT, Nakai Y, Fujimoto S et al (2016) Super-resolution mapping of neuronal circuitry with an index-optimized clearing agent. Cell Rep 14:2718–2732. https://doi.org/10.1016/j.celrep.2016.02.057
50. Rios AC, Capaldo BD, Vaillant F et al (2019) Intraclonal plasticity in mammary tumors revealed through large-scale single-cell resolution 3D imaging. Cancer Cell 35:618–632.e6. https://doi.org/10.1016/J.CCELL.2019.02.010

Chapter 3

Single-Cell Transcriptomic and Epigenetic Analyses of Mouse Mammary Development Starting with the Embryo

Zhibo Ma, Nikki K. Lytle, Cynthia Ramos, Razia F. Naeem, and Geoffrey M. Wahl

Abstract

Cancers are caricatures of normal development. Yet, for most organs we are only beginning to learn about the molecular events underlying the embryonic antecedents of organogenesis and when differentiation into the cell types found in the adult actually begins. Here, we will focus on the powerful single-cell RNA sequencing and Assay for Transposase Accessible DNA by DNA sequencing (ATAC-seq) that we and others have been using to decipher the key regulators and signal transduction pathways involved in normal mammary development. We will first describe the techniques we use to identify, dissect, and isolate embryonic mammary rudiments and their constituent cells. We then describe the methods we have employed to perform single-cell RNA-seq and single-nucleus ATAC-seq using the small number of cells obtainable from mouse embryos. Finally, we will discuss the bioinformatic techniques we have used to interpret the vast amount of data obtained with these methods.

Key words Fetal mammary gland, Mammary rudiments, Single-cell RNA-seq, Single-nucleus ATAC-seq, Integrated analysis, Bioinformatic analysis

1 Introduction

Cancer has been referred to as a "caricature of normal development" and the "ever-healing wound that never heals" [1, 2]. Studies over the past decade have strengthened these ideas by adding some molecular detail to these descriptions. This has required a deeper knowledge of the epigenetic, transcriptomic and resulting downstream signaling events underpinning the development of the involved organs. Here, we will focus on how we have used single-cell transcriptomics and epigenetics to study mammary development in the mouse to determine whether breast cancers do in fact reprise some of the molecular pathways employed to generate

Zhibo Ma and Nikki K. Lytle contributed equally with all other contributors.

Maria dM. Vivanco (ed.), *Mammary Stem Cells: Methods and Protocols*,
Methods in Molecular Biology, vol. 2471, https://doi.org/10.1007/978-1-0716-2193-6_3,

bipotential and unipotential progenitors. This is an important and understudied area given that breast cancer is the most common cancer in women, the second leading cause of women's death due to cancer [3], and analyses of the earliest events in mammary development have largely been inaccessible to analysis until only recently.

Mammary glands arose late in evolution, likely as derivatives of skin glands that produced nutritious secretions [4]. Mesenchyme-ectoderm interactions generate a placode, which then thickens and sinks into the underlying dermis to form a mammary bud [5]. The mesenchyme plays a significant role in determining the form and function of the resulting structure. For example, mixing embryonic day 13 (E13) ventral mesenchyme with dorsal ectoderm induces mammary bud formation, a stunning result as mammary buds don't develop in dorsal ectoderm [6]. Subsequent events generate primitive mammary ducts. Importantly, mammary placodes and bud enlargement result from migration of ectodermal cells along one or more Wnt signaling gradients associated with a mesodermal ridge often called the "milk line" and not from local proliferation [7]. Proliferation is undetectable between ~E11.5–14.5, a time corresponding to the formation of the mammary mesenchyme [5, 7]. Proliferation begins ~E15.5 leading to rudimentary sprouting at E16. Recently, intra-amniotic injection of barcoded lentiviruses along with other lineage tracing studies confirm that bipotent mammary cells are present in the embryo [8–11]. However, lineage tracing approaches are insufficient to reveal when bipotency arises, and they cannot reveal the molecular programs leading to bipotency and subsequent commitment to differentiation.

Analysis of early-stage mammary rudiments is complicated by the fragility of the developing skin, difficulty of identifying and separating mammary rudiments, and the small number of cells within mammary rudiments up until about E15. Some of these challenges were overcome in an early bulk transcriptomic analysis of E12.5 rudiments by B. Howard and her colleagues who used microsurgical techniques to separate the mesenchyme from the epithelium and then used microarrays to ascertain the transcriptomes of these early-stage cells [12]. This study was shortly followed by studies from our lab that used both microarrays and the first single-cell analysis using the Fluidigm Biomark microfluidic strategy to analyze the expression of genes in 96 individual cells isolated from E18.5 embryos [13, 14]. This time point was chosen since it was the earliest time in development at which we were able to find evidence of mammary cell bipotency using in vivo limiting dilution transplantation and clonal mammosphere formation as functional assays and able to obtain sufficient cells for single-cell sequencing [14]. Our functional data were also consistent with lineage tracing studies performed at E17 [8]. Recently, single-cell transcriptomes from 69 E14 and 690 (E16) mammary cells were reported [13, 15]. Taken together, these studies provided evidence that most of the individual embryonic mammary cells co-express genes typically found in adult luminal or basal cells. These data support the bipotential nature of cells in the embryo. Importantly, we and others have also found that the transcriptomes of the embryonic mammary cells, which the bulk of current data indicate contain the only bipotent progenitors of the mammary gland, are enriched in human basal-like breast cancers [13, 14, 16].

These early studies begged the question of when mammary embryonic bipotency arises and the molecular pathways that enable mammary to be derived from its skin progenitor. These questions ultimately involve understanding how the transcriptome of the skin changes to generate a mammary transcriptome. The answer to this fundamental question almost certainly requires understanding how the skin epigenome is altered to generate the epigenome programmed to express both luminal and basal mammary transcriptional regulators. These epigenetic changes likely involve the generation of instructive cues from the mesenchyme surrounding the mammary epithelium. The solution to these problems ultimately requires the implementation of single-cell epigenetic and transcriptomic strategies. The remainder of this chapter will describe the methods needed to obtain the cells needed to perform such studies, the molecular strategies employed, and the bioinformatic tools needed to analyze the data obtained.

2 Materials

1. Male and female mice for timed pregnancies, age 8–15 weeks (*see* **Note 1**).
2. Gross anatomy probe (Fine Science Tools, cat. no. 10088-15).
3. Curved blunt forceps (Fine Science Tools, cat. no. 11051-10).
4. Curved blunt scissors (Fine Science Tools, cat. no. 14011-15).
5. Angled spring scissors (Fine Science Tools, cat. no. 15006-09).
6. Dumont #5 Forceps (Fine Science Tools, cat. no. 11251-20).
7. 10 and 6 cm culture dishes.
8. DPBS (Dulbecco's Phosphate-Buffered Saline).
9. 70% Ethanol.
10. Disposable absorbent underpad.
11. Dissecting microscope with upper and lower lighting.
12. Fluorescent dissecting microscope with upper and lower lighting.
13. Low binding 1.5 mL micro tubes.

14. Dissociation Medium: EpiCult-B Basal Medium, 5% FBS, 1× Penicillin-Streptomycin, 1× Amphotericin B.
15. 10× Collagenase/Hyaluronidase (Stemcell Technologies, Catalog #07912).
16. Hydrocortisone: 1 mg/mL in 100% ethanol.
17. Shaking incubator set to 37 °C.
18. 5 U/mL Dispase.
19. DNase I Solution (Stemcell Technologies, Catalog #07900).
20. HF: Hanks' Balanced Salt Solution (HBSS), 2% FBS.
21. 0.8% Ammonium Chloride.
22. CD16/CD32 Rat anti-Mouse, Mouse BD FC Block, Clone 2.4G2.
23. Antibodies for flow cytometry (*see* **Note 8**).
24. Falcon Round-Bottom Polystyrene Test Tubes with Cell Strainer Snap Cap.
25. Optional: Pyrex® 9 Depression Glass Spot Plate (Corning, Catalog #7220-85).

3 Methods

3.1 Production of Timed Pregnant Females

See ref. 17 for suggestions on how to optimize the likelihood of successful timed pregnancies.

1. Combine 1–2 adult females with single-housed males in the evening. For the experiments below, we utilized Sox10-Venus and K14-TdTomato mice that have been backcrossed at least ten generations to CD-1 mice (*see* **Note 1**). To ensure 100% of embryos carry the reporter alleles, we suggest crossing males that are homozygous for both reporter alleles with wild-type CD-1 females. All embryos will be heterozygous for both reporter alleles. Heterozygous and homozygous reporter alleles are easily detected using a dissecting microscope equipped for fluorescence detection (below).
2. Check the females for vaginal plugs the next morning (preferentially 8–12 h after introduction to the males). Separate the females from the males for the remainder of the day.
3. Females with vaginal plugs are considered days post coitum (dpc) 0.5. They are likely pregnant with embryonic day 0.5 (E0.5) embryos. Embryonic time points are calculated from this point on.
4. The presence of a vaginal plug does not guarantee the female is pregnant. To ensure embryos are the age that corresponds to the plug date, females with vaginal plugs are not used in a timed

mating again for at least 14 days, or until it is clear that pregnancy was unsuccessful.

5. Females without a vaginal plug can be used in a timed mating immediately.

3.2 Embryo Collection and Identification of Females

1. At the embryonic time point of interest, euthanize the pregnant dam according to approved protocols. Place the dam on an absorbent disposable pad ventral side up and spray thoroughly with 70% ethanol.
2. Use blunt forceps and curved blunt dissecting scissors to open the skin and body wall over the belly region. Take care not to cut through any of the embryos by pulling up on the skin to create space between the body wall and uterine horn.
3. Use blunt forceps to carefully pull out the mouse uterine horn containing embryos. The uterus will be extensively attached to the dam via mesometrium and a blood vessel network. Carefully release the uterus by cutting away these attachments (*see* Fig. 2 of [18]). Place the uterus (containing embryos) in a 10 cm dish on ice containing ice-cold PBS.
4. Using fine forceps, carefully tear away at the uterine wall and muscle to release each individual embryo, then tear away the placenta from the embryo. Once released, embryos should be carefully handled with large blunt forceps. Early-stage (<E15.5) embryos are particularly fragile and should be scooped, rather than pinched, with blunt forceps. *See* **Note 2** and Figs. 1, 2, and 3, if Sox10-Venus; K14-TdTomato reporter mice were used for timed pregnancy.
5. Embryo sex can be rapidly determined by genital anatomy as early as E16.5 [19]. For later timepoints (E16.5, E17.5, and E18.5), collect the embryos into an empty 10 cm dish. It is helpful to line them up along corresponding sagittal planes, which allows for rapid comparison of genital anatomies between embryos. Place the 10 cm dish containing embryos under a dissecting microscope and use blunt forceps to screen genitalia. Male embryos have larger ano-genital distances and a bulge between the genitals and anus from the developing testis. Male genitalia shape is also distinct from female genitalia and males have a visible urethral seam while females have a urethral meatus [19].
6. For earlier timepoints (prior to E16.5), gender identification by visual inspection can be challenging and often inaccurate. For E15.5 and earlier, embryo sex can be determined by genotyping for the *Sry* gene [20], or by using a X- or Y-chromosome linked fluorescent reporter [21] (*see* **Note 3**).
7. Place all female embryos in a cell culture dish on ice containing ice-cold PBS. Euthanize male embryos by decapitation.

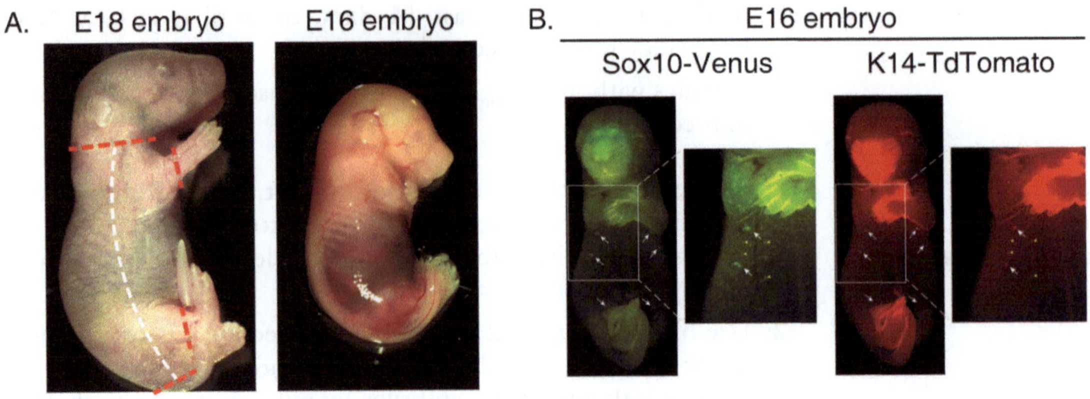

Fig. 1 Late-stage embryo preparation. (**a**) Example brightfield images of E18 and E16 female embryos (lateral view). For mammary rudiment isolation of E15–E18 embryos, flanks are first removed from the rest of the body (*see* Subheading 3.3). The head, tail, and limbs are removed with dissection scissors (red dotted lines), then the skin is carefully cut with fine spring scissors up the lateral sides (white dotted line). The flank is carefully dissected away. (**b**) Example fluorescent images of E16 embryos from Sox10-Venus and K14-TdTomato reporters (ventral view). Both reporters are visible in late-stage embryos and mammary rudiments can be identified in female embryos through the skin (white arrows marking second, third, and fourth mammary rudiments on the left and right sides; first and fifth mammary rudiments are also visible but covered by the limbs in these images). Hair follicles appear as smaller fluorescent puncta in Sox10-Venus and K14-TdTomato reporters (examples indicated by yellow arrowheads)

3.3 Isolation of Late (E15.5 to E18.5) Mammary Rudiments

1. Carefully remove the flank region of each female (Fig. 1, *see* **Notes 4** and **5**). First decapitate the female using large dissection scissors and cut off the tail at the point that the tail meets the body. Remove each limb by cutting in the middle of the humerus or femur for forelimbs and hindlimbs, respectively. Using blunt forceps to stabilize the embryo, carefully cut the skin with fine spring scissors from the opening at the base of embryo (where the tail was removed) up the right lateral side to the top opening (where the head was removed). Repeat on the left lateral side. Using fine forceps, carefully tease the skin flank away from the body wall. Be sure to keep all fat depots attached to the flank, as these will include early mammary rudiments. Place flanks in 6 cm dish on ice containing ice-cold PBS.
2. Lay a flank skin-side down in an empty 6 cm dish bottom (*see* **Note 5**).
3. Using the lower lighting function on the dissecting microscope, identify the location of each of the ten mammary rudiments (Fig. 2).
4. Using fine forceps first release fat depots containing mammary rudiments from body wall and/or muscle so that the mammary rudiment and fat pad are only attached at the nipple. Pinch at

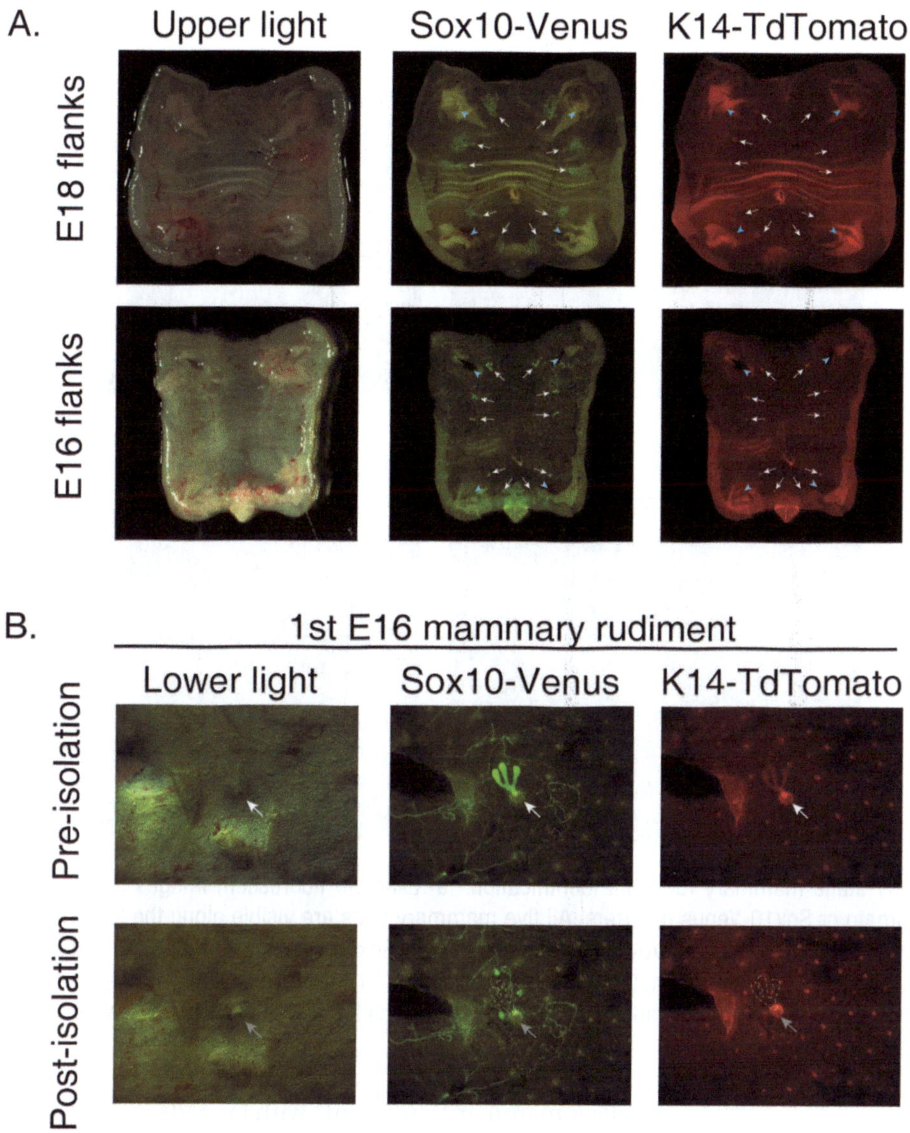

Fig. 2 Late-stage mammary rudiment identification and isolation. (**a**) Example flank images from E18 (top) and E16 (bottom) female embryos: brightfield (left), Sox10-Venus fluorescence (middle), and K14-TdTomato fluorescence (right). Flank is laid on a flat surface with skin facing downward. All 10 mammary rudiments are visible in both reporters in E15–E18 embryos (white arrows). Blue arrowheads indicate limb positions. (**b**) Example high magnification images of first mammary rudiment from E16 flank. In the absence of reporter expression, rudiments can be identified using the lower light function on the dissecting microscope, creating a translucent outline of mammary rudiments (top left, white arrow). However, rudiment identification is expedited by the use of Sox10-Venus (top middle, white arrow) or K14-TdTomato reporters (top right, white arrow). Bottom panels show lack of mammary rudiment after isolation, with only nipple remaining (gray arrows; white dashed lines in lower panels indicate rudiment position before isolation)

the base of the mammary rudiment (right at the nipple) and carefully pull up to release (Fig. 2).

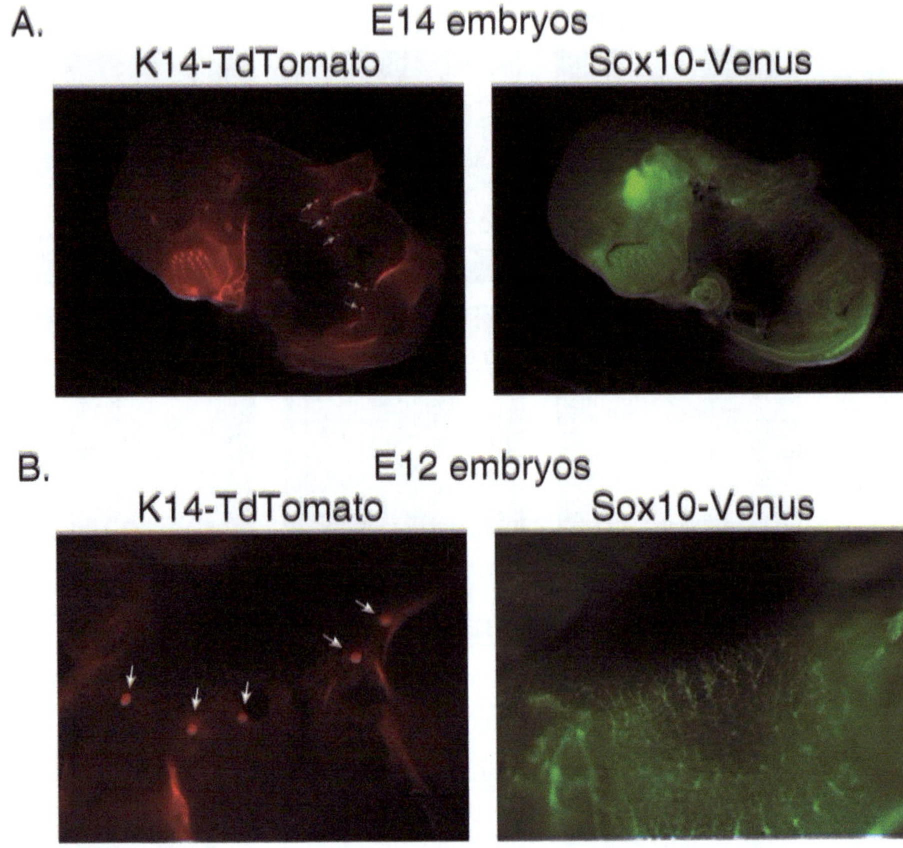

Fig. 3 Early-stage mammary rudiment identification. (**a**) Example fluorescent images of E14 embryos from K14-TdTomato or Sox10-Venus reporters. All five mammary buds are visible along the left lateral side of the K14-TdTomato embryo (white arrows) but absent on the Sox10-Venus embryo. Sox10-Venus expression becomes visible at E15. (**b**) Example fluorescent images of E12 embryos. Similar to E14 embryos, all five mammary buds are fluorescently labeled along each side of the K14-TdTomato embryo (white arrows) but not on the Sox10-Venus embryo

5. Collect mammary rudiments with fat pads in a low bind 1.5 mL tube containing 900 μL ice-cold dissociation media on ice. For E15.5 and E16.5 embryos, collect up to 40 rudiments per tube. For E17.5 and E18.5, collect up to 30 rudiments per tube.

3.4 Isolation of Early (E12.5 to E14.5) Mammary Rudiments

Due to the early developmental stage of the skin, dissecting off the flanks (described in Subheading 3.3, **step 1**) of embryos prior to E15.5 is not feasible. We have developed two alternative methods for collecting rudiments from embryos at these early developmental stages; one approach describes collection through the outer layer of the skin (**steps 1–4** below), and the second describes collection through the inside body wall (**steps 5–10** below).

1. Method 1: mammary bud collection through the outer skin layer (Fig. 3).
2. Move an embryo to a 6 cm dish (*see* **Note 5**). Using curved scissors and blunt forceps, decapitate the embryo and remove the limbs.
3. With the embryo positioned on its side, use the DsRed filter on the fluorescent dissecting microscope to identify the K14-TdTomato+ mammary rudiments. Pinch at the base of each red rudiment with fine forceps and carefully pull up to release. Collect rudiments in low bind 1.5 mL tube containing 900 μL ice-cold dissociation media on ice.
4. For E12.5–14.5 embryos, collect up to 50 rudiments per tube.
5. Method 2: mammary bud collection through the inside body wall.
6. Move an embryo to a 6 cm dish (*see* **Note 5**). Using curved blunt scissors and blunt forceps, decapitate the embryo and cut along the coronal plane from the tail to neck on both sides.
7. Gently hold onto the limbs with blunt forceps. Use fine forceps to remove most of the flesh inside the peritoneum until the body is hollow.
8. Remove the tail and the limbs and lay the hollowed-out carcass outer skin-side down on an empty 6 cm dish (*see* **Note 5**).
9. Using a fluorescent dissecting microscope, identify the K14-TdTomato fluorescent rudiments with the DsRed filter. While keeping the upper red fluorescent light on, turn on the lower brightfield light (illuminating the sample from underneath) to help with visible tracking of forceps position (*see* **Note 6**).
10. Using fine forceps, carefully pick the red rudiments and collect them in low bind 1.5 mL tube containing 900 μL ice-cold dissociation media on ice.

3.5 Mammary Rudiment Dissociation into Single-Cell Suspension

1. After all rudiments have been collected, add 100 μL 10× Collagenase/Hyaluronidase and 1 μL Hydrocortisone per tube.
2. Invert to ensure all rudiments are suspended, then tape tubes on their side in a shaking incubator set at 37 °C.
3. Shake at 120 rpm for 1.5 h for late rudiments and 30 min for early rudiments. Halfway through incubation, invert tubes to ensure rudiments are suspended.
4. Remove tubes from incubator. Carefully triturate 10× with a P1000 set at 1000 μL.

Table 1
Suggested antibody panel used for flow cytometry

Population	Antibody/dye	Other name	Clone	Company	Catalog #	Dilution
Epithelial cells	Alexa Fluor 647 anti-mouse CD326	Ep-CAM	G8.8	Biolegend	118212	1:500
Epithelial cells	APC/Cyanine7 anti-human/mouse CD49f	ITGA6	GoH3	Biolegend	313628	1:500
Immune cells	Brilliant Violet 421 anti-mouse CD45		30-F11	Biolegend	103134	1:500
Erythroid cells	Brilliant Violet 421 anti-mouse TER-119		TER-119	Biolegend	116234	1:500
Endothelial cells	Brilliant Violet 421 anti-mouse CD31	PECAM-1	390	Biolegend	102424	1:500
Dead cells	4′,6′-Diamidino-2-Phenylindole, Dihydrochloride	DAPI		Thermo Fisher	D1306	1:1000

This antibody panel is conducive with mammary rudiments isolated from both K14-TdTomato and Sox10-Venus reporter mice. *See* Fig. 4 for an example gating scheme. Some, but not all, of non-epithelial cell lineages are CD45+, TER-119+, or CD31+. This combination of markers can be stained with overlapping fluorophores (e.g., BV421) for exclusion, and combined with a dead cell exclusion dye in the same channel (e.g., DAPI). Propidium Iodide (Thermo Fisher, P3566; used at 1:1000 dilution) is another common dead cell exclusion dye, but is not optimal for use with TdTomato+ cells. Epithelial cells are Ep-CAM+. We observe evidence of early lineage specification of E18.5 mammary epithelial cells, with luminal-leaning cells expressing Ep-CAM^{hi}/CD49f^{mid} and basal-leaning cells Ep-CAM^{mid}/CD49f^{hi} (*see* Fig. 4). This distinction becomes distinct in adult mammary epithelium. Some stromal cells also express CD49f but are Ep-CAM-negative

5. Centrifuge tubes at 600 RCF for 4 min at 4 °C. A large pellet should be visible. Carefully remove supernatant with a P1000 pipette and discard (*see* **Note** 7).
6. Resuspend pellet in 200 μL dispase containing 15 μL DNase I pre-warmed to 37 °C. Triturate for 3 min at room temperature. Add 800 μL cold HF. Spin 600 RCF for 4 min at 4 °C. Carefully remove supernatant using P1000 and discard.
7. Resuspend pellet in 100 μL HF. Add 400 μL ammonium chloride, briefly vortex, and incubate on ice for 2 min. Add 500 μL HF and spin at 600 RCF for 4 min at 4 °C. Discard supernatant. If specific cell populations will be isolated by FACS prior to sequencing, proceed to **step 8**. If all cells will be used for sequencing, proceed to **step 9**.
8. For cell sorting: Resuspend pellet in 250 μL HF containing 0.5 μL FC block and incubate on ice for 5 min. Add antibodies of interest directly to the tube containing cells and incubate for 20 min on ice protected from light (*see* **Note 8**, Table 1). Wash with 1 mL cold HF and spin at 600 RCF for 4 min at 4 °C. Carefully remove supernatant. Resuspend in 500 μL HF

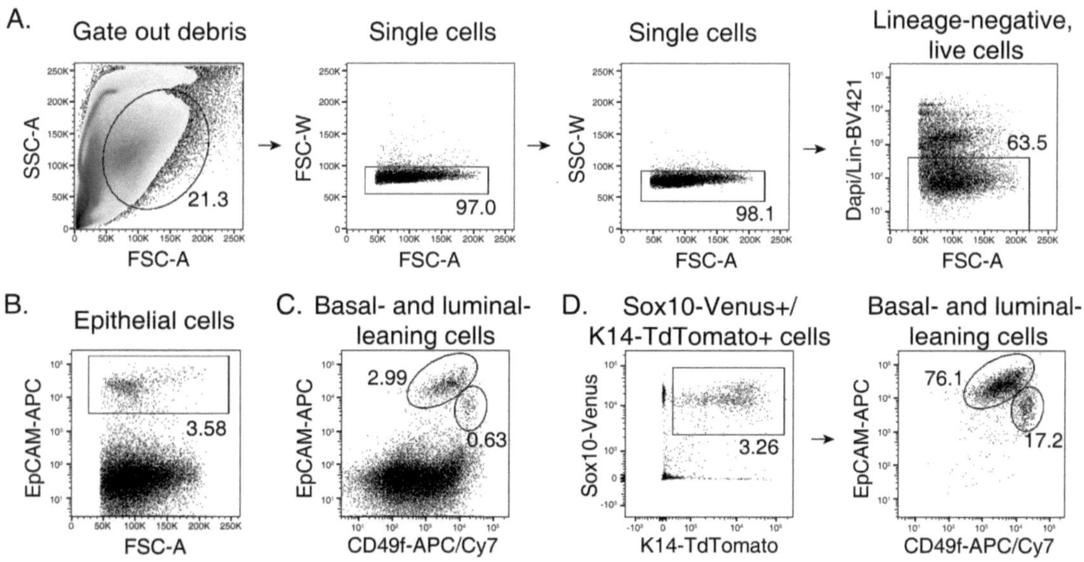

Fig. 4 Example gating scheme for isolation of embryonic mammary epithelium by fluorescent activated cell sorting (FACS). (**a**) Gates to sequentially exclude debris (FSC-A/SSC-A), doublets (FSC-A/FSC-W and FSC-A/SSC-W), and dead or lineage+ (CD45/Ter-119/CD31) cells (FSC-A/BV421). From the lineage-negative, live cell gate, mammary epithelium can be identified by gates in (**b**)–(**d**): (**b**) gate for all Ep-CAM+ epithelial cells, (**c**) gates for luminal-leaning (Ep-CAMhi/CD49fmid) or basal-leaning (Ep-CAMmid/CD49fhi) epithelial cells, or (**d**) gate for Sox10-Venus+/K14-TdTomato+ cells (left), which further enriches for luminal- or basal-leaning epithelial cells (right)

containing dead cell stain (*see* Table 1) and filter through a Falcon round-bottom polystyrene tube with 70 μM cell strainer cap. Cells are now ready for isolation by flow cytometry (Fig. 4, *see* **Note 9**).

9. Resuspend pellet in appropriate buffer/volume according to intended single-cell sequencing platform (*see* Subheading 3.6, Table 2, and **Note 10**).

3.6 Performing RNA/ATAC-seq Using a Preferred Method

With the rapid development of single-cell sequencing technologies, there are many options available for sequencing dissociated single-cells/nuclei. Selection of single-cell sequencing platforms often depends on the actual experimental question, budget, and instrument availability. To aid the selection of sequencing method according to your actual experimental question, we briefly summarize the key features of the two most widely used commercial platforms: the Droplet-based 10× Chromium system and microfluidic plate-based Fluidigm C1 SMARTer-Seq system (*see* Table 2). The 10× Chromium system also supports single-nucleus ATAC-seq and joint profiling of RNA and ATAC in the same cell. Each method has specific requirements for reagents, solutions, instruments, and workflow. Please follow the manufacturer recommended protocol of the selected method for library preparation (*see* **Note 11**).

Table 2
Comparison of 10× Genomics Chromium system and Fluidigm C1 system

	10× Genomics Chromium	Fluidigm C1 (SMARTer-Seq)
Transcript coverage	3′ or 5′ end mRNA	Full length mRNA
Cells captured per run	500–10k (usually 4-6k)	Up to 96
nReads/nUMI per cell	2–5k UMI per cell on average for RNA	500k reads per cell on average[a]
nGenes per cell	500–6000 (vary largely between cells)	4000–9000
RNA selection	polyA	polyA
UMI	Yes	No
Instrument	Chromium Controller	Fluidigm C1
Strength	High cell throughput, could potentially detect rare cell populations	Full length, higher sensitivity on low expression genes
Limitation	Low sensitivity, high dropout	Limited throughput; high cost per cell
Cell size limitation	Up to 30 μm	5–25 μm based on IFC of selection
Visual quality check	No	Microscope exam
Cost	+	++++
Modality	RNA/ATAC/RNA + ATAC	RNA

[a] Largely depends on the sequencing depth

3.7 Processing of scRNA-seq Data

Sequencing results typically come back in a fastq file format. This needs to be aligned to the reference transcriptome and then analyzed for the reads per gene to construct the gene-cell expression matrix for subsequent bioinformatic analyses. Many options are available for processing the raw data into an expression matrix. To help those readers lacking experience in this area, we provide examples of how to start analysis with the primary single-cell RNA-seq data by using the primary 10× scRNAseq-seq data we generated previously [13].

3.7.1 FastQC

After obtaining sequencing data, check the quality of the sequenced reads, for example using FastQC, the most commonly used tool (https://www.bioinformatics.babraham.ac.uk/projects/fastqc/). FastQC provides a modular set of analyses that provide a quick impression of data quality before doing any further analysis.

```
# create a directory to store the results from FastQC:
mkdir fastqc_results
# run FastQC for all fastq files. Run "fastqc -h" for help
fastqc -o fastqc_results -t 4 *.fastq.gz
```

Once the command is complete, you will get two files for each fastq file as output. The html file contains the report.

3.7.2 Running 10× Cell Ranger Pipeline

Fastq files from the 10× Genomics platform can be directly processed using the Cell Ranger pipeline, which uses STAR for read alignment, filtering, barcode counting, UMI counting, feature-barcode matrices construction, and optional secondary analysis. Install the latest version of Cell Ranger and download the latest genome reference from 10× Genomics website (https://support.10xgenomics.com/single-cell-gene-expression/software/pipelines/latest/what-is-cell-ranger). Use cellranger count –help to find out the detailed usage and other available optional arguments.

```
# align four samples from different developmental stages
for sampleName in E16 E18 P4 Adults
do
     cellranger count --id=$sampleName \
     --transcriptome=/opt/refdata-gex-GRCh38-2020-A \
     --fastqs=/directory-to-fastqs/ \
     --sample= $sampleName \
     --localcores=16 \
     --localmem=128
Done
```

Fastq files downloaded from the database usually require renaming before running Cell Ranger (*see* **Note 12**).

3.7.3 Quality Control in Seurat

Quality control is essential for data from 10× cellranger pipeline. Usually the filtered feature-barcode matrix from cellranger is further QCed to remove cells with low quality and then analyzed using other packages, such as Seurat [22], Scater [23], scran [24] in R, and Scanpy [25] in python. We use the Seurat package in R.

1. Import the 10× data into Seurat

```
     library(Seurat)
     library(ggplot2)
     setwd("/path-to-your-working-directory")
     samples <- c("E16","E18","P4","Adults")
     Path <- paste0("path-to-cellranger-output-results-direc-
tory/", samples,"/outs/filtered_feature_bc_matrix")
     # import the filtered expression matrices from cellranger
```

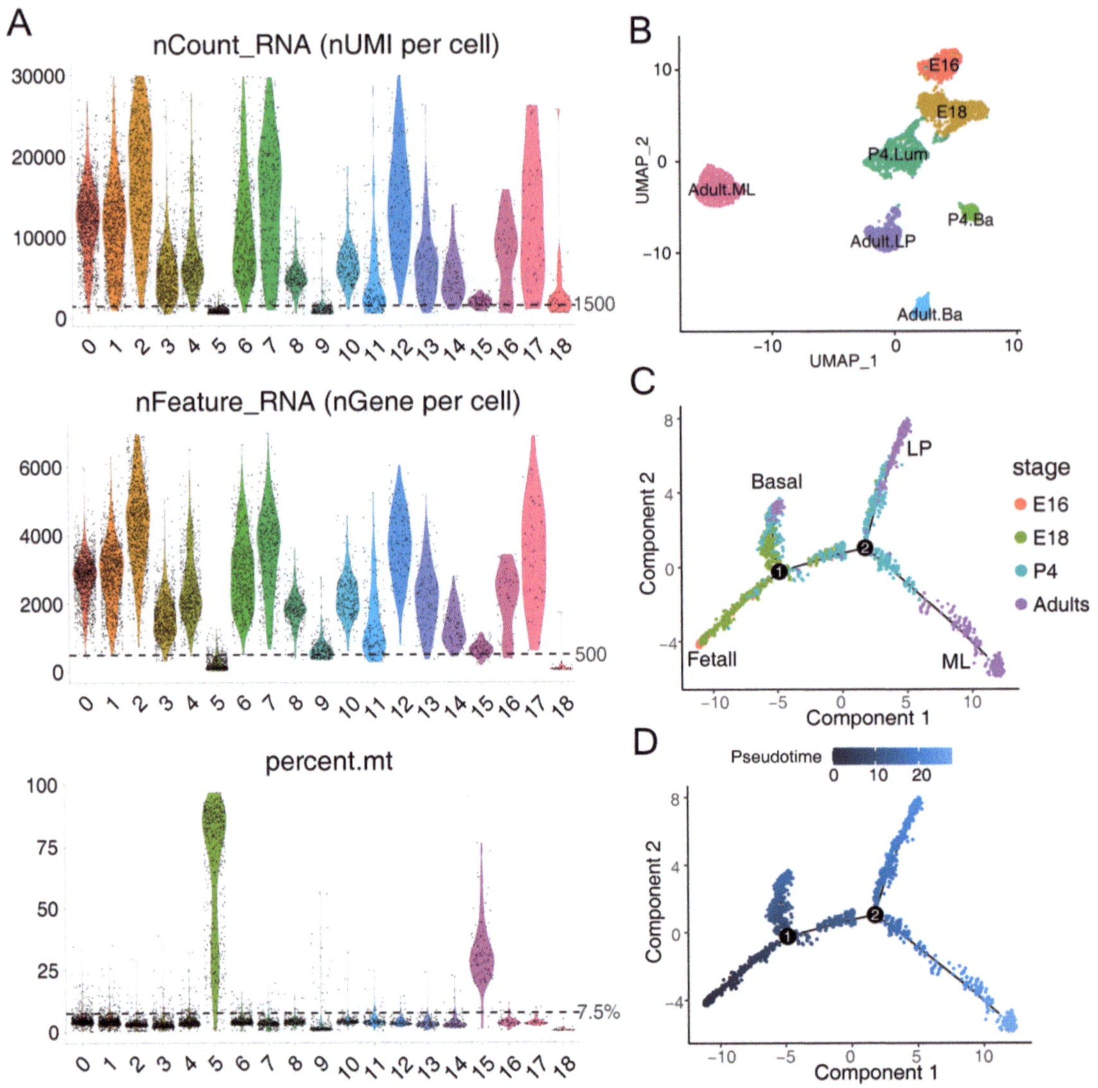

Fig. 5 Example plots for scRNA-seq analysis in Seurat. (**a**) Violin plot of nCount_RNA(number of UMI count per cell), nFeature_RNA(number of genes detected per cell), and percent.mt (percent of mitochondrial reads) across 18 Seurat clusters. The selected QC cutoffs are marked by the dashed line in each plot. (**b**) UMAP plot of all the cells colored by the cluster cell type labels. (**c**) Monocle2 trajectory plot of all cells colored by the developmental stages of samples. (**d**) Monocle2 trajectory plot of all cells overlaid by the value of pseudotime. Trajectory tree is drawn as solid black line and the branching points are marked in the trajectory plot (**c**)–(**d**)

```
ma.list <- lapply(Path, FUN=function(x){
        x <- Read10X(data.dir = x)
        x <- CreateSeuratObject(counts = x
})
# merge the samples to a single Seurat object
   all  <-  merge(ma.list[[1]],  ma.list[2:4],  add.cell.
ids=samples)
```

```
# label the samples in the merged dataset
all$stage <- factor(gsub("_[ATCG]+-1$" ,"" ,colnames(all)), levels = samples)
# add information about mitochondrial reads proportion for QC step
all[["percent.mt"]] <- PercentageFeatureSet(all, pattern = "^mt-")
```

2. Quality control

Three common measures of cell quality are library size, number of genes detected in each cell, and proportion of reads from mitochondrial genes in each cell. Low number of reads/genes and high percentage of mitochondrial reads usually indicate low cell quality [26] (*see* **Note 13**). If the cells have been FACS-selected, then another part of the QC step is removing cell contamination due to imperfect sorting. Removing contaminated cells improves the accuracy of downstream analysis, such as differential gene expression, trajectory inferences, etc.

```
library(dplyr)
# first, cluster the raw data to determine the QC threshold by considering the difference between cell types/clusters. (see Note 14)
all <- NormalizeData(all) %>% FindVariableFeatures() %>% ScaleData() %>% RunPCA()
# plot the Elbowplot to select the significant PCs
ElbowPlot(all, ndims = 50)
all <- RunUMAP(all, dims = 1:40)#use the first 40 PCs
all <- FindNeighbors(all, dims = 1:40) %>% FindClusters()
# examine the cell embedding and clustering on UMAP
DimPlot(all, label = T)
DimPlot(all, label = T, group.by = "stage")
# examine maker genes/features on UMAP to annotate clusters with major cell types
FeaturePlot(all, features = c("percent.mt", "nCount_RNA", "nFeature_RNA", "Epcam", "Trp63", "Elf5", "Foxa1", "Mki67", "Ptprc", "Col1a1", "Rgs5", "Pecam1"), pt.size = 0.1, label = T)
```

Cell types and states can be bioinformatically distinguished based on their expression of marker genes. In this example, the marker genes/features patterns indicate that cluster 11 is an immune cell; 13 is fibroblast + pericyte; 16 is endothelial cell; 17 is embryonic melanocyte; 18 is erythrocyte contamination; cluster 5, 9, and 15 are potential low-quality cells, the rest are mammary epithelium cells to be kept (*see* **Note 15** for the annotation of marker genes/features).

```
# determine the QC thresholds for library size, number of
genes, and proportion of mitochondrial reads (Fig. 5a)
p1 <- VlnPlot(all, features = "nCount_RNA", y.max = 30000)
+geom_hline(yintercept = 1500)
p2 <- VlnPlot(all, features = "nFeature_RNA", y.max =
7000)+geom_hline(yintercept = 500)
p3 <- VlnPlot(all, features = "percent.mt")+geom_hline
(yintercept = 7.5)
ggpubr::ggarrange(p1,p2,p3, ncol=1, nrow=3)
# Subset the data to remove low quality cells (see Note
16).
all <- subset(all, subset = nCount_RNA >=1500 &
nFeature_RNA >= 500 & percent.mt <= 7.5)
# reprocess the data to get new embedding and clustering
(see Note 17).
all <- FindVariableFeatures(all) %>% ScaleData() %>%
RunPCA() %>% RunUMAP(dims=1:30, n.neighbors = 20) %>% Find-
Neighbors(dims=1:30, force.recalc=T, k.param=15) %>% FindClus-
ters()
# plot marker genes/features again to get the cell type
identify of each cluster (not shown). Subset to remove the
stromal cells.
all.epi <- subset(all, idents = c(2,1,6,9,4,5,8,7,3,12,0))
# Plot cells to confirm the accuracy of the cell selection
(see Note 18)
DimPlot(all.epi, cells.highlight = WhichCells(all.epi,
expression = Pmel>0 & Pax3 >0))
# remove the remaining melenocytes
all.epi <- subset(all.epi, cells = WhichCells(all.epi,
expression = Pmel>0 & Pax3 >0, invert = T))
```

3. Reprocess the final QCed mammary epithelium subset

```
all.epi <- FindVariableFeatures(all.epi) %>% ScaleData() %
>% RunPCA() %>% RunUMAP(dims = 1:25) %>% FindNeighbors(dims=1:
25) %>% FindClusters()
# visualize the final embedding and clustering results
(Fig. 5)
DimPlot(all.epi, label = T)+theme(legend.position =
"none")
DimPlot(all.epi, group.by = "stage")
FeaturePlot(all.epi, features = c("Trp63","Elf5","Fox-
a1","Mki67"), pt.size = 0.1, label = T)
```

Cell clusters are subsequently marked by cell type labels according to the marker gene expression patterns (Fig. 5b). This final QCed dataset is now ready for many downstream analyses to address specific experimental questions. Commonly performed downstream analysis include differential gene expression, developmental trajectory inference [27, 28], regulon analysis [29], signaling pathway activity analysis [30], and cell-cell communication analysis [31, 32]. The final QCed raw count-cell matrix and normalized expression matrix could be extracted from the processed Seurat object (raw: all.epi@assays $RNA@counts, normalized: all.epi@assays$RNA@data) and directly used in many other packages.

3.7.4 Marker Gene Identification and Differential Gene Analysis

One routine procedure in single-cell RNA-seq data analysis involves identifying the marker genes of cells belonging to the same cluster. We can conveniently identify markers for all the clusters in Seurat (*see* **Note 19**).

```
# identify marker genes for all the clusters
markers.clusters <- FindAllMarkers(all.epi, logfc.threshold =
1, min.pct = 0.2, min.diff.pct = 0.3, return.thresh = 0.01,
only.pos = T)
```

Differential gene expression analysis can be conveniently done by using the FindMarker() function.

```
# get the markers of two clusters of E18 mammary cells
marker.c6vsc1 <- FindMarkers(all.epi, ident.1 = 6, ident.2 =
1, logfc.threshold = 1, min.pct = 0.2, min.diff.pct = 0.3,
only.pos = T)
```

3.7.5 Pseudotime Trajectory Inference Using monocle2 Package

Our mammary epithelium dataset includes a set of samples collected at different developmental time points from fetal to adult. A developmental trajectory can be constructed from pseudotime ordering of single-cells to help decipher the transcriptomic underpinnings of cell state changes. Pseudotime trajectory analysis algorithms, like monocle2, learn the sequence of gene expression changes that each cell must pass through as part of a dynamic biological process, taking advantage of an individual cell's asynchronous progression [27].

```
# prepare the input data for creating monocle2 cds object (see
Note 20)
counts <- all.epi@assays$RNA@counts
pd <- new("AnnotatedDataFrame", data = all.epi@meta.data)
fd <- new("AnnotatedDataFrame", data = data.frame(gene_short_=
name=rownames(all.epi)))
rownames(fd) <- rownames(all.epi)
# create monocle 2 cds object
cds <- newCellDataSet(cellData = as.matrix(counts), phenoData
```

```
= pd, featureData = fd)
# Estimate size factors and dispersions
cds <- estimateSizeFactors(cds)
cds <- estimateDispersions(cds)
# identify the top 1000 variable genes for pseudotime ordering
(see Note 21).
cds <- detectGenes(cds)
genes.expr <- row.names(subset(fData(cds), num_cells_ex-
pressed >= 10))
DEGs <- differentialGeneTest(cds[genes.expr,], fullModelFor-
mulaStr = "~seurat_clusters",cores = 8)
ordering_genes <- row.names(DEGs)[order(DEGs$qval)][1:1000]
# set the genes for pseudotime ordering in the monocle2 cds
cds <-setOrderingFilter(cds, ordering_genes = ordering_genes)
# build the DDRTree embedding (see Note 22)
cds <- reduceDimension(cds, max_components = 3, method =
"DDRTree",norm_method = "log")
plot_cell_trajectory(cds)
# Visualize sample stage labels on DDRTree embedding
plot_cell_trajectory(cds, color_by = "stage")#(Fig. 5c)
# order cells in pseudotime
cds <- orderCells(cds)
plot_cell_trajectory(cds, color_by = "Pseudotime")
# manually set root state based on actual developmental stage
(see Note 23)
cds <- orderCells(cds, root_state=4)
plot_cell_trajectory(cds, color_by = "Pseudotime")#(Fig. 5d)
```

3.8 Analysis of snATAC-seq Data

Recently a few packages, including ArchR [33], snapATAC [34], Signac [35], cisTopic [36], and cicero [37], have been developed for analyzing single-nucleus ATAC-seq data. We will use ArchR (v1.0.1) to demonstrate the analysis starting from the raw fastq files using our previously generated combinatorial barcoding snATAC-seq data [38] (*see* **Note 24**).

ArchR takes two formats of aligned data generated from different single-nucleus ATAC-seq experiments. One is in the format of fragment files, such as the fragments.tsv file generated by the 10× Genomics cellranger-atac pipeline. It can be generated from BAM files as well; the other is the BAM format, which is the common output from most aligners.

3.8.1 Import Data into ArchR Object (See Note 25)

1. FastQC and reads trimming

   ATAC libraries are sequenced in paired-ends. Similar to scRNA-seq, one should first QC sequencing reads to get an impression of the read quality before any further analysis. FastQC is often used in this step and can be performed in the same way as described in Subheading 3.7.1. Depending on the actual read quality and selected aligner, trimming of Illumina adaptor sequences and low-quality sequences from 3′ end is often performed prior to the alignment (*see* **Note 26**). Recommended tools in their dedicated paired-end modes include: Trimmomatic [39], Cutadapt [40], and Skewer [41].

2. Paired-end reads alignment using bowtie2

```
# create a directory to store aligned bam files
mkdir bam
# align the reads and add the cell barcode in a Tag in bam
for files in $(ls *.R1.fastq.gz)
do
    sampleName <- $(basename($files, .R1.fatq.gz))
   bowtie2 -p 16 -t -X 2000 --no-mixed --no-discordant \
   -x /Path-to-Bowtie2Index \
   -1 ./$[1].R1.fastq.gz \ -2 ./${sampleName}.R2.fastq.gz \
   -S ./bam/${sampleName}.sam
   # store cell barcode from the reads name to a CB tag (see
Note 27)
   samtools view -S ./bam/${sampleName}.sam |\
    awk '/^@/ {print;next} {split($1,bc,":");print $0 "\tCB:
Z:"bc[1]}'|\
   samtools view -bS > ./bam/${filename}.CBTag.bam
   rm ./bam/${sampleName}.sam
done
# sort the bam file by position
  find . -name "*.bam" | parallel -j 0 "samtools sort {} -o
{.}.sorted.bam"
```

3. Create ArchR object

```
library(ArchR)
addArchRThreads(threads = 16)
addArchRGenome("mm10")
 # creates arrow files, this also does pre-QC inspections
(see Note 28)
  bam <- list.files(path = "./bam/",pattern="CBTag",full.
names = T)
  inputFiles <- c(fetal1=bam[3], fetal2=bam[4], adult1=bam
[1],adult2=bam[2])
  ArrowFiles <- createArrowFiles(inputFiles = inputFiles,
sampleNames = names(inputFiles), filterTSS = 4,filterFrags =
1000,addTileMat = TRUE, addGeneScoreMat = TRUE, bcTag = "CB")
 #create the archr object
 projMammary <- ArchRProject(ArrowFiles = ArrowFiles, out-
putDirectory = "Mammary_snATAC", copyArrows = TRUE)
```

Now the data has been imported into ArchR, ready for QC and subsequent analysis.

3.8.2 Quality Control

Similar to the scRNA-seq data, snATAC-seq data also includes many low-quality cells. We often use unique fragments and TSS enrichment score as the major selection criteria. Cells with very few fragments captured will not provide enough data for robust interpretations and are therefore discarded. The TSSEnrichment score has been used as standard workflow for determining signal-to-background ratio (for example, the ENCODE ATACseq pipeline). A very low TSS enrichment often indicates failed ATAC experiments. We also expect to see a nucleosomal periodicity in the distribution of fragment sizes for a successful ATAC-seq experiment.

At Subheading 3.8.1, ArchR generated several plots for QC matrices. Examine and determine the cutoff for each sample (*see* **Note 29**).

```
# subset to remove low quality cells
projMammary <- subsetArchRProject(ArchRProj = projMammary,
cells = projMammary$cellNames[which(projMammary$nFrags >
2000)], outputDirectory = "Mammary_snATAC_QCed",dropCells = T)
```

3.8.3 Dimensionality Reduction with ArchR

Dimensionality reduction with scATAC-seq data is challenging due to the extreme ***sparsity and binary nature*** of the data. In scATAC-seq, a particular site can either be accessible on one allele, both alleles, or neither. With current chemistry, even the high-quality scATAC-seq data fails to capture the majority of accessible regions. This sparsity makes standard dimensionality reduction using a technique such as PCA unsuitable. Latent Semantic Indexing (LSI) has been previously found to be helpful in processing the sparse snATAC-seq data [42]. ArchR stores the accessibility profile of each cell in a binary TileMatrix counted at 500 bp genomic bins. To embed the cells in low dimensions, ArchR implements an iterative version of LSI to remedy the low reproducibility issue when integrating multiple samples [33].

```
# perform LSI, cluster cells, and embed cells in UMAP (see Note
30)
projMammary <- addIterativeLSI(ArchRProj = projMammary, useMa-
trix = "TileMatrix", iterations = 2, clusterParams = list
(resolution = 0.8))
# cluster cells using default Seurat method, optimize resolu-
tion as needed
projMammary<- addClusters(projMammary, resolution = 0.8)
# embedding of cells in UMAP using the default parameters
projMammary <- addUMAP(projMammary)
```

```
# visualize sample and cluster annotations overlaid on UMAP
plotEmbedding(projMammary, colorBy = "cellColData", name =
"Sample", embedding = "UMAP", pal=c("darkgreen", "dark-
green","darkorange", "darkorange"))# (Fig. 6a)
plotEmbedding(ArchRProj = projMammary, colorBy = "cellColDa-
ta", name = "Clusters", embedding = "UMAP")# (Fig. 6b)
```

3.8.4 Explore the Clusters Identify Using ArchR Gene Score

Different from scRNA-seq data, snATAC-seq data captures DNA fragments that are not directly labeled with gene names. To associate our knowledge in cell state associated genes expression with chromatin accessibility, different models of gene activity estimation using chromatin accessibility data have been proposed. ArchR estimates gene activity score using a custom distance-weighted accessibility model (*see* **Note 31**).

```
# identify the marker genes using ArchR gene score
markersGS <- getMarkerFeatures(projMammary, useMatrix = "Gen-
eScoreMatrix", groupBy = "Clusters", bias = c("TSSEnrichment",
"log10(nFrags)"))
markerList <- getMarkers(markersGS, cutOff = "FDR <= 0.01 &
Log2FC >= 1.25")
```

Based on the marker gene activity of each cluster, you may now label the clusters. The dataset may be further filtered to remove cell contaminations, such as the stromal cells (Cluster C3, marked by fibroblast associated genes) in this dataset.

Because of the sparsity of snATAC-seq data, gene scores of many genes often appear quite variable. We can use MAGIC to impute gene scores by smoothing signal across nearby similar cells.

```
projMammary <- addImputeWeights(projMammary)
# visualize MAGIC imputed gene score
plotEmbedding(projMammary, colorBy = "GeneScoreMatrix", name =
"Foxa1", imputeWeights = getImputeWeights(projMammary))
```

3.8.5 Generate Cluster Aggregated ATAC Signal Tracks

In addition to the gene score estimation, we can also browse and compare local chromatin accessibility at marker genes on a per cluster basis in ArchR. To do this, we use the plotBrowserTrack() function.

```
markerGenes <- c("Krt14","Elf5","Kit","Krt8","Fox-
a1","Esr1","Col1a2","Ddr2")
p.tracks <- plotBrowserTrack(projMammary, groupBy = "Clus-
ters", geneSymbol = markerGenes)
# visualize the track plot for a selected gene
grid::grid.newpage() & grid::grid.draw(p.tracks$Krt14)
```

3.8.6 Pseudo-Bulk Replicates and Peak Calling

Because of the binary nature of snATAC-seq data, it is not possible to perform analyses directly on a single-cell. In addition, many statistical tests require replicates to measure significance. Under the assumption that cells in the same cluster are similar enough that as to not have significant underlying differences, ArchR generates pseudo-bulk replicates on a per cluster basis through the addGroupCoverages() function.

```
projMammary <- addGroupCoverages(projMammary, groupBy = "Clus-
ters")
```

3.8.7 Call Peaks in Each Cluster

With these pseudo-bulk replicates generated, we can now call reproducible peaks in our data in a cluster-specific fashion. ArchR provides a wrapper function to use the most commonly used masc2 tool.

```
projMammary <- addReproduciblePeakSet(projMammary, groupBy =
"Clusters"
projMammary <- addPeakMatrix(projMammary)
```

3.8.8 Explore Transcription Motif Deviations Using chromVar TF Activity Score

TF motif enrichments can help us predict which regulatory factors are most active in our cell type of interest. ChromVAR was previously created to estimate TF motif enrichment in each single-cell while accounting for both Tn5 insertion sequence bias and mean accessibility [43]. It uses a deviation *z*-score to show the relative enrichment of TF motif in peak regions in comparison to matched background. The motif activity score can be efficiently calculated in ArchR using the addDeviationsMatrix() function.

add the TF motif annotation using TFs from the CIS-BP database

```
projMammary <- addMotifAnnotations(projMammary, motifSet =
"cisbp", name = "Motif")
# add backgrounds peaks, calculates TF z-score(chromVAR TF
activity score)
projMammary <- addBgdPeaks(projMammary,force = T)
projMammary <- addDeviationsMatrix(projMammary, peakAnnota-
tion = "Motif")
# check activity of known lineage specific TFs
markerTFs <- c("Trp63","Elf5","Foxa1","Sox11)
markerMotifs <- getFeatures(projMammary, select = paste(mar-
kerTFs, collapse="|"), useMatrix = "MotifMatrix")
p <- plotEmbedding(projMammary, colorBy = "MotifMatrix", name
= grep("z:",markerMotifs, value = T), imputeWeights = NULL)#
figure ##-#
```

This creates a TF-cell activity score matrix which can be extracted through getMatrixFromProject(projMammary, useMatrix = "MotifMatrix") function and used in further analysis in other packages, such as pseudotime trajectory construction in monocle2 [27].

3.8.9 Enrichment Analysis of Custom Genome Annotations

Similar to the motif enrichment deviation, we can also check the enrichment of other types of genome annotation, such as ChIP-seq peaks and signature ATAC peaks. Using the unique accessible regions generated from bulk ATAC-seq of different sorted adult mammary lineages, we've successfully identified the early epigenetic priming of E18 mammary cells towards basal-luminal bifurcation (Fig. 6c) [38].

```
# download UAR files, convert to bed format, specify the paths
UARs <- list.files(path = "./UARURR/", pattern = "UAR", full.
names = T)
names(UARs) <- c("Adult.Ba","fMaSC","Adult.LP","Adult.ML")
projMammary <- addPeakAnnotations(projMammary, regions = UARs,
name = "UAR")
projMammary <- addDeviationsMatrix(projMammary, peakAnnota-
tion = "UAR")
# visualize the relative enrichment of UARs
p <- plotEmbedding(projMammary, colorBy = "UARMatrix", name
=paste0("z:",names(UARs), imputeWeights = NULL, quantCut = c
(0.1,0.9), pal = c("grey90","darkmagenta"))
```

3.8.10 Peak-Peak Co-accessibility

All previous analyses of the snATAC-seq data focused either on estimated gene score or solely on peak regions. Majority of the peak regions are located in the intergenic regions, which makes the interpretation of the data challenging. Bulk ATAC-seq usually assigns intergenic peaks to the nearest genes, which is suboptimal considering the long-range interaction properties of distal enhancers. The snATAC-seq data offers a unique opportunity to link distal peaks to its target gene by co-accessibility, which was first introduced in the cicero package [37]. ArchR adopted this approach and create the addCoAccessibility() to compute peak co-accessibility.

```
projMammary <- addCoAccessibility(projMammary)
# visualize the co-accessbility loops in browswer track
p<- plotBrowserTrack(projMammary,groupBy = "Clusters", gene-
Symbol = "Sox10", loops = getCoAccessibility(projMammary,
corCutoff=0.5))
grid::grid.newpage() & grid::grid.draw(p$Sox10)
```

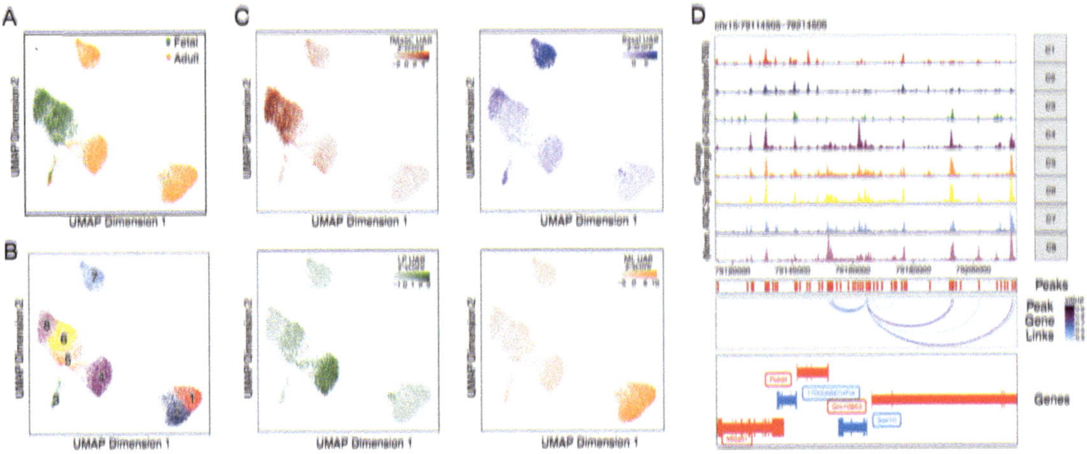

Fig. 6 Example plots for snATAC-seq analysis in ArchR. (**a**) UMAP plot of all cells colored by the developmental stages. (**b**) UMAP plot of all cells colored by the clusters. (**c**) UMAP plot of all cells overlaid with the deviation *z*-score of fMaSC, adult basal, adult luminal progenitor (LP), and adult mature luminal (ML) uniquely accessible regions (UARs). (**d**) Track browser plot in the Sox10 region showing the normalized cluster aggregated ATAC signal, combined peaks, peak-to-gene links, and gene modules

3.9 Integration of scRNA-seq and snATAC-seq Data

Although snATAC-seq data provides a unique opportunity to infer the linkage of a distal regulatory region to its target promoter, this does not necessarily mean there is a regulatory relationship between these peaks because open chromatin is necessary but not sufficient for transcription. In the case where matched scRNA-seq and snATAC-seq are both available, the peak-to-gene linkage could be inferred by leveraging the peak accessibility and gene expression. This peak-to-gene expression linkage may be more relevant to the cis-regulation of genes.

ArchR package also builds a function addPeak2GeneLinks() to conveniently infer such linkages.

1. Add pseudo-scRNA-seq profiles for each cell in scATAC-seq

   ArchR adopted the Seurat label transfer method for integrating scATAC-seq data with scRNA-seq data. By finding anchors between the GeneScore from snATAC-seq data and RNA expression from the scRNA-seq data, each cell in snATAC-seq dataset gets an imputed pseudo-scRNA-seq profile based on the most similar cells from the scRNA expression profile. ArchR supports the integration through one function addGeneIntegrationMatrix().

```
# import processed matching scRNA-seq data then define a developmental stage matched integration
groupList <- SimpleList(E18 = SimpleList(ATAC = projMammary$cellNames[projMammary$Sample %in% c("fetal1","fetal2")], RNA = grep("E18", colnames(all), value = T)), Adults = SimpleList(ATAC = projMammary$cellNames[projMammary$Sample %in% c
```

```
("adult1","adult2")], RNA = grep("Adult", colnames(all), value
= T)))
     # add integration (see Note 32)
       projMammary <- addGeneIntegrationMatrix(projMammary,
useMatrix = "GeneScoreMatrix", matrixName = "GeneIntegration-
Matrix", seRNA = all, groupList = groupList, groupRNA =
"celltype", nameCell = "predictedCell", nameGroup = "predic-
tedGroup", nameScore = "predictedScore")
     # visualize the imputed gene expression
     plotEmbedding(projMammary, colorBy = "GeneIntegrationMa-
trix", name = "Trp63", imputeWeights = NULL)# optional to plot
MAGIC imputed value
     # visualize the predicted cell labels
     plotEmbedding(projMammary, colorBy = "GeneIntegrationMa-
trix", name = "predictedGroup")
```

Optional: You may also use the predictedGroup as a new source of inference for labeling the cells in the snATAC-seq dataset.

2. Add peak-to-gene expression linkage

```
projMammary <- addPeak2GeneLinks(projMammary)
```

This peak-to-gene linkage loop could be directly visualized in tracks plot (Fig. 6d), similar to Subheading 3.8.10.

```
p <- plotBrowserTrack(projMammary,groupBy = "Clusters", gene-
Symbol = "Sox10", loops = getPeak2GeneLinks(projMammary, cor-
Cutoff=0.5))
grid::grid.newpage() & grid::grid.draw(p$Sox10)
```

3. Identification of positive TF regulators using peak-to-gene linkages

One of the challenges in TF motif-based analysis is the similarity of DNA motifs between closely related TFs. The integrated snATAC-seq and scRNA-seq dataset could help identify TFs whose gene expression is positively correlated to changes in the accessibility of their corresponding motif, which are the potential positive regulators. ArchR does this by examining the correlation between imputed pseudo-scRNA expression with chromVAR TF deviation *z*-scores (*see* **Note 33**).

```
# identify deviant TF motifs
seGroupMotif <- getGroupSE(projMammary, useMatrix = "MotifMa-
trix", groupBy = "Clusters")
seZ <- seGroupMotif[rowData(seGroupMotif)$seqnames=="z",]
# identify the max change of TF z-score between all clusters
```

```
rowData(seZ)$maxDelta <- lapply(seq_len(ncol(seZ)), function
(x){
rowMaxs(assay(seZ) - assay(seZ)[,x])
}) %>% Reduce("cbind", .) %>% rowMaxs
# identify correlated TF motifs and TF expression
corGIM_MM <- correlateMatrices(projMammary, useMatrix1 =
GeneIntegrationMatrix", useMatrix = "MotifMatrix")
```

There is no consensus best option in identifying the cell type specific positive regulators. For a demonstration purpose in this example, we consider positive regulators as those TFs whose correlation between motif and imputed gene expression is greater than 0.5 with an adjusted p-value less than 0.01 and a maximum inter-cluster difference of z-score in the first quartile.

```
# identify the positive TF regulators.
corGIM_MM$maxDelta <- rowData(seZ)[match(cor-
GIM_MM$MotifMatrix_name, rowData(seZ)$name), "maxDelta"]
corGIM_MM <- corGIM_MM[order(abs(corGIM_MM$cor), decreas-
ing = TRUE), ]
corGIM_MM <- corGIM_MM[which(!duplicated(gsub("\cr-.*",
"", corGIM_MM[,"MotifMatrix_name"]))),]
corGIM_MM$TFRegulator <- "NO"
corGIM_MM$TFRegulator[which(corGIM_MM$cor > 0.5 & cor-
GIM_MM$padj < 0.01 & corGIM_MM$maxDelta > quantile(cor-
GIM_MM$maxDelta, 0.75))] <- "YES"
regulators <- sort(corGIM_MM[corGIM_MM$TFRegulator=="-
YES",1])
# plot the candidate positive regulators on UMAP
p <- plotEmbedding(projMammary, colorBy = "GeneIntegra-
tionMatrix", name=regulators, imputeWeights = NULL, pal=c
("grey90","darkmagenta"))
```

4 Notes

1. We recommended using CD-1 mice (Charles River, Strain 022) due to their high success rate of timed pregnancy and large litter size. We will be using Sox10-Venus [44] and Keratin14-TdTomato (K14-TdTomato, unpublished) reporter mice that have been backcrossed to CD-1 mice for at least ten generations. Reporter expression aids in early mammary rudiment identification and isolation. Sox10-Venus fluorescence is detectable at E15.5 and K14-TdTomato is detectable as early as E12.5. Both reporters are detectable through adulthood. K14-TdTomato mice were engineered by CRISPR-mediated knock-in of the TdTomato gene into the 3′end of Keratin14

locus. The precise description of this mouse model will appear in a forthcoming manuscript. As an alternative, s-Ship-GFP mice (Jackson Laboratory, Strain 024808) have been previously used for isolation of mammary rudiments as early as E12.0 [45].

2. If homozygous reporter mice were used for timed pregnancy (*see* Subheading 3.1, **step 1**), all embryos will be heterozygous for the Sox10-Venus and K14-TdTomato alleles. If heterozygous reporter mice were used for timed pregnancy, the embryos need to first be screened for reporter expression. This can be done rapidly by collecting embryos in a 10 cm dish and checking whole-embryo fluorescence under low magnification using the fluorescence dissecting scope. *See* Figs. 2 and 3 for examples of embryos expressing each reporter.
3. Male and female mammary gland development is anatomically indistinguishable prior to E14.5, but diverges dramatically in response to a testosterone surge in male embryos on E14.5 [46, 47]. To our knowledge there has been no transcriptomic or epigenetic studies comparing male and female mammary epithelium throughout development, so it is possible there are molecular differences prior to E14.5 and likely substantial differences starting at E14.5. Therefore, depending on the experimental question and approach, it is important to consider if embryo sex needs to be determined so that male and female rudiments can be isolated and analyzed separately. Alternatively, male and female epithelium can be pooled for single-cell analysis then later bioinformatically segregated using Y-chromosome-specific transcripts.
4. E15.5 flanks are extremely fragile but become more durable with every additional day in development. E18.5 flanks are relatively tough.
5. While it is possible to perform flanking and rudiment isolation using 6 cm dishes, it can be helpful to instead use a Pyrex® depression glass spot plate. Placing the embryos in a depression can help stabilize the torso during flanking. Additionally, flanks can be laid on a flat section of the glass plate during rudiment identification and isolation.
6. The rudiments are easier to identify under fluorescent light in a dark room. It is optimal to keep the brightfield light (illuminating under the sample) on a low setting.
7. It is not critical to remove all supernatant during rudiment dissociation. To ensure cells are NOT removed, leave ~50 μL at the bottom of the tube after each spin.
8. In our experience, most antibodies for cell sorting can be used at 1:500 dilution. It is possible some antibodies may require optimization.

9. To minimize cell stress, it is best to sort cells using a 100 μM nozzle and low pressure (20 psi or less).
10. If a plate-based method is chosen for single-cell sequencing, then you can sort directly into the plates.
11. FACS enrichment of cells of interest is often paired with low throughput SMART-Seq methods. Several improved C1 SMART-Seq protocols, including SMART-seq2 and SMART-seq v4, are available at Fluidigm Script Hub (https://www.fluidigm.com/c1openapp/scripthub). When determining the targeted number of cells per sample for droplet-based high throughput methods, it is important to consider whether one is interested in detecting a rare cell population to ensure analyzing sufficient cells. We try to ensure using a large enough sample to obtain at least 300–500 cells per major cell type. This typically requires a total of 5–6k cells per run to avoid a high percentage of barcode collision events.
12. Cell Ranger requires a specific format in fastq file naming. Make sure it is like "SampleName_S#_L###_R1_001.fastq.gz". This is only relevant if your data is downloaded from databases such as SRA, rather than directly generated from running cellranger mkfastq. Check (https://support.10xgenomics.com/single-cell-gene-expression/software/pipelines/latest/using/fastq-input) for details.
13. High proportion of mitochondrial reads does not always mean low cell quality. Some cell types/states have higher percent of mitochondrial reads than others. This has been observed in other tissues, such as gastric epithelium, and some tumor samples. It is important to double check with other matrices such as library size. It is often helpful to QC by checking each sample separately. Sometimes there are sample-to-sample variations in some QC matrices. A threshold should be selected based on the distribution of matrices across cell types or samples. Different thresholds may be applied between different cell types or between different samples if necessary.
14. Some cell types naturally contain less RNA or fewer transcripts when compared to other cell types, such as lymphocytes. Assessing the low threshold without considering each cell type may eliminate a cell type from the data.
15. nCount_RNA: number of UMI; nFeature_RNA: number of genes; Epcam: Epithelium; *Trp63*: Basal; *Elf5*: Luminal Progenitor; *Foxa1*: Mature luminal; *Mki67*: Proliferation; *Pax3*, *Pmel*: melanocytes; *Ptprc*: aka *Cd45*, Immune; *Col1a1*: Fibroblast; *Rgs5*: Pericytes; *Pecam1*: Endothelium. Additional markers from uncertain clusters were further examined by using the FindMarkers() function on specific clusters. Usually, the

top marker genes from a cluster are good indicators of the cell type.

16. Based on the examination, nCount_RNA>=1500; nFeature_RNA>=500; percent.mt<7.5 seems to be a reasonable combination. It's optional to further filter the maximum reads per cell to remove potential doublets. We did not filter the upper range because there aren't many outliers. In addition, cells in this dataset cluster both by cell type and developmental stage. It is unnecessary to separately examine the QC threshold per sample as we would normally recommend.
17. Lowering the n.neighbors in RunUMAP() and k.param in FindNeighbors() helps better label rare cell populations. Tweak to find the optimal combination.
18. Cluster labels are often less accurate on rare cell populations. Plot the marker genes/features again to check if there are significant number of stromal cells remaining. In this dataset, some melanocytes still remain (data not shown). Further remove using marker genes in the following step.
19. Several arguments could be adjusted to control the specificity of the resulting marker geneset, including logfc.threshold, min.pct, min.diff.pct, and return.thresh. There are also many test methods available with this function. Use ?FindAllMarkers() to find the detailed usage documentation. In general, the default Wilcoxon Rank Sum test works well for clusters with a sufficient cell number (while there is no clear cutoff, we are usually more cautious about ascribing a specific cluster when there are less than 50 cells). Starting in Seurat v4, the default base for log fold change is set to 2 instead of the natural base *e* that was used as default in previous versions. Use the base = exp(1) argument if you prefer the behavior in the older Seurat versions.
20. Monocle2 does not support importCDS() from the Seurat newer than version 2. DDRTree trajectory structure works well with mammary gland development. However, due to slow computational speed in large datasets, DDRTree algorithm is not included in the newer monocle3, which supports conversion from Seurat 3 or 4 object.
21. There are many different approaches in determining the genes for pseudotime ordering of cells. Here we use the top 1000 variable genes across the Seurat clusters. Genes expressed in at least ten cells are considered for DEG test. Depending on the actual experiments, this filter could be adjusted with the consideration of both background noise and rare cell population of interest.
22. Good QC is critical in building the pseudotime trajectory. In our experience, removal of non-related cell contamination and

doublets often greatly helps with the trajectory stability. We also find that increasing the embedding max_components to 3 sometimes better preserves the variations and avoids over-simplification of the tree.

23. Inferring the root state in pseudotime trajectory is challenging. It is often manually set in monocle2 based on the known biology of the system. Root state can only be manually identified after the cells have been ordered with orderCells() function.
24. Among the currently available packages, ArchR seems to be the most comprehensive and user-friendly for data analysis and visualization purposes. We also noticed that the gene activity score calculated in ArchR is suboptimal when it is compared to the cicero gene accessibility score.
25. For data from the widely available 10× platform, the fragments.tsv.gz file generated by the 10× cellranger-atac pipeline can be directly used to create arrow files and build ArchR object. Run ?createArrowFiles() for help.
26. ATAC-seq contains many short fragments, which could lead to the presence of Illumina adaptor sequences at the 3′ end. This adapter contamination may prevent the reads from aligning to the reference genome if you are using an aligner without soft-clipping/local alignment. We generally apply adapter trimming in combination with an optional quality trimming with threshold quality score of Q15.
27. The demultiplexed cell barcodes are often stored in the fastq headers. Bam files generated from different pipelines may store the cell barcode at different fields/tags. In this example dataset, the barcode is stored together with the read names. We had to manually add the cell barcode into a tag named "CB". It is important to properly add the barcode into the bam file so that the ArchR could extract the cell barcode information. By default, ArchR assumes qname file stores the barcode information. The createArrowFiles() function supports custom specification of the actual cell barcode field via the bcTag argument. This is critical for loading bam files into ArchR project.
28. Do not set the filtering thresholds (filterTSS and filter Frags) too high in this very first step as you can always further filter later in QC after fully inspecting the data.
29. In this specific dataset, it looks like the default TSSEnrichment score cutoff at 4 is working well. In practice, we often try with different cutoffs and compare the results of cell embedding and clustering. If there is significant sample-to-sample variation in QC matrices, we set the threshold per sample based on the distribution of QC matrices to only keep the majority of the

high-quality cells. We then do subsequent analyses with caution.

30. Cluster resolution needs to be selected based on the heterogeneity of the actual data. This could be selected by running addIterativeLSI() with only 1 iteration, then plot UMAP labeled with clusters calculated from different cluster resolution in addClusters() step. Usually, 2-iterations performs well. Increase the iteration number and/or drop the cluster resolution if you see subtle batch effects downstream. Using a too small resolution value may mask rare cell populations. This step usually needs multiple rounds of optimization in our experience.
31. Another popular gene activity estimation is the gene accessibility score from the cicero package [37]. It uses a different model, which is based on the accessibility of gene promoter region and its co-accessible distal (enhancer) regions.
32. This integration step has different options. Depending on the actual experiments, we could either do unconstrained cell-cell similarity matching or constrained integration by matching known cell groups or samples. In this example, we matched cells from the same developmental stages. In practice, we often do both constrained and unconstrained integration and compare the results. Constrained integration could give better result when the groups are accurately annotated.
33. TFs expressed at very low levels will not work well due to the high dropout rate in most high throughput scRNA-seq experiments. The accuracy of such TFs should be treated with caution and further examined. In the case where scRNA-seq is not available, one can use the estimated ArchR gene score in the place of the imputed pseudo-scRNA expression.

Acknowledgments

This work was supported by a Cancer Center Core Grant (5 P30CA014195), NIH/National Cancer Institute (R35 CA197687), the Susan G. Komen Foundation (SAC110036), and the Breast Cancer Research Foundation (BCRF). NKL was supported by a T32 Cancer Training Grant (5 T32CA9370), Salk Women & Science Research Awards, and the Hope Funds for Cancer Research Postdoctoral Fellowship (HFCR-20-03-03).

References

1. Dvorak HF (1986) Tumors: wounds that do not heal. Similarities between tumor stroma generation and wound healing. N Engl J Med 315(26):1650–1659. https://doi.org/10.1056/NEJM198612253152606

2. Pierce GB, Speers WC (1988) Tumors as caricatures of the process of tissue renewal: prospects for therapy by directing differentiation. Cancer Res 48(8):1996–2004
3. CDC (2020) Basic information about breast cancer. https://www.cdc.gov/cancer/breast/basic_info/index.htm. Accessed May 3 2021
4. Oftedal OT (2002) The mammary gland and its origin during synapsid evolution. J Mammary Gland Biol Neoplasia 7(3):225–252. https://doi.org/10.1023/a:1022896515287
5. Veltmaat JM (2013) Investigating molecular mechanisms of embryonic mammary gland development by bead-implantation in embryonic flank explant cultures - a protocol. J Mammary Gland Biol Neoplasia 18(2):247–252. https://doi.org/10.1007/s10911-013-9297-1
6. Cunha GR, Hom YK (1996) Role of mesenchymal-epithelial interactions in mammary gland development. J Mammary Gland Biol Neoplasia 1(1):21–35
7. Propper AY, Howard BA, Veltmaat JM (2013) Prenatal morphogenesis of mammary glands in mouse and rabbit. J Mammary Gland Biol Neoplasia 18(2):93–104. https://doi.org/10.1007/s10911-013-9298-0
8. Lloyd-Lewis B, Davis FM, Harris OB, Hitchcock JR, Watson CJ (2018) Neutral lineage tracing of proliferative embryonic and adult mammary stem/progenitor cells. Development 145(14):dev164079. https://doi.org/10.1242/dev.164079
9. Lloyd-Lewis B, Harris OB, Watson CJ, Davis FM (2017) Mammary stem cells: premise, properties, and perspectives. Trends Cell Biol 27(8):556–567. https://doi.org/10.1016/j.tcb.2017.04.001
10. Van Keymeulen A, Rocha AS, Ousset M, Beck B, Bouvencourt G, Rock J, Sharma N, Dekoninck S, Blanpain C (2011) Distinct stem cells contribute to mammary gland development and maintenance. Nature 479(7372):189–193. https://doi.org/10.1038/nature10573
11. Ying Z, Beronja S (2020) Embryonic barcoding of equipotent mammary progenitors functionally identifies breast cancer drivers. Cell Stem Cell 26(3):403–419.e404. https://doi.org/10.1016/j.stem.2020.01.009
12. Wansbury O, Mackay A, Kogata N, Mitsopoulos C, Kendrick H, Davidson K, Ruhrberg C, Reis-Filho JS, Smalley MJ, Zvelebil M, Howard BA (2011) Transcriptome analysis of embryonic mammary cells reveals insights into mammary lineage establishment. Breast Cancer Res 13(4):R79. https://doi.org/10.1186/bcr2928
13. Giraddi RR, Chung CY, Heinz RE, Balcioglu O, Novotny M, Trejo CL, Dravis C, Hagos BM, Mehrabad EM, Rodewald LW, Hwang JY, Fan C, Lasken R, Varley KE, Perou CM, Wahl GM, Spike BT (2018) Single-cell transcriptomes distinguish stem cell state changes and lineage specification programs in early mammary gland development. Cell Rep 24(6):1653–1666.e1657. https://doi.org/10.1016/j.celrep.2018.07.025
14. Spike BT, Engle DD, Lin JC, Cheung SK, La J, Wahl GM (2012) A mammary stem cell population identified and characterized in late embryogenesis reveals similarities to human breast cancer. Cell Stem Cell 10(2):183–197. https://doi.org/10.1016/j.stem.2011.12.018
15. Wuidart A, Sifrim A, Fioramonti M, Matsumura S, Brisebarre A, Brown D, Centonze A, Dannau A, Dubois C, Van Keymeulen A, Voet T, Blanpain C (2018) Early lineage segregation of multipotent embryonic mammary gland progenitors. Nat Cell Biol 20(6):666–676. https://doi.org/10.1038/s41556-018-0095-2
16. Zvelebil M, Oliemuller E, Gao Q, Wansbury O, Mackay A, Kendrick H, Smalley MJ, Reis-Filho JS, Howard BA (2013) Embryonic mammary signature subsets are activated in Brca1-/- and basal-like breast cancers. Breast Cancer Res 15(2):R25. https://doi.org/10.1186/bcr3403
17. Yeadon J (2014) 6 Steps for setting up timed pregnant mice. https://www.jax.org/news-and-insights/jax-blog/2014/september/six-steps-for-setting-up-timed-pregnant-mice. Accessed May 3 2021
18. Barak H, Boyle SC (2011) Organ culture and immunostaining of mouse embryonic kidneys. Cold Spring Harb Protoc 2011(1):pdb prot5558. https://doi.org/10.1101/pdb.prot5558
19. Murdaugh LB, Mendoza-Romero HN, Fish EW, Parnell SE (2018) A novel method for determining sex in late term gestational mice based on the external genitalia. PLoS One 13(4):e0194767. https://doi.org/10.1371/journal.pone.0194767
20. Kunieda T, Xian M, Kobayashi E, Imamichi T, Moriwaki K, Toyoda Y (1992) Sexing of mouse preimplantation embryos by detection of Y chromosome-specific sequences using polymerase chain reaction. Biol Reprod 46(4):692–697. https://doi.org/10.1095/biolreprod46.4.692

21. Hadjantonakis AK, Gertsenstein M, Ikawa M, Okabe M, Nagy A (1998) Non-invasive sexing of preimplantation stage mammalian embryos. Nat Genet 19(3):220–222. https://doi.org/10.1038/893
22. Stuart T, Butler A, Hoffman P, Hafemeister C, Papalexi E, Mauck WM III, Hao Y, Stoeckius M, Smibert P, Satija R (2019) Comprehensive integration of single-cell data. Cell 177(7):1888–1902.e1821. https://doi.org/10.1016/j.cell.2019.05.031
23. McCarthy DJ, Campbell KR, Lun AT, Wills QF (2017) Scater: pre-processing, quality control, normalization and visualization of single-cell RNA-seq data in R. Bioinformatics 33(8):1179–1186. https://doi.org/10.1093/bioinformatics/btw777
24. Lun AT, McCarthy DJ, Marioni JC (2016) A step-by-step workflow for low-level analysis of single-cell RNA-seq data with Bioconductor. F1000Res 5:2122. https://doi.org/10.12688/f1000research.9501.2
25. Wolf FA, Angerer P, Theis FJ (2018) SCANPY: large-scale single-cell gene expression data analysis. Genome Biol 19(1):15. https://doi.org/10.1186/s13059-017-1382-0
26. Ilicic T, Kim JK, Kolodziejczyk AA, Bagger FO, McCarthy DJ, Marioni JC, Teichmann SA (2016) Classification of low quality cells from single-cell RNA-seq data. Genome Biol 17:29. https://doi.org/10.1186/s13059-016-0888-1
27. Qiu X, Mao Q, Tang Y, Wang L, Chawla R, Pliner HA, Trapnell C (2017) Reversed graph embedding resolves complex single-cell trajectories. Nat Methods 14(10):979–982. https://doi.org/10.1038/nmeth.4402
28. Street K, Risso D, Fletcher RB, Das D, Ngai J, Yosef N, Purdom E, Dudoit S (2018) Slingshot: cell lineage and pseudotime inference for single-cell transcriptomics. BMC Genomics 19(1):477. https://doi.org/10.1186/s12864-018-4772-0
29. Aibar S, Gonzalez-Blas CB, Moerman T, Huynh-Thu VA, Imrichova H, Hulselmans G, Rambow F, Marine JC, Geurts P, Aerts J, van den Oord J, Atak ZK, Wouters J, Aerts S (2017) SCENIC: single-cell regulatory network inference and clustering. Nat Methods 14(11):1083–1086. https://doi.org/10.1038/nmeth.4463
30. Zhang Y, Zhang Y, Hu J, Zhang J, Guo F, Zhou M, Zhang G, Yu F, Su J (2020) scTPA: a web tool for single-cell transcriptome analysis of pathway activation signatures. Bioinformatics 36(14):4217–4219. https://doi.org/10.1093/bioinformatics/btaa532
31. Efremova M, Vento-Tormo M, Teichmann SA, Vento-Tormo R (2020) CellPhoneDB: inferring cell-cell communication from combined expression of multi-subunit ligand-receptor complexes. Nat Protoc 15(4):1484–1506. https://doi.org/10.1038/s41596-020-0292-x
32. Jin S, Guerrero-Juarez CF, Zhang L, Chang I, Ramos R, Kuan CH, Myung P, Plikus MV, Nie Q (2021) Inference and analysis of cell-cell communication using CellChat. Nat Commun 12(1):1088. https://doi.org/10.1038/s41467-021-21246-9
33. Granja JM, Corces MR, Pierce SE, Bagdatli ST, Choudhry H, Chang HY, Greenleaf WJ (2021) ArchR is a scalable software package for integrative single-cell chromatin accessibility analysis. Nat Genet 53(3):403–411. https://doi.org/10.1038/s41588-021-00790-6
34. Fang R, Preissl S, Li Y, Hou X, Lucero J, Wang X, Motamedi A, Shiau AK, Zhou X, Xie F, Mukamel EA, Zhang K, Zhang Y, Behrens MM, Ecker JR, Ren B (2021) Comprehensive analysis of single cell ATAC-seq data with SnapATAC. Nat Commun 12(1):1337. https://doi.org/10.1038/s41467-021-21583-9
35. Stuart T, Srivastava A, Lareau C, Satija R (2020) Multimodal single-cell chromatin analysis with Signac. bioRxiv:373613. https://doi.org/10.1101/2020.11.09.373613
36. Bravo Gonzalez-Blas C, Minnoye L, Papasokrati D, Aibar S, Hulselmans G, Christiaens V, Davie K, Wouters J, Aerts S (2019) cisTopic: cis-regulatory topic modeling on single-cell ATAC-seq data. Nat Methods 16(5):397–400. https://doi.org/10.1038/s41592-019-0367-1
37. Pliner HA, Packer JS, McFaline-Figueroa JL, Cusanovich DA, Daza RM, Aghamirzaie D, Srivatsan S, Qiu X, Jackson D, Minkina A, Adey AC, Steemers FJ, Shendure J, Trapnell C (2018) Cicero predicts cis-regulatory DNA interactions from single-cell chromatin accessibility data. Mol Cell 71(5):858–871.e858. https://doi.org/10.1016/j.molcel.2018.06.044
38. Chung CY, Ma Z, Dravis C, Preissl S, Poirion O, Luna G, Hou X, Giraddi RR, Ren B, Wahl GM (2019) Single-cell chromatin analysis of mammary gland development reveals cell-state transcriptional regulators and lineage relationships. Cell Rep 29(2):495–510.e496. https://doi.org/10.1016/j.celrep.2019.08.089
39. Bolger AM, Lohse M, Usadel B (2014) Trimmomatic: a flexible trimmer for Illumina sequence data. Bioinformatics 30(15):

2114–2120. https://doi.org/10.1093/bioinformatics/btu170

40. Martin M (2011) Cutadapt removes adapter sequences from high-throughput sequencing reads. EMBnet J 17(1):10–12. https://doi.org/10.14806/ej.17.1.200

41. Jiang H, Lei R, Ding SW, Zhu S (2014) Skewer: a fast and accurate adapter trimmer for next-generation sequencing paired-end reads. BMC Bioinformatics 15:182. https://doi.org/10.1186/1471-2105-15-182

42. Cusanovich DA, Daza R, Adey A, Pliner HA, Christiansen L, Gunderson KL, Steemers FJ, Trapnell C, Shendure J (2015) Multiplex single cell profiling of chromatin accessibility by combinatorial cellular indexing. Science 348(6237):910–914. https://doi.org/10.1126/science.aab1601

43. Schep AN, Wu B, Buenrostro JD, Greenleaf WJ (2017) chromVAR: inferring transcription-factor-associated accessibility from single-cell epigenomic data. Nat Methods 14(10):975–978. https://doi.org/10.1038/nmeth.4401

44. Shibata S, Yasuda A, Renault-Mihara F, Suyama S, Katoh H, Inoue T, Inoue YU, Nagoshi N, Sato M, Nakamura M, Akazawa C, Okano H (2010) Sox10-Venus-mice: a new tool for real-time labeling of neural crest lineage cells and oligodendrocytes. Mol Brain 3:31. https://doi.org/10.1186/1756-6606-3-31

45. Kogata N, Bland P, Tsang M, Oliemuller E, Lowe A, Howard BA (2018) Sox9 regulates cell state and activity of embryonic mouse mammary progenitor cells. Commun Biol 1:228. https://doi.org/10.1038/s42003-018-0215-3

46. Hens JR, Wysolmerski JJ (2005) Key stages of mammary gland development: molecular mechanisms involved in the formation of the embryonic mammary gland. Breast Cancer Res 7(5):220–224. https://doi.org/10.1186/bcr1306

47. Kratochwil K (1977) Development and loss of androgen responsiveness in the embryonic rudiment of the mouse mammary gland. Dev Biol 61(2):358–365. https://doi.org/10.1016/0012-1606(77)90305-0

Chapter 4

Computational Methods to Identify Cell-Fate Determinants, Identity Transcription Factors, and Niche-Induced Signaling Pathways for Stem Cell Research

Muhammad Ali, Mariana Messias Ribeiro, and Antonio del Sol

Abstract

The large-scale development of high-throughput sequencing technologies has not only allowed the generation of reliable omics data related to various regulatory layers but also the development of novel computational models in the field of stem cell research. These computational approaches have enabled the disentangling of a complex interplay between these interrelated layers of regulation by interpreting large quantities of biomedical data in a systematic way. In the context of stem cell research, network modeling of complex gene–gene interactions has been successfully used for understanding the mechanisms underlying stem cell differentiation and cellular conversion. Notably, it has proven helpful for predicting cell-fate determinants and signaling molecules controlling such processes. This chapter will provide an overview of various computational approaches that rely on single-cell and/or bulk RNA sequencing data for elucidating the molecular underpinnings of cell subpopulation identities, lineage specification, and the process of cell-fate decisions. Furthermore, we discuss how these computational methods provide the right framework for computational modeling of biological systems in order to address long-standing challenges in the stem cell field by guiding experimental efforts in stem cell research and regenerative medicine.

Key words Gene regulatory networks, Stem cell research, Cellular reprogramming, Cell-fate determinants, Lineage specifier, Core identity TFs, Systems biology

1 Introduction

The human body comprises different organs, constituted by different cell types, that are supposed to perform dedicated tasks. These cells can be divided into different categories based on, e.g., their ability to differentiate and their function in the body. For example, cells that have the ability to renew themselves and differentiate into any kind of cells are called stem cells—a prominent example is the mammary stem cells (MaSCs) that are known to form the two main

Muhammad Ali and Mariana Messias Ribeiro contributed equally.

Maria dM. Vivanco (ed.), *Mammary Stem Cells: Methods and Protocols*,
Methods in Molecular Biology, vol. 2471, https://doi.org/10.1007/978-1-0716-2193-6_4,

cellular lineages of the human mammary gland [1, 2]. Stem cells further give rise to progenitor cells that have a limited spectrum of differentiation as they can only differentiate into particular specialized cell types. For instance, basal and luminal progenitor cells differentiate into basal/myoepithelial cells that exhibit contractile capacity and luminal cells capable of producing milk, respectively [2]. Although underlying genetic material is the same in each cell of the body, it is the cross-talk between the epigenetic and transcriptional regulatory machinery that controls the identity of different cell types by only allowing specific regions of the genome to be accessible for the transcriptional machinery and being transcribed. Cell identity specification is considered to be determined by cell-specific gene-expression programs that are tightly controlled at the chromatin level. It is widely understood that cell identity is mostly regulated by the action of transcription factors (TFs) that are very particular in their tendency to recognize and bind to specific genomic sequences and regulate transcription [3, 4]. In order for a cell-specific gene-expression program to be expressed, the genomic regions corresponding to the cell type-specific TFs and their distal regulatory regions must be in euchromatin state, harboring active chromatin modifications. Although the number of TFs that are generally expressed in a cell type is considered to be the half of all known TFs [5], a small group of them, commonly known as core TFs, are suggested to be critical for defining the cell identities [6–9].

In some particular cases during cellular differentiation, the expression levels of two inter-dependent genes or TFs determine the fate of the cell through a toggle switch. A toggle switch consists of two genes or TFs that repress each other in a mutual fashion. Usually, this regulatory motif is active during differentiation of a stem/progenitor cell (parent state) into two different lineages (daughter states) and is thought to act as a memory device, being able to choose and maintain cell-fate decisions [10]. In this biomolecular battle, overexpression of each TF corresponds to one of the two mutually exclusive daughter cell fates, where one TF inhibits the other TF and subsequently activates its lineage-determining target TFs and genes. In contrast, a stabilized expression of both TFs maintains the stem/progenitor state [11, 12]. These parent and daughter cellular states are characterized by stable gene-expression programs, determined by underlying gene regulatory networks (GRNs) and the constituent subnetworks (GRN motifs) that are functionally important. A toggle switch is in fact a classical GRN motif that constitutes a molecular mechanism determining the cell-fate decisions and providing stability to transcriptional programs of binary cell-fate choices. A well-known example is the mutual inhibition of an erythroid determinant *GATA1* and a myeloid determinant *SPI1*, two TFs responsible for the development of erythroid and myeloid blood cells from common myeloid progenitors in the hematopoietic stem cell (HSC) system [13, 14].

The natural ability of stem cells to generate more specialized cell types in a coordinated manner and accurately regulate their activity makes them indispensable for maintaining the tissue homeostasis [15]. This intricate decision of either self-renewing or differentiation into different cell fates is controlled by multiple cell-intrinsic and extrinsic factors. One such important factor is the interactions between stem cells and their in vivo microenvironment, also known as a niche. In their niche, stem cells receive the stimuli that determine their behavior, such as maintaining the dormant state (quiescent state), or either induce self-renewal or differentiate into a particular cell fate (active state) [16]. These stimuli include external cues such as cell–cell and cell–matrix interactions which are translated by the niche into intracellular signaling events that activate and/or repress genes and transcriptional programs. The hypothesis of a specialized stem cell niche was postulated by Schofield in 1978 in his description of the hematopoiesis process [17]. He hypothesized that a cellular niche has a defined anatomical location where a stem cell must be associated with other cells in its environment. This environment determines the stem cell behavior and losing its association with the cellular niche results in the differentiation of stem cells. To this end, molecular signaling pathways in the niche are recognized as important modulators of stem cell maintenance and function. These signaling pathways are redundant in different niches but have different roles according to the specific niches. For example, integrins, heterodimeric transmembrane extracellular matrix (ECM) receptors, consisting of various α and β subunits, have been implicated in the control of stem cell maintenance and progenitor cell differentiation in various embryonic and adult tissues [18, 19]. In particular, α2-, α6-, β1-, and β3-integrin subunits have been shown to serve as surface markers for the isolation of MaSCs populations from basal and luminal mammary epithelial layers, suggesting the important role of ECM in MaSCs microenvironment [20, 21]. Therefore, elucidation of molecular signaling pathways that maintain different stem cell niches, cellular phenotypes, and integrity of the organism is of broad interest and paramount to the design of effective treatments for various human pathologies that are associated with defects in the normal cellular differentiation process.

Challenges exist to identify these properties and single-cell technologies can help with the modeling for addressing these challenges. During the last decade, various experimental techniques have enabled the large-scale generation of high-throughput biological data across different regulatory levels (genomics, epigenomics, and transcriptomics) that led to the development of computational approaches in the field of systems biology and stem cell research. The development of these genome-wide transcriptomic and epigenetic profiling techniques has enabled more generic endeavors to predict candidate TFs that control cell identity. In this regard, computational methods have been developed to perform genome-wide transcriptomic and/or epigenomic analysis across multiple cell types to identify core TFs critical for cell identity TFs [3, 7, 22–26]. For example, D'Alessio et al. [3] presented a computational approach that relies on transcriptomic data for predicting cell identity TFs and generated an atlas of candidate core TFs for a variety of different human cell types and tissues. Briefly, this computational method searches for the two fundamental characteristics of core TFs: cell-type-specificity and relatively high expression levels. The algorithm they proposed quantifies both these properties by using an entropy-based measure of Jensen–Shannon divergence [27]. They also demonstrated the experimental validation of their core TFs predictions made for reprogramming human fibroblasts into functional retinal pigment epithelial (RPE) cells that possess the morphological and gene-expression features similar to those derived from healthy individuals. Although many computational methods have addressed this challenge of identifying cell identity TFs and several of them have also experimentally verified their predictive power, they all share some important limitations. Foremost, they are unable to systematically integrate the regulatory information from the epigenetic and transcriptomic levels, which has been shown to mediate cell-type-specific gene-expression programs via intricate and interconnected regulatory links as well as controlling the homeostasis of differentiated or pluripotent cells [28, 29]. Most of these methods rely only on transcriptomic data and ignore the fact that cell-type-specificity is also determined by the epigenetic program which is characterized by accessible chromatin regions, active enhancers, and differential binding of regulators [30–32]. Furthermore, these global attempts are broad in their scope as they assess their predictive power using scalable methods and do not systematically evaluate whether predicted factors are sufficient to establish cell identity. In addition, most of the existing methods focus mainly on quantifying the differences between different cell identities and less on the direct identification of TFs and co-factors controlling cell identity.

Similarly, several computational approaches use transcriptomics data for predicting cell-fate determinants during the cellular differentiation process [6, 22, 26, 33, 34]. For example, Okawa et al. presented a computational framework that models the cell differentiation process in the form of GRNs and predicts cell-fate determinants and their GRN motifs [34]. The proposed tool predicted the overexpression of *Esr1* and *Runx2* for the induction of neuronal and astrocytic lineages, respectively, from the mouse neural stem cells (NSCs). Their in silico predictions were also experimentally verified, showing the application of the tool in stem cell research and regenerative medicine. However, similar to the computational methods aiming for predicting cell identity TFs, these methods are also limited in their usage of bulk transcriptome data while undermining the crucial information from the epigenetic layer of regulation. In addition, none of these methods systematically addresses the problem of cell differentiation efficiency and fidelity, a long-standing challenge in stem cell biology. Unlike most of these approaches that rely on bulk transcriptomic data and large quantities of background or training data sets, only a couple of these methods are able to predict cell-fate determinants at an increased resolution of different cell subpopulations using the single-cell transcriptomic data [35, 36].

The advancements in single-cell sequencing techniques have offered the quantification of expression levels of genes and TFs at single-cell resolution, allowing more reliable measurements of absolute gene-expression levels for a given cell type and its subpopulations. This revolutionary technique has opened new gateways for the development of computational methods that can elucidate complex molecular interaction networks and predict lineage specifiers within a heterogeneous cell population. Similarly, the identification of conserved signaling pathways that provide an accurate description of individual cell behavior while being in a heterogeneous niche requires the accurate characterization of all cellular subpopulations. To this end, sensitive full-length single-cell transcriptome profiling will serve as a basis for the development of single cell-based computational frameworks that can identify niche determinants of different stem cell systems. As the key role of the cell niche in health as well as in several infectious [37] and degenerative diseases [38] is evident, the identification of niche determinants holds the potential to further advance our understanding of underlying disease mechanisms and it can possibly aid in the development of novel therapeutic strategies. Moreover, modeling core GRNs using single-cell data could allow the identification of subpopulations with the highest conversion propensity, thus helping in overcoming the barrier of limited efficiency in directed cellular conversion experiments. Furthermore, single-cell data can help in designing novel experimental strategies to achieve more efficient cellular conversions by predicting cell identity TFs that can initially prime a cell population and then subsequently induce the desired cell type conversion. Thus, guiding experimental attempts for achieving effective in vivo cellular transitions, where limited conversion efficiency is a critical barrier for its application in regenerative medicine.

An overview of currently available computational tools designed to address the above-mentioned three major challenges in stem cell biology is provided in Table 1. Also, a detailed description of state-of-the-art single cell-based computational methods

Table 1
An overview of currently available computational tools designed to use bulk and single-cell transcriptomic and/or epigenetic datasets for addressing challenges in stem cell research, such as predicting cell-fate determinants, lineage specifiers, and identity TFs required for cellular conversions

Method name	Input data	Description	Reference
CellNet	Gene-expression (microarray-based) data from 56 published reports and predefined GRNs	A network biology platform designed to assess the fidelity of cellular engineering and generate hypotheses for improving cell derivations.	[7]
Mogrify	Gene-expression data of 173 human cell types and 134 tissues	A computational framework that combines gene-expression data with regulatory network information to predict the reprogramming factors necessary to induce desired cellular conversion.	[6]
D'Alessio et al.	504 gene-expression profiles, representing 106 cell and tissues types	A computational approach for identifying candidate TFs that control cell identity.	[3]
Crespo et al.	Gene-expression data and predefined GRNs	A computational tool that generalizes the concept of TF cross-repression to predict core TFs for cell conversion.	[39]
Davis et al.	Gene-expression data and chromatin modification ChIP-seq profiles	A computational method for predicting TFs that convert adult cell identity.	[25]
SLICE	Single-cell RNA-seq (scRNA-seq) data	A computational tool that quantitatively measures cellular differentiation states based on single-cell entropy and predicts cell differentiation lineages.	[36]
SeesawPred	Gene-expression data and predefined GRNs	A web application for predicting cell-fate determinants from transcriptomics data.	[33]
Okawa et al.	Gene-expression data and predefined GRNs	A computational algorithm for predicting cell-fate determinants and their GRN motifs.	[34]
Okawa et al.	scRNA-seq data and predefined transcriptional regulatory network (TRNs)	A computational tool for predicting lineage specifiers for different cell subpopulations in binary-fate differentiation events.	[35]
Ravichandran et al.	Gene-expression data, predefined TRNs, and signaling interactome network	A computational approach for identifying niche determinants of cellular phenotypes.	[40]
TransSyn	Single-cell RNA-seq (scRNA-seq) data	A computational platform for the identification of cell population	[41]

(continued)

Table 1
(continued)

Method name	Input data	Description	Reference
		identities by defining their synergistic transcriptional cores.	
SigHotSpotter	Single-cell RNA-seq (scRNA-seq) data and signaling interactome network	A computational web application that predicts key signaling molecules (hotspots) responsible for controlling cell phenotype.	[42]

developed to address these challenges is provided below with their schematic workflow described in Fig. 1.

Cellular differentiation is an intricate process where a less specialized cell type (a stem/progenitor cell, e.g., myoepithelial progenitor cell) evolves into another more specialized cell type (a specific cell, e.g., myoepithelial cell). Despite the recent advances in transcriptomic profiling techniques and increasing research efforts, identification of TFs that determine cell fates during stem cell differentiation (cell-fate determinants) remains a challenge. Identification of cell-fate determinants becomes more challenging for closely related daughter cell types that originate from a common progenitor state such as differentiation of luminal progenitor cells to ductal and alveolar cells [43]. Furthermore, the existence of different cell subpopulations within a particular cell population also hampers our understanding of cellular differentiation processes because of closely related but different gene-expression programs and underlying transcriptional regulatory networks (TRNs) that characterize these states. Therefore, the development of appropriate computational tools is absolutely necessary to utilize the single-cell transcriptomic data for characterizing subpopulation-specific TRNs and identifying TFs that determine specific lineage commitment.

The increasing importance of single-cell gene-expression data in stem cell biology has triggered the development of computational approaches that provide TRN-based modeling of stem cell differentiation at the subpopulation-specific levels and identify lineage specifiers for different cell subpopulations. One prominent example of such a computational framework is the SeesawPred (*see* **Note 1**) [33, 35], a web application that models the cell differentiation process in the form of gene regulatory networks (GRN) and predicts cell-fate determinants. One important feature of this computational method is its flexibility to take into account the single-cell as well as the bulk transcriptomic data for predicting the cell-fate determinants. The application of this computational method to three different stem cell systems predicted already known and experimentally validated lineage specifiers. In addition,

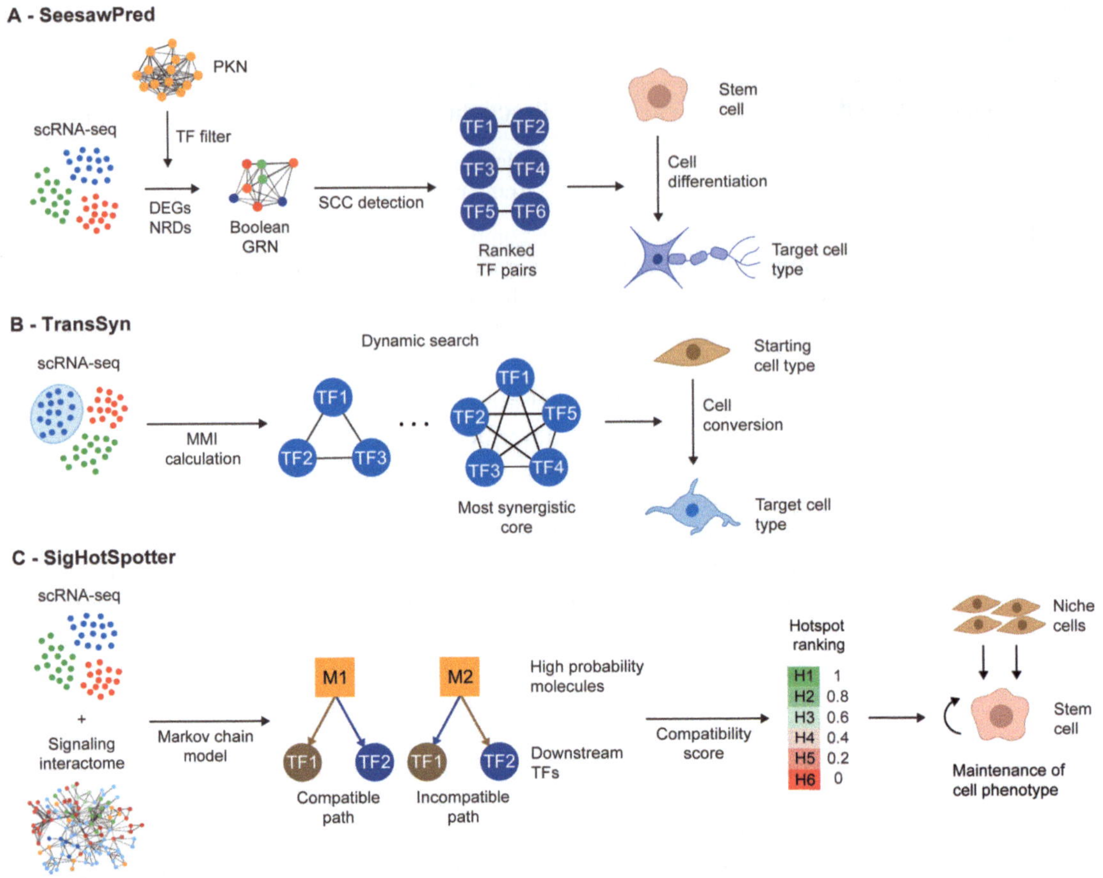

Fig. 1 Workflow of single cell-based computational methods for addressing current challenges in stem cell research. (**a**) SeesawPred workflow overview. This method uses scRNA-seq and a prior knowledge network (PKN) to predict cell-fate determinants, based on a GRN model of cell differentiation. Differentially expressed TFs (DEGs) are identified and the normalized ratio difference (NRD) is calculated for each TF pair. TFs are filtered based on the NRD score and then subdivided based on the provided PKN. The predicted network is pruned using the Boolean GRN formalism and the SCCs identified. Based on the identified SCCs, the final score for each TF pair is calculated and the TF pairs are ranked. The highest-ranking TF pairs are classified as cell-fate determinants and can then be applied in cell differentiation protocols. (**b**) TransSyn workflow overview. This method identifies the most synergistic transcriptional cores of a given population based solely on scRNA-seq data. Based on multivariate mutual information (MMI), TransSyn performs a dynamic search for the most synergistic core by progressively adding TFs to the core, one at a time. The search finishes when the MMI no longer decreases when adding a new TF to the current combination (there is no more increase in the synergy). This last TF combination is identified as the most synergistic core and it can then be applied to cell conversion protocols. (**c**) SigHotSpotter workflow overview. This method predicts key signaling molecules (hotspots) by integrating scRNA-seq data and a pre-compiled signaling interactome in a probabilistic Markov chain model. Based on the compatibility score of the predicted high probability molecules with the respective downstream TFs, SigHotSpotter identifies and ranks signaling hotspots that can then be used to maintain and control a cell phenotype

the experimental validation of two predicted cell-fate determinants confirmed that overexpression of *Runx2* and *Esr1* in mouse NSCs induces astrocyte and neuronal differentiation, respectively [34]. Furthermore, this tool was also employed to interrogate the differentiation of cardiac progenitor cells to more specialized cell types during normal and abnormal cardiogenesis to study the devastating consequences of disruption in specific cellular subpopulations [44]. The experimental validation of its in silico predictions shows the potential of SeesawPred to guide differentiation experiments in stem cell research and regenerative medicine. The version of SeesawPred that relies on single-cell transcriptomic data is described here [35, 44], whereas the one that uses the bulk transcriptomic data is freely available at https://seesaw.lcsb.uni.lu/.

SeesawPred has been applied to multiple scRNA-seq and bulk transcriptomic datasets, predicting cell-fate determinants and their GRN motifs in multiple stem cell differentiation systems [34, 35, 44]. The application of this approach in stem cell research with therapeutic potential has been demonstrated in cardiogenesis to reveal the basis for organ level developmental defects. The presented computational method revealed the key lineage-specifying TF *Hand2* as a specifier of outflow tract (OFT) cells. Furthermore, it predicted that *Hand2*-null OFT-fated cells have disrupted specification, which was subsequently experimentally verified [44]. In addition, SeesawPred was also used to predict the cell-fate determinants for astrocytic and neuronal lineages from the mouse NSCs. SeesawPred predicted the master regulator *Runx2* and *Esr1* overexpression for astrocytic and neuronal fate, respectively [34]. Moreover, the application of this computational framework to three different stem cell systems predicted already known and experimentally validated lineage specifiers in stem cell subpopulations [35]. For this purpose, three binary-fate stem cell differentiation systems were selected for which high-quality single-cell gene-expression data were available. Precisely, the lineage specifiers were predicted for inner cell mass (ICM), hematopoietic stem cells (HSC), multipotent progenitor (MPP), common myeloid progenitor (CMP), and lung alveolar bipotential progenitor (BP) differentiations. Briefly, SeesawPred predicted the Gata6 and Klf2 to be cell-fate determinants for the primitive endoderm (PE) and epiblast (EPI) differentiation from the inner cell mass (ICM), which is in full agreement with independent experimental observations [45, 46]. Similarly, in line with existing experimental reports, *Hes1* [47] and *Pou6f1* [48] were predicted as lineage specifiers for the alveolar type 1 (AT1) and alveolar type 2 (AT2) subpopulations from the differentiating lung alveolar BP. Taking these facts together, the presented approach addresses crucial challenges in stem cell research as it has the ability to guide experimentalists in the design of new strategies for stem cell therapies and treatment with potential application in regenerative medicine.

On the other hand, recent advances in single-cell RNA sequencing (scRNA-seq) has provided us with a large compilation of high-resolution datasets that reflect the cellular heterogeneity within a cell population. The high resolution and statistical power of this data has significantly advanced our ability to classify cells into distinct subpopulations, based on their gene-expression profiles. The identity of these cell subpopulations can vary from well-known cell types, subtypes, to uncharacterized subpopulations that, without the scRNA-seq technology, would have remained unknown. Maintaining the identity of different cell populations has been linked to specific sets of TFs [9]. Moreover, the scRNA-seq advent opened doors for the characterization of TFs that act synergistically in order to induce cell transitions between cell populations [41]. Thus, identification of such core TFs is of utmost importance for characterizing and converting any cell population. The quest to advance our knowledge of core TFs identification and characterization of their crucial role in cellular conversions has prompted the development of new computational methods [6–9] that aim at deciphering the underlying regulatory mechanism and interactions responsible for cell identity and transitions.

One such state-of-the-art computational platform is TransSyn (*see* **Note 2**), a tool for the identification of cell population identities by defining their synergistic transcriptional cores [41]. This method is based on the assumption that a cell population identity arises from the synergistic activity of specific TFs that stabilize the expression levels of genes characteristic of that cell population. TransSyn method does not rely on GRNs nor depends on any prior knowledge, and it only requires a scRNA-seq dataset of distinct populations as input. Hence, TransSyn determines cell population identities and promotes the design of experimental strategies by means of predicting core TFs necessary for converting the cell population under consideration to different other cell populations.

As a proof of concept, TransSyn predicted synergistic transcriptional cores recapitulated known identity TFs in 85% of the cases and known synergistic TF interactions related to cell identity [41]. Furthermore, TransSyn cell subpopulation-specific synergistic transcriptional core successfully drove the experimental conversion of human hindbrain neuroepithelial cells (hNES) into medial floor plate midbrain progenitors (hProgFPM), which later differentiated into midbrain dopaminergic (mDA) neurons [41]. TransSyn is freely available at https://sourceforge.net/projects/transsyn/.

TransSyn has been applied to multiple scRNA-seq datasets, predicting synergistic transcriptional cores of 186 cell populations [41]. The predicted synergistic transcriptional cores frequently contained TFs known to maintain the respective cell population identities. For instance, the pluripotency factors *Pou5f1*, *Sox2*, and *Nanog* were predicted as the most synergistic transcriptional core in human embryonic stem cells (ESCs). Indeed, these TFs are known to maintain the ESC phenotype and were described to act synergistically through enhancer clusters [49]. Also, *Phox2a* and *Isl1*, which have been described to act synergistically to define cranial motor neurons from mouse ESCs [50], were predicted as part of the synergistic core of human fetal oculomotor and trochlear nucleus population. Finally, *Tal1, Runx1, Gata2, and Fli1*, identified as the synergistic transcriptional core of blood cell progenitors, have been shown to act synergistically by interacting at a protein level, stabilizing their cofactor binding to DNA and controlling the cell population identity [51, 52].

TransSyn predictions not only capture the synergy between TFs but also recapitulate several known TFs interactions that control cell population identities. For instance, *Eomes, Otx2, Zic3, Foxa2,* and *Hnf4a* were identified in the synergistic core of embryonic visceral endoderm subpopulation. These TFs have been described to regulate each other and multiple downstream targets specific to that cell population [53, 54]. In addition, *Gata1* and *Ikzf1*, known to functionally regulate each other and maintain the identity of embryonic blood cells [55, 56], were identified in the synergistic TF core of the embryonic erythrocyte population [52]. Differentiation towards vascular endothelial cell fate is regulated by *Id3* and members of the Krüppel-like family of TFs [57, 58], which were included on the predicted synergistic transcriptional core for this population. For mouse enteroendocrine cells, the synergistic TF core consisted of *Foxa1, Foxa2, Insm1, Lmx1a, Neurog1, Neurog3, Nkx2.2*, and *Pax4*, all known to play an important role in the functioning of this cell type [59–62]. Finally, the synergistic transcriptional core obtained for fetal dopaminergic neurons contained *Nurr1* and *Foxa1*, both known for controlling mDA identity and neurogenesis [63, 64].

In order to demonstrate TransSyn efficacy in cell conversion protocols, this computational platform was used to predict the synergistic core of hProgFPM [41]. The results included *Foxa2, Otx2*, and *Lmx1a*, TFs which were previously described as important for mouse dopaminergic neuron (mDA) development [64–66]. Since *Otx2* is known to induce *Lmx1a* expression [67], an experiment was designed to shift the identity of hindbrain hNES cell line towards hProgFPM targeting only *Foxa2* and *Otx2* [41]. These two TFs were induced by treatment of hNES with two small molecules: smoothened agonist and Dickoppf1 for *Otx2* and *Foxa2*, respectively. Anteriorization and acquisition of midbrain identity were shown by an increase in the number of *Otx2* positive cells and the increased ratio of *Otx2/Gbx2* expression. In addition, increased levels of *Foxa2* and decreased levels of *Pax6* and *Irx3* were observed, which reveals efficient ventralization. Furthermore, cells obtained by targeting these two TFs generated a significant increase in both Th gene expression and Th positive cells, which were also positive for *Lmx1a, Nurr1*, and *Pbx1*, known markers of midbrain identity [63, 66, 68]. Th positive cells also expressed *Map2*, a mature neuronal marker, and were observed to acquire long processes and varicosities, usually found in mDA neurons. Taken together, such an extensive validation of TransSyn proves its ability to facilitate the design of novel strategies for conversion of cell subpopulation identities with potential applications in regenerative medicine.

Finally, cell rejuvenation strategies are essential to prevent aged and disease cell niches to impair normal cellular function [69–71]. Characterizing and controlling these cell subpopulation phenotypes has great potential for developing novel regenerative medicine strategies. The statistical power and increasing number of scRNA-seq datasets have been helping us to profile distinct cell subpopulations. However, the lack of computational methods that use scRNA-seq data to identify key factors involved in the maintenance of a specific cell phenotype and controlling cell rejuvenation remains a challenge.

In order to bridge this gap in the literature, researchers have introduced SigHotSpotter, a computational web application that integrates scRNA-seq with a probabilistic Markov chain model of signal transduction to predict key signaling molecules (hotspots) that are responsible for the control of a cell phenotype [42]. This method is based on the assumption that in steady-state, single-cell gene expression can be representative of the corresponding protein level [72]. SigHotSpotter integrates signaling and transcriptional networks in order to identify specific signaling hotspots that are involved in sustaining the transmission of external niche signals. These hotspots are predicted based on the highest sustained signal flux instead of their interference in the dysregulation of the entire signaling pathway. In fact, signaling molecules matching these characteristics have been shown to be more likely to transmit stable signals involved in niche maintenance, rather than transient but strong signals, normally related to changes in cell phenotype [73]. Furthermore, SigHotSpotter ranks the predicted hotspots according to their compatibility towards the downstream TFs. This will help experimentalists in prioritizing their targets for further studies and prompt the development of cell rejuvenation strategies for counteracting the loss of activation potential in niche cells due to aging or disease.

SigHotSpotter was applied in silico to four different cell systems where it accurately recapitulated signaling molecules known to be involved in maintaining specific cell phenotypes [42]. Namely, SigHotSpotter predicted that the inhibition of Gsk3β and Map2k1 is responsible for the maintenance of the mouse ESCs in a naive pluripotency state. SigHotSpotter also identified specific signaling pathways involved in fibroblast switching from neonatal to mature state, a process that was shown to prompt the maturation of cardiomyocytes in vivo [74]. Furthermore, the Markov chain applied in this method identified Spr5 as the signaling hotspot mediating a switch from canonical to non-canonical Wnt activity, inducing quiescence in the aging mice brain [75]. SigHotSpotter is available at https://SigHotSpotter.lcsb.uni.lu and its source code can be accessed at https://gitlab.com/srikanth.ravichandran/sighotspotter.

SigHotSpotter has been applied in four different cell systems based on scRNA-seq datasets from mouse ESCs, hair-follicle stem cells (HFSCs), hematopoietic stem cells (HSCs), and oligodendrocyte progenitor cells [42]. The method correctly predicted signaling hotspots with known experimental validation in those cell systems. For instance, SigHotSpotter accurately predicted the inhibition of Gsk3β and Map2k1 in ESCs under 2i culture conditions. Indeed, different culture conditions are known to maintain ESCs in different phenotypic states. In particular, 2i culture media, which contains selective inhibitors for Gsk3β and Map2k1, is known to maintain ESCs in a naive pluripotency state while leukemia inhibitory factor (LIF) alone is known to maintain ESCs in a primed pluripotency state [76]. In addition, Gsk3β and Bmpr1a, components of, respectively, the Wnt and BMP signaling pathways were identified as inhibited hotspots in HFSCs. Accordingly, Wnt and BMP signaling pathways were reported to play a role in the activation of quiescent HFSCs to generate specialized mesenchymal cells necessary to induce the hair cycle [77]. In young long-term HSCs, Map2k1 and Map3k1, proteins of the ERK signaling pathway, were predicted as activated signaling hotspots together with Gsk3β, a protein regulated by the PI3K signaling pathway. The other two members of the PI3K signaling pathway, Akt1 and Icam1, were identified as inhibited in the same phenotype. Notably, long-term HSCs have been described to be the main cell type involved in the production of blood cells, a process where ERK and PI3K pathways were identified as key regulators of the balance between HSCs dormancy and activation [78–81]. Finally, Notch1 and Gsk3β, that have been reported to inhibit oligodendrocyte differentiation and myelination as inhibited, were also predicted as inhibited in OPCs, while Fyn, a protein reported to be involved in oligodendrocyte migration, was predicted as activated in this cell type [82–84].

The Markov chain applied in this method was used to predict niche signals responsible for age-dependent changes in NSCs [75]. The results predicted Sfrp5, an antagonist of the non-canonical Wnt signaling pathway, as a key signaling intermediate for old quiescent NSCs. Sfrp5 was inhibited in vivo by administering a neutralizing antibody for 14 days. Sfrp5 inhibition significantly decreased the number of proliferating cells in an old mice brain when compared to IgG-treated control old mice. This result indicates a reduced NSC activation, showing that Sfrp5 antagonization leads to increased Wnt canonical activity and, thus, quiescence in the aging mice brain. Furthermore, this shows that NSCs activation is regulated by a switch from canonical to non-canonical Wnt activity, a new finding that highlights the importance of this method in characterizing different cell phenotypes.

Recently, SigHotSpotter was also applied to further understand the signaling pathways involved in fibroblast identity in mouse heart development [74]. The results identified Foxo, mTOR, and VEGF signaling pathways as specific for neonatal fibroblasts while cell adhesion, estrogen, and TGFβ signaling pathways were identified only in mature fibroblasts. These results suggest that fibroblast switching from neonatal to mature state prompts the in vivo maturation of cardiomyocytes.

In conclusion, in recent years stem cell therapy has become a very promising and important scientific research topic. Advances in this area allowed for a deeper understanding of human disease and prompted the design of novel strategies for tissue regeneration, namely in cell transplantation. The advent of single-cell technology continues to expand our knowledge about molecular mechanisms underlying cell conversion and differentiation, the characterization of cell phenotypes in different conditions, and intrinsic molecular dysregulations in different pathologies. By applying single-cell technology to stem cell research, multiple computational methods were developed to accurately identify key factors in cell conversion, explain differences between cell phenotypes, and predict signaling molecules controlling these processes. To sum up, the discussed computational models provide mechanistic insights into biological processes and generate new predictions that guide experimental research. However, stem cell therapy still has challenges to overcome. For instance, cells transplanted from a different tissue or obtained from in vitro experiments are not always successfully integrated into the patient's tissue [85]. In addition, most in vitro reprogramming and differentiation protocols often have a low conversion efficiency, making it more expensive and time-consuming to collect enough target cells for clinical use or further research. Furthermore, in vitro cell conversion protocols often generate non-functional and immature variants of target cells, failing to obtain the desired cell phenotype and functionalities.

Computational modeling can help address these limitations by developing network-based models that combine GRNs with the hierarchical organization of cell identity TFs, in order to predict specific cell type and subtype identity TFs that lead to in vitro generation of functionally mature target cells [85, 86]. Moreover, these methods can be further used to elucidate mechanisms underlying the dysregulation of cell differentiation associated with, for instance, congenital disorders [44]. Integrating different types of single-cell data in multiscale computer models would allow for a more thorough characterization of a biological system and help experimentalists in designing novel strategies for stem cell therapy.

For instance, combining scRNA-seq and single-cell sequencing assay for transposase-accessible chromatin (scATAC-seq) would provide a better characterization of novel cell types and phenotypes, which would lead to the development of more accurate models for predicting cell-fate determinants.

Modeling biological systems at the tissue level, based on cell–cell or receptor–ligand interactions, could help elucidate the basic processes involved in tissue regeneration and homeostasis and lead to novel predictions of cell–cell interactions that sustain tissue regeneration. For instance, recent scRNA-seq computational methods focused on modeling stem cell niche interactions by integrating scRNA-seq with signaling and transcriptional networks [42, 87]. These approaches acknowledge the fact that predicting key signaling molecules that mediate niche signals responsible for maintaining cell phenotypes is of the utmost importance for stem cell rejuvenation therapies that focus on reverting impaired cell function and improve tissue repair processes in degenerative and age-related diseases. Furthermore, modeling cell-cell communication networks based on scRNA-seq data has helped in predicting key cell–cell interactions involved in the regulation of tissue homeostasis [88–90]. By comparing these reference cell-cell interactomes with the networks of injured or diseased tissues, a computational model could identify which interactions are dysregulated and help develop novel strategies for reverting tissue degeneration. Moreover, multiscale modeling can improve the prediction of relevant cell–cell interactions in tissue homeostasis. To this end, integrating phosphoproteomics data with scRNA-seq would help to identify signaling pathways that maintain stem cell phenotypes while combining scRNA-seq with imaging data could further improve our understanding of cell-cell communication.

In summary, stem cell research can greatly benefit from the increased resolution of scRNA-seq and its integration with new single-cell technologies, such as proteomes, epigenomes, and spatial information in order to have a more comprehensive classification and characterization of cell types, their interactions, and function. Closer collaboration between computational researchers and experimentalists can further stimulate advances in stem cell research. For example, by joining efforts they could come up with a more generic and integrative model that will provide an accurate and deeper insight into the molecular mechanisms underlying the regulation and control of stem cell homeostasis and its dysregulation in disease, therefore, accelerating the translation of our biological knowledge to clinical application.

2 Materials

2.1 SeesawPred: Cell-Fate Determinants

1. SeesawPred requires as an input the gene-expression profiles of the stem/progenitor and two daughter cell types.
2. A prior knowledge network (PKN) that contains interactions between the TFs (*see* **Notes 3** and **4**).

2.2 TransSyn: Cell Identity TFs

1. scRNA-seq gene-expression matrix with column names labeled according to which population each cell belongs, and the rows contain gene-expression values (*see* **Note 5**).
2. List of all known TFs of the species from which the gene-expression matrix was obtained. TFs should be labeled with the same nomenclature used in the gene-expression matrix mentioned in **item 1** (*see* **Note 6**).

2.3 SigHotSpotter: Niche-Induced Signaling Pathways

1. Two scRNA-seq gene-expression matrices from the same cell type in different conditions, such as ESCs cultured in 2i and LIF culture media (condition 1 and condition 2). scRNA-seq gene-expression matrices should have columns representing cell data or replicates and row names labeled according to Gene Symbols nomenclature. Gene-expression matrices data should be normalized or in read counts (TPM/FPKM).
2. Differential expression file containing only the significantly differentially expressed TFs, where "1" denotes upregulated TFs in condition 1 and "−1" denotes upregulated TFs in condition 2. The differential expression file should have Gene Symbols on the first column and differentially expression information on the second column (*see* **Note 7**).
3. Pre-compiled, species-specific signaling interactome network, including receptor/ligand information, a list of TFs and TF interactions (*see* **Notes 8** and **9**).

3 Methods

3.1 SeesawPred: Cell-Fate Determinants

1. The computation of differentially expressed TFs between two daughter cells.
2. The reconstruction of Boolean TRNs from differentially expressed TFs by retrieving interactions from the MetaCore (Clarivate Analytics) database and then prune this network by discarding interactions incompatible with Booleanized gene-expression data of two daughter cells (*see* **Note 10**).
3. The computation of statistically significant normalized ratio difference (NRD) TF pairs to identify pairs of TFs whose

expression ratios showed a significant change in daughter cells in comparison with the stem/progenitor cells (*see* **Note 11**).

4. The identification of strongly connected components (SCC) in the TRN of parental cell subpopulation.
5. TF pairs that are present in the parental SCC and whose TFs are differentially expressed in daughter cell subpopulations are considered candidate opposing lineage specifier pairs (*see* **Note12**).
6. The NRD TF pairs are then tested for significance and filtered. The criteria for keeping a TF pair is that it has to be: (1) differentially expressed between the two daughter cell types, (2) constitute significant NRD TF pair on significance test among all TFs, and (3) show high absolute NRD values in both daughter cell types ($|NRD| > 0.5$) (*see* **Note 13**).
7. Finally, SCCs with significant NRD TF pairs which satisfy the presented criteria are considered the final predictions.

3.2 TransSyn: Cell Identity TFs

1. Discard unclassified populations or populations with less than three cells in the dataset.
2. Filter gene-expression matrix for TFs-only based on the TF list in **item 2** of the Subheading 2.
3. Classify TFs as expressed if their expression values are higher than or equal to 1 in RNA-seq FPKM/RPKM/TPM values, higher than or equal to 10 in normalized read counts, or higher than or equal to 1 in UMI counts. TFs that are below these expression cut-offs are to be considered not expressed (*see* **Note 14**).
4. Select target cell populations for transcriptional synergistic core identification.
5. Calculate the fraction of cells expressing each TF per selected population. Keep the top 10% most frequently expressed TFs that are also expressed in more than 70% of the cells. If the number of TFs is more than 150, calculate the coefficient of variation and select the top 150 TFs (*see* **Note 15**).
6. Convert the zero gene-expression values into one. Log 10-transform the gene-expression values. Convert the gene-expression values with mean equal to zero into one. Discretize obtained gene-expression values using the Freedman–Diaconis rule (*see* **Note 16**). For FPKM/RPKM/TPM values and normalized read counts, set the input for the Freedman–Diaconis rule to the number of cells plus 1. For UMI counts, set it to the number of cells plus 6. Set the range of gene-expression values to be between 0 and the maximum value of each cell population.

7. Compute Shannon's entropy of each TF based on the discretized values and normalize these values by the theoretical maximum entropy (*see* **Note 17**).
8. Calculate multivariate mutual information (MMI) [91] for all combinations of three TFs using the total number of cells in each population. Select the top 1% lowest MMI combinations (*see* **Note 18**). Rank these combinations by total correlation (TC) and select the top 1% highest combinations.
9. Use the selected TC combinations as the initial seeds for the dynamic heuristic search of high-level synergistic TF cores. Add new TFs to each seed combination, one by one, and compute the new MMI. Select the combinations that show lower MMI (less than 0.05) than the seed and compute TC. Select the top ten best TC combinations as seed for the next iteration. The heuristic search is finished when no new combination has a lower MMI (more synergistic) than the seed or when the number of TFs within a combination reaches 15 (*see* **Note 19**).
10. Compute the TC of the final combinations obtained in **step 9**. Calculate MMI for the 20 best TC combinations. If MMI is lower, rank new combinations by TC. Otherwise, rank combinations from **step 9** by TC to obtain the final synergistic transcriptional cores. If there is a tie between combinations, they are ranked according to the highest summed mean gene expression and the top three combinations are kept as the final synergistic transcriptional cores.
11. Identify the cell conversion TFs by calculating the mean gene expression for the TFs in each synergistic transcriptional core and ranking them based on the fold change between the target cell population and the starting cell population (remaining populations in the dataset).

3.3 SigHotSpotter: Niche-Induced Signaling Pathways

1. Model the signal transduction process from the niche to intracellular signaling pathways as a finite discrete time-homogeneous Markov chain, where the signal originates from the niche and propagates successively through a finite set of signaling molecules (*see* **Note 20**).
2. Use the mass action principle to construct the state transition probability matrix, assuming that the probability of interaction between two molecules is proportional to the dot product of their corresponding gene-expression values (*see* **Note 4**). Transition probability values should only be calculated from interactions present in the signaling interactome (*see* **Note 21**).
3. Establish a cut-off parameter that determines the percentage of cells for which both interacting molecules must be expressed for the interaction to be considered in the Markov chain model. For instance, if the cut-off is set to 30%, then interactions

among two molecules are only considered if both genes are simultaneously expressed in at least 30% of the cells in the scRNA-seq.

4. Calculate the stationary distribution of the transition probability matrix by finding the eigenvector of the transposed transition probability matrix with an eigenvalue equal to 1 [92]. This computation gives the steady-state probability distribution of where the signal will be present, at any time point (*see* **Note 22**).
5. Establish a percentile parameter that determines the top fraction of molecules classified as high probability signaling molecules, which will be used for the following calculations.
6. Classify TFs as interface TFs and non-interface TFs. Non-interface TFs are defined as TFs that do not have an incoming edge from a signaling molecule. Interface TFs are defined as TFs that have an incoming edge from a signaling molecule. Filter this list for differentially expressed non-interface TFs (ntDETFs).
7. Evaluate the concordance between the overall effect of the high probability signaling molecules (activation or inhibition) on the expression status of the downstream ntDETFs (up- or downregulated). Trace all directed shortest paths from the high probability signaling molecules to the ntDETFs. Based on these results, calculate the weight for each shortest path, which is obtained from the product of the steady-state probabilities of the nodes contained in the path and its sign. If the number of inhibitor edges is even and the ntDETF is upregulated, the sign of the shortest path is equal to 1. However, if the ntDETF is downregulated, the sign is equal to −1. If the number of inhibitor edges is odd and the ntDETF is downregulated, the sign of the shortest path is equal to 1. On the other hand, if the ntDETF is upregulated, the sign is equal to −1 (*see* **Note 23**).
8. Calculate the compatibility score for a given signaling molecule and a target ntDETF based on the shortest path weight obtained on the previous step. The compatibility score is defined as the fraction of the sum of positive edge weights relative to the sum of absolute edge weights. This score provides a quantitative measurement between a high probability signaling molecule and a target ntDETF. Calculate the overall compatibility score of a signaling molecule for a given phenotype/condition by computing the mean of the compatibility scores over all ntDETFs.
9. Based on the compatibility scores obtained on the previous step, identify the active and inactive signaling hotspots (*see* **Note 24**). Construct the signaling network around these

hotspots by computing all the shortest paths from the niche-node to the predicted signaling hotspots, and then from the hotspots to the ntDETFs (*see* **Notes 25** and **26**).

4 Notes

1. The SeesawPred algorithm was written in R and the web interface has been developed using the Shiny web technology.
2. AnimalTFDB [93] database can be a useful resource to obtain the species-specific TF list.
3. A literature-based manually curated PKN can be retrieved from the MetaCore (Clarivate Analytics) database or any other source of user's choice.
4. TransSyn algorithm was written in C++ and wrapped in R using Rcpp and gtools R packages.
5. It is recommended to use the same population classification and to not reprocess the raw data and gene-expression values defined in the original dataset.
6. AnimalTFDB [93] database can be a useful resource to obtain the species-specific TF list.
7. AnimalTFDB [93] annotation can be useful to define TFs. Differentially expressed TFs can be identified by using a *t*-test. Shortlist the set of obtained differentially expressed TFs by using Benjamini–Hochberg correction and a cut-off for the adjusted *p*-value lower than 0.05.
8. We recommend using Omnipath [94] and ReactomeFI [95] databases to construct the signaling interactome network, since they contain information about the directionality of the interactions and their regulatory nature. For example, given a gene A and a gene B, we can know if A and B interact with each other or if it is a unidirectional interaction, and if these interactions are activations or inhibitions. Gene Ontology classification of the plasma membrane (GO:0005886) and receptor activity (GO:0004872) can be used to compile a list of receptors/ ligands. Transcriptional interactions can be obtained from Zaffaroni et al. [96] which contains a high number of manually curated TF interactions from MetaCore (Clarivate Analytics).
9. In order to account for the niche influence on the intracellular signaling and to model sustained signaling, it is recommended to introduce an external niche-node which should be connected to all receptors and ligands as well as all the TFs nodes in the signaling interactome. This step ensures a continuous signal transmission from niche to TFs through signaling intermediates, based on the assumption that, once a signal

reaches a TF, this signal starts once again from the external niche-node. Also, it is recommended to remove non-receptor/ligand nodes that have zero in-degree and non-TF nodes that have a zero out-degree since these nodes do not contribute to sustain the signal transduction.

10. The PKN retrieved from MetaCore contains direct interactions between the TFs obtained from different interaction categories, such as binding, transcriptional regulation, and influence on expression.
11. The required input of SeesawPred web application is a tab-separated value file where columns represent gene-expression (microarray or RNA-Seq) replicates of the stem/progenitor cell type and the two daughter cell types labeled as "Progenitor," "Daughter1," "Daughter2," respectively, and the rows are labeled according to the TF symbols.
12. The PKN retrieved from MetaCore or defined by the user must also be a tab-separated value file containing the list of TF–TF interactions that serve as potential links in the network.
13. Example input files for two cellular differentiation systems (mouse NSCs differentiation into neurons and astrocytes, and Mouse Hematopoietic Stem Cell (HSC) differentiation to erythroid and myeloid) are provided in the online web application of SeesawPred.
14. If for a given dataset setting the cut-offs to the recommended values results in too many expressed TFs, making the following computation steps impracticable, the expression cut-off can be set to 10 instead. If the input dataset is binarized, TFs with mean counts lower than 1 can be discarded.
15. The maximum number of TFs used for the subsequent computations are suggested to be set to 150 since running the algorithm becomes infeasible for a standard desktop computer if this number is higher than that.
16. The Freedman–Diaconis rule was implemented using the R nclass.FD function.
17. Normalizing the TF entropy using the theoretical maximum entropy enables a direct comparison between different TF entropies.
18. MMI measures the information gained by adding a variable (in this case, a TF), which cannot be explained by the sum of the information given by the subsets of variables. Thus, when MMI is negative it means that the TFs are synergistically interacting with each other because the information given by the TFs together is higher than the sum of the information given by the TFs separately. MMI is calculated with all cells in each

population in the dataset, except for the ones in the population for which MMI is being computed.

19. The suggested maximum number of TFs per synergistic core is 15 because continuing the computations with higher numbers is often very demanding and it has been observed that, at that point, most TFs are shared among different combinations.
20. According to this model, the probability of a given signal to propagate from molecule A to molecule B on the next time step only depends on where the signal is currently present and not on where it was on the previous time steps. Thus, the transition probability of a signal to propagate from molecule A to B is defined by a transition probability matrix, calculated based on the scRNA-seq expression matrix.
21. The transition probability value between two molecules is calculated by dividing the dot product of their corresponding single-cell gene-expression vectors (defined as interaction weight) by the sum of the product of the interaction weights between their neighboring molecules.
22. The transition probability matrix will be stochastic since only the interactions present in the signaling interactome will have a non-zero transition probability calculated from the data. The others will have zero probabilities. Since the interaction probability for any two molecules is proportional to the scalar product of their gene-expression values, this probability will be higher only when both the molecules are highly expressed in the same cell and expressed in a large number of cells for that population.
23. The compatibility of the shortest path from a high probability signaling molecule to a ntDETF is determined by the sign. If the sign of the shortest path is negative (-1), it means that the signaling molecule and ntDETF are incompatible and so the weight will be negative. If the sign of the shortest path is positive (1), the signaling molecule and ntDETF are compatible, therefore the weight will be positive.
24. The compatibility score is equal to 1 if all shortest paths between a signaling molecule and a target ntDETF are compatible, making this signaling molecule an active signaling hotspot. If the net effect of the signaling molecule is entirely incompatible with the expression status of the downstream TFs, the compatibility score is equal to 0 and the signaling molecule classified as an inactive signaling hotspot.
25. We recommend using the Igraph implementation of Dijkstra's algorithm [97] to calculate the shortest paths network. This network will show how the signaling hotspots meditate the transmission of the signal from the external niche-node to the ntDETFs.

26. We recommend to output a list of the top ten active and inactive signaling hotspots for each phenotype and a complete list of all ranked predictions, together with the shortest path network of the predicted signaling hotspots in SIF format so it can be easily visualized in, for instance, Cytoscape [98].

References

1. Fu NY, Nolan E, Lindeman GJ, Visvader JE (2020) Stem cells and the differentiation hierarchy in mammary gland development. Physiol Rev 100(2):489–523. https://doi.org/10.1152/physrev.00040.2018
2. Visvader JE, Clevers H (2016) Tissue-specific designs of stem cell hierarchies. Nat Cell Biol 18(4):349–355
3. D'Alessio AC, Fan ZP, Wert KJ et al (2015) A systematic approach to identify candidate transcription factors that control cell identity. Stem Cell Rep 5(5):763–775. https://doi.org/10.1016/j.stemcr.2015.09.016
4. Whyte WA, Orlando DA, Hnisz D et al (2013) Master transcription factors and mediator establish super-enhancers at key cell identity genes. Cell 153(2):307–319. https://doi.org/10.1016/j.cell.2013.03.035
5. Vaquerizas JM, Kummerfeld SK, Teichmann SA, Luscombe NM (2009) A census of human transcription factors: function, expression and evolution. Nat Rev Genet 10(4): 252–263
6. Rackham OJL, Firas J, Fang H et al (2016) A predictive computational framework for direct reprogramming between human cell types. Nat Genet 48(3):331–335. https://doi.org/10.1038/ng.3487
7. Cahan P, Li H, Morris SA et al (2014) CellNet: network biology applied to stem cell engineering. Cell 158(4):903–915. https://doi.org/10.1016/j.cell.2014.07.020
8. Buganim Y, Faddah DA, Jaenisch R (2013) Mechanisms and models of somatic cell reprogramming. Nat Rev Genet 14(6):427–439
9. Morris SA, Daley GQ (2013) A blueprint for engineering cell fate: current technologies to reprogram cell identity. Cell Res 23(1):33–48
10. Strasser M, Theis FJ, Marr C (2012) Stability and multiattractor dynamics of a toggle switch based on a two-stage model of stochastic gene expression. Biophys J 102(1):19–29. https://doi.org/10.1016/j.bpj.2011.11.4000
11. Roeder I, Glauche I (2006) Towards an understanding of lineage specification in hematopoietic stem cells: a mathematical model for the interaction of transcription factors GATA-1 and PU.1. J Theor Biol 241(4):852–865. https://doi.org/10.1016/j.jtbi.2006.01.021
12. Huang S, Guo YP, May G, Enver T (2007) Bifurcation dynamics in lineage-commitment in bipotent progenitor cells. Dev Biol 305(2): 695–713. https://doi.org/10.1016/j.ydbio.2007.02.036
13. Zhang P, Behre G, Pan J et al (1999) Negative cross-talk between hematopoietic regulators: GATA proteins repress PU.1. Proc Natl Acad Sci U S A 96(15):8705–8710. https://doi.org/10.1073/pnas.96.15.8705
14. Arinobu Y, Mizuno S-i, Chong Y et al (2007) Reciprocal activation of GATA-1 and PU.1 marks initial specification of hematopoietic stem cells into myeloerythroid and myelolymphoid lineages. Cell Stem Cell 1(4):416–427. https://doi.org/10.1016/j.stem.2007.07.004
15. Biteau B, Hochmuth CE, Jasper H (2011) Maintaining tissue homeostasis: dynamic control of somatic stem cell activity. Cell Stem Cell 9(5):402–411
16. Ferraro F, Lo Celso C, Scadden D (2010) Adult stem cells and their niches. Adv Exp Med Biol 695:155–168
17. Schofield R (1978) The relationship between the spleen colony-forming cell and the haemopoietic stem cell. A hypothesis. Blood Cells 4(1–2):7–25
18. Glukhova MA, Streuli CH (2013) How integrins control breast biology. Curr Opin Cell Biol 25(5):633–641
19. Brizzi MF, Tarone G, Defilippi P (2012) Extracellular matrix, integrins, and growth factors as tailors of the stem cell niche. Curr Opin Cell Biol 24(5):645–651
20. Shehata M, Teschendorff A, Sharp G et al (2012) Phenotypic and functional characterisation of the luminal cell hierarchy of the mammary gland. Breast Cancer Res 14(5):R134. https://doi.org/10.1186/bcr3334
21. Stingl J, Eirew P, Ricketson I et al (2006) Purification and unique properties of mammary epithelial stem cells. Nature 439(7079): 993–997. https://doi.org/10.1038/nature04496

22. Heinäniemi M, Nykter M, Kramer R et al (2013) Gene-pair expression signatures reveal lineage control. Nat Methods 10(6):577–583. https://doi.org/10.1038/nmeth.2445
23. Lang AH, Li H, Collins JJ, Mehta P (2014) Epigenetic landscapes explain partially reprogrammed cells and identify key reprogramming genes. PLoS Comput Biol 10(8):e1003734. https://doi.org/10.1371/journal.pcbi.1003734
24. Roost MS, Van Iperen L, Ariyurek Y et al (2015) KeyGenes, a tool to probe tissue differentiation using a human fetal transcriptional atlas. Stem Cell Rep 4(6):1112–1124. https://doi.org/10.1016/j.stemcr.2015.05.002
25. Davis FP, Eddy SR (2013) Transcription factors that convert adult cell identity are differentially polycomb repressed. PLoS One 8(5): e63407. https://doi.org/10.1371/journal.pone.0063407
26. Morris SA, Cahan P, Li H et al (2014) Dissecting engineered cell types and enhancing cell fate conversion via cellnet. Cell 158(4): 889–902. https://doi.org/10.1016/j.cell.2014.07.021
27. Cabili M, Trapnell C, Goff L et al (2011) Integrative annotation of human large intergenic noncoding RNAs reveals global properties and specific subclasses. Genes Dev 25(18): 1915–1927. https://doi.org/10.1101/gad.17446611
28. Beerman I, Rossi DJ (2015) Epigenetic control of stem cell potential during homeostasis, aging, and disease. Cell Stem Cell 16(6): 613–625
29. Avgustinova A, Benitah SA (2016) Epigenetic control of adult stem cell function. Nat Rev Mol Cell Biol 17(10):643–658
30. Neph S, Stergachis AB, Reynolds A et al (2012) Circuitry and dynamics of human transcription factor regulatory networks. Cell 150(6): 1274–1286. https://doi.org/10.1016/j.cell.2012.04.040
31. Huang M, Chen Y, Yang M et al (2018) DbCoRC: a database of core transcriptional regulatory circuitries modeled by H3K27ac ChIP-seq signals. Nucleic Acids Res 46(D1): D71–D77. https://doi.org/10.1093/nar/gkx796
32. Saint-André V, Federation AJ, Lin CY et al (2016) Models of human core transcriptional regulatory circuitries. Genome Res 26(3): 385–396. https://doi.org/10.1101/gr.197590.115
33. Hartmann A, Okawa S, Zaffaroni G, del Sol A (2018) SeesawPred: a web application for predicting cell-fate determinants in cell differentiation. Sci Rep 8(1):13355. https://doi.org/10.1038/s41598-018-31688-9
34. Okawa S, Nicklas S, Zickenrott S et al (2016) A generalized gene-regulatory network model of stem cell differentiation for predicting lineage specifiers. Stem Cell Rep 7(3):307–315. https://doi.org/10.1016/j.stemcr.2016.07.014
35. Okawa S, del Sol A (2015) A computational strategy for predicting lineage specifiers in stem cell subpopulations. Stem Cell Res 15(2): 427–434. https://doi.org/10.1016/j.scr.2015.08.006
36. Guo M, Bao EL, Wagner M et al (2017) SLICE: determining cell differentiation and lineage based on single cell entropy. Nucleic Acids Res 45(7):e54. https://doi.org/10.1093/nar/gkw1278
37. Feingold BJ, Vegosen L, Davis M et al (2010) A niche for infectious disease in environmental health: rethinking the toxicological paradigm. Environ Health Perspect 118(8):1165–1172
38. Decimo I, Bifari F, Krampera M, Fumagalli G (2012) Neural stem cell niches in health and diseases. Curr Pharm Des 18(13):1755–1783. https://doi.org/10.2174/138161212799859611
39. Crespo I, Del Sol A (2013) A general strategy for cellular reprogramming: the importance of transcription factor cross-repression. Stem Cells 31(10):2127–2135. https://doi.org/10.1002/stem.1473
40. Ravichandran S, Okawa S, Martínez Arbas S, del Sol A (2016) A systems biology approach to identify niche determinants of cellular phenotypes. Stem Cell Res 17(2):406–412. https://doi.org/10.1016/j.scr.2016.09.006
41. Okawa S, Saltó C, Ravichandran S et al (2018) Transcriptional synergy as an emergent property defining cell subpopulation identity enables population shift. Nat Commun 9(1): 2595. https://doi.org/10.1038/s41467-018-05016-8
42. Ravichandran S, Hartmann A, del Sol A (2019) SigHotSpotter: scRNA-seq-based computational tool to control cell subpopulation phenotypes for cellular rejuvenation strategies. Bioinformatics 36(6):1963–1965. https://doi.org/10.1093/bioinformatics/btz827
43. Lee E, Piranlioglu R, Wicha MS, Korkaya H (2019) Plasticity and potency of mammary stem cell subsets during mammary gland development. Int J Mol Sci 20(9):2357. https://doi.org/10.3390/ijms20092357
44. de Soysa TY, Ranade SS, Okawa S et al (2019) Single-cell analysis of cardiogenesis reveals

basis for organ-level developmental defects. Nature 572(7767):120–124. https://doi.org/10.1038/s41586-019-1414-x

45. Fujikura J, Yamato E, Yonemura S et al (2002) Differentiation of embryonic stem cells is induced by GATA factors. Genes Dev 16(7): 784–789. https://doi.org/10.1101/gad.968802
46. Yeo JC, Jiang J, Tan ZY et al (2014) Klf2 is an essential factor that sustains ground state pluripotency. Cell Stem Cell 14(6):864–872. https://doi.org/10.1016/j.stem.2014.04.015
47. Ito T, Udaka N, Yazawa T et al (2000) Basic helix-loop-helix transcription factors regulate the neuroendocrine differentiation of fetal mouse pulmonary epithelium. Development 127(18):3913–3921
48. Sandbo N, Kregel S, Taurin S et al (2009) Critical role of serum response factor in pulmonary myofibroblast differentiation induced by TGF-β. Am J Respir Cell Mol Biol 41(3): 332–338. https://doi.org/10.1165/rcmb.2008-0288OC
49. Hnisz D, Abraham BJ, Lee TI et al (2013) Super-enhancers in the control of cell identity and disease. Cell 155(4):934–947. https://doi.org/10.1016/j.cell.2013.09.053
50. Mazzoni EO, Mahony S, Closser M et al (2013) Synergistic binding of transcription factors to cell-specific enhancers programs motor neuron identity. Nat Neurosci 16(9): 1219–1227. https://doi.org/10.1038/nn.3467
51. Wilson NK, Foster SD, Wang X et al (2010) Combinatorial transcriptional control in blood stem/progenitor cells: genome-wide analysis of ten major transcriptional regulators. Cell Stem Cell 7(4):532–544. https://doi.org/10.1016/j.stem.2010.07.016
52. Scialdone A, Tanaka Y, Jawaid W et al (2016) Resolving early mesoderm diversification through single-cell expression profiling. Nature 535(7611):289–293. https://doi.org/10.1038/nature18633
53. Chen WS, Manova K, Weinstein DC et al (1994) Disruption of the HNF-4 gene, expressed in visceral endoderm, leads to cell death in embryonic ectoderm and impaired gastrulation of mouse embryos. Genes Dev 8(20):2466–2477. https://doi.org/10.1101/gad.8.20.2466
54. Coffinier C, Thépot D, Babinet C et al (1999) Essential role for the homeoprotein vHNF1/HNF1β in visceral endoderm differentiation. Development 126(21):4785–4794
55. Vassen L, Beauchemin H, Lemsaddek W et al (2014) Growth factor independence 1b (Gfi1b) is important for the maturation of erythroid cells and the regulation of embryonic globin expression. PLoS One 9(5):e96636. https://doi.org/10.1371/journal.pone.0096636
56. Ross J, Mavoungou L, Bresnick EH, Milot E (2012) GATA-1 utilizes ikaros and polycomb repressive complex 2 to suppress Hes1 and to promote erythropoiesis. Mol Cell Biol 32(18): 3624–3638. https://doi.org/10.1128/mcb.00163-12
57. Das JK, Voelkel NF, Felty Q (2015) ID3 contributes to the acquisition of molecular stem cell-like signature in microvascular endothelial cells: its implication for understanding microvascular diseases. Microvasc Res 98:126–138. https://doi.org/10.1016/j.mvr.2015.01.006
58. Suzuki T, Aizawa K, Matsumura T, Nagai R (2005) Vascular implications of the Krüppel-like family of transcription factors. Arterioscler Thromb Vasc Biol 25(6):1135–1141
59. Gierl MS, Karoulias N, Wende H et al (2006) The zinc-finger factor Insm1 (IA-1) is essential for the development of pancreatic β cells and intestinal endocrine cells. Genes Dev 20(17): 2465–2478. https://doi.org/10.1101/gad.381806
60. Ye DZ, Kaestner KH (2009) Foxa1 and Foxa2 control the differentiation of goblet and enteroendocrine L- and D-cells in mice. Gastroenterology 137(6):2052–2062. https://doi.org/10.1053/j.gastro.2009.08.059
61. Larsson LI, St-Onge L, Hougaard DM et al (1998) Pax 4 and 6 regulate gastrointestinal endocrine cell development. Mech Dev 79(1–2):153–159. https://doi.org/10.1016/S0925-4773(98)00182-8
62. Gross S, Garofalo DC, Balderes DA et al (2016) The novel enterochromaffin marker Lmx1a regulates serotonin biosynthesis in enteroendocrine cell lineages downstream of Nkx2.2. Development 143(14):2616–2628. https://doi.org/10.1242/dev.130682
63. Zetterström RH, Solomin L, Jansson L et al (1997) Dopamine neuron agenesis in Nurr1-deficient mice. Science 276(5310):248–250. https://doi.org/10.1126/science.276.5310.248
64. Ferri ALM, Lin W, Mavromatakis YE et al (2007) Foxa1 and Foxa2 regulate multiple phases of midbrain dopaminergic neuron development in a dosage-dependent manner. Development 134(15):2761–2769. https://doi.org/10.1242/dev.000141

65. La Manno G, Gyllborg D, Codeluppi S et al (2016) Molecular diversity of midbrain development in mouse, human, and stem cells. Cell 167(2):566–580.e19. https://doi.org/10.1016/j.cell.2016.09.027
66. Andersson E, Tryggvason U, Deng Q et al (2006) Identification of intrinsic determinants of midbrain dopamine neurons. Cell 124(2): 393–405. https://doi.org/10.1016/j.cell.2005.10.037
67. Ono Y, Nakatani T, Sakamoto Y et al (2007) Differences in neurogenic potential in floor plate cells along an anteroposterior location: midbrain dopaminergic neurons originate from mesencephalic floor plate cells. Development 134(17):3213–3225. https://doi.org/10.1242/dev.02879
68. Villaescusa JC, Li B, Toledo EM et al (2016) A PBX1 transcriptional network controls dopaminergic neuron development and is impaired in Parkinson's disease. EMBO J 35(18): 1963–1978. https://doi.org/10.15252/embj.201593725
69. del Sol A, Okawa S, Ravichandran S (2019) Computational strategies for niche-dependent cell conversion to assist stem cell therapy. Trends Biotechnol 37(7):687–696
70. Lane SW, Williams DA, Watt FM (2014) Modulating the stem cell niche for tissue regeneration. Nat Biotechnol 32(8):795–803. https://doi.org/10.1038/nbt.2978
71. Neves J, Sousa-Victor P, Jasper H (2017) Rejuvenating strategies for stem cell-based therapies in aging. Cell Stem Cell 20(2):161–175
72. Liu Y, Beyer A, Aebersold R (2016) On the dependency of cellular protein levels on mRNA abundance. Cell 165(3):535–550
73. Wang LD, Wagers AJ (2011) Dynamic niches in the origination and differentiation of haematopoietic stem cells. Nat Rev Mol Cell Biol 12(10):643–655
74. Wang Y, Yao F, Wang L et al (2020) Single-cell analysis of murine fibroblasts identifies neonatal to adult switching that regulates cardiomyocyte maturation. Nat Commun 11(1):2585. https://doi.org/10.1038/s41467-020-16204-w
75. Kalamakis G, Brüne D, Ravichandran S et al (2019) Quiescence modulates stem cell maintenance and regenerative capacity in the aging brain. Cell 176(6):1407–1419.e14. https://doi.org/10.1016/j.cell.2019.01.040
76. Ying QL, Wray J, Nichols J et al (2008) The ground state of embryonic stem cell self-renewal. Nature 453(7194):519–523. https://doi.org/10.1038/nature06968
77. Yang H, Adam RC, Ge Y et al (2017) Epithelial-mesenchymal micro-niches govern stem cell lineage choices. Cell 169(3): 483–496.e13. https://doi.org/10.1016/j.cell.2017.03.038
78. Baumgartner C, Toifl S, Farlik M et al (2018) An ERK-dependent feedback mechanism prevents hematopoietic stem cell exhaustion. Cell Stem Cell 22(6):879–892.e6. https://doi.org/10.1016/j.stem.2018.05.003
79. Huang J, Zhang Y, Bersenev A et al (2009) Pivotal role for glycogen synthase kinase-3 in hematopoietic stem cell homeostasis in mice. J Clin Invest 119(12):3519–3529. https://doi.org/10.1172/JCI40572
80. Liu Y-F, Zhang S-Y, Chen Y-Y et al (2018) ICAM-1 deficiency in the bone marrow niche impairs quiescence and repopulation of hematopoietic stem cells. Stem Cell Rep 11(1): 258–273. https://doi.org/10.1016/j.stemcr.2018.05.016
81. Kowalczyk MS, Tirosh I, Heckl D et al (2015) Single-cell RNA-seq reveals changes in cell cycle and differentiation programs upon aging of hematopoietic stem cells. Genome Res 25(12):1860–1872. https://doi.org/10.1101/gr.192237.115
82. Liu X, Lu Y, Zhang Y et al (2012) Slit2 regulates the dispersal of oligodendrocyte precursor cells via Fyn/RhoA signaling. J Biol Chem 287(21):17503–17516. https://doi.org/10.1074/jbc.M111.317610
83. Zhang Y, Argaw AT, Gurfein BT et al (2009) Notch1 signaling plays a role in regulating precursor differentiation during CNS remyelination. Proc Natl Acad Sci U S A 106(45): 19162–19167. https://doi.org/10.1073/pnas.0902834106
84. Azim K, Butt AM (2011) GSK3β negatively regulates oligodendrocyte differentiation and myelination in vivo. Glia 59(4):540–553. https://doi.org/10.1002/glia.21122
85. Xu J, Du Y, Deng H (2015) Direct lineage reprogramming: strategies, mechanisms, and applications. Cell Stem Cell 16(2):119–134
86. Berneman-Zeitouni D, Molakandov K, Elgart M et al (2014) The temporal and hierarchical control of transcription factors-induced liver to pancreas transdifferentiation. PLoS One 9(2): e87812. https://doi.org/10.1371/journal.pone.0087812
87. Browaeys R, Saelens W, Saeys Y (2020) NicheNet: modeling intercellular communication by linking ligands to target genes. Nat Methods 17(2):159–162. https://doi.org/10.1038/s41592-019-0667-5

88. Camp JG, Sekine K, Gerber T et al (2017) Multilineage communication regulates human liver bud development from pluripotency. Nature 546(7659):533–538. https://doi.org/10.1038/nature22796
89. Skelly DA, Squiers GT, McLellan MA et al (2018) Single-cell transcriptional profiling reveals cellular diversity and intercommunication in the mouse heart. Cell Rep 22(3): 600–610. https://doi.org/10.1016/j.celrep.2017.12.072
90. Raredon MSB, Adams TS, Suhail Y et al (2019) Single-cell connectomic analysis of adult mammalian lungs. Sci Adv 5(12):eaaw3851. https://doi.org/10.1126/sciadv.aaw3851
91. Bell AJ (2003) Co-information lattice. In: 4th int symp indep compon anal blind source
92. Stewart JW (1994) An introduction to the numerical solution of Markov chains. Princeton University Press, New Jersey
93. Zhang HM, Chen H, Liu W et al (2012) AnimalTFDB: a comprehensive animal transcription factor database. Nucleic Acids Res 40 (Database issue):D144–D149. https://doi.org/10.1093/nar/gkr965
94. Türei D, Korcsmáros T, Saez-Rodriguez J (2016) OmniPath: guidelines and gateway for literature-curated signaling pathway resources. Nat Methods 13(12):966–967
95. Wu G, Feng X, Stein L (2010) A human functional protein interaction network and its application to cancer data analysis. Genome Biol 11(5):R53. https://doi.org/10.1186/gb-2010-11-5-r53
96. Zaffaroni G, Okawa S, Morales-Ruiz M, Del Sol A (2019) An integrative method to predict signalling perturbations for cellular transitions. Nucleic Acids Res 47(12):e72. https://doi.org/10.1093/nar/gkz232
97. Csardi G, Nepusz T (2006) The igraph software package for complex network research. InterJournal Complex Syst 1695(5):1–9
98. Shannon P, Markiel A, Ozier O et al (2003) Cytoscape: a software environment for integrated models of biomolecular interaction networks. Genome Res 13(11):2498–2504. https://doi.org/10.1101/gr.1239303

Chapter 5

Differential Proteomic Analysis of Complex Mixtures by Label-Free nLC MS/MS

Iraide Escobés, Mikel Azkargorta, Ibon Iloro, and Felix Elortza

Abstract

After over two decades of constant evolution, proteomics can be truly considered nowadays as a high-throughput technique. Latest advances performed in sample preparation, instrumentation, and data analysis tools enable proteome-wide detection and quantification of proteins in complex samples.

Label-free quantification by nanoscale liquid chromatography coupled online to tandem mass spectrometry (nLC MS/MS) is a straightforward procedure for relative protein quantification. This approach allows to get deeper insights of what molecular changes are involved in the biological system we want to study in an unbiased manner.

This chapter describes methods for sample preparation prior to mass spectrometry analysis. Besides, we describe a standard acquisition method, and some common bioinformatics analyses that help extracting biologically relevant information out of the achieved data.

Key words Mass spectrometry, Proteome, Proteomics, Label-free quantification, Relative protein quantification, Biomarker discovery, Precision-medicine

1 Introduction

The proteome is the expression of a genome in any given moment. It may vary on time and in response to different stimulus. Besides, molecular changes associated to any disease will be mirrored on the proteome. Proteins are the major effectors of molecular activity within cell and organs, and they are the targets to most of the drugs used in clinics. Thus, the interest on proteomics is attracting increasing attention. The proteome is a complex entity and the analytical challenges that its analysis faces are high due to different reasons. Having approximately 20,000 genes, a mammalian cell can display around 10,000–12,000 proteins and they are expressed showing a high dynamic range of protein concentration. Moreover, there is a large number of different existing proteoforms (e.g., proteins generated from same gene but showing different aminoacid sequence due to alternative splicing) without mentioning the

Maria dM. Vivanco (ed.), *Mammary Stem Cells: Methods and Protocols*,
Methods in Molecular Biology, vol. 2471, https://doi.org/10.1007/978-1-0716-2193-6_5,

existing different post-translational modifications (e.g., phosphorylation, acetylation, methylation, glycosylation, ubiquitylation, etc.) that proteins may suffer during their life time having a direct impact on their activity, localization, etc. [1, 2].

In a typical proteomic analysis, proteins extracted from biological samples, such as tissues, cells, or biofluids, are enzymatically digested into peptides. These peptides are the analytes for the so-called shot-gun proteomics and they are resolved in nanoscale reverse-phase liquid chromatography and coupled online by electrospray to a tandem mass spectrometer (nLC MS/MS). Advances in chromatography and mass spectrometry instrumentation enable the analysis of unfractionated samples with an up to now unforeseen sensitivity, throughput, and depth. Label-free nLC MS/MS quantification approach is lately getting momentum since it is rather straightforward and inexpensive comparing to other methods (e.g., isobaric labeling methods). Following the nLC MS/MS data acquisition, dedicated software is used for protein identification and quantification (Fig. 1).

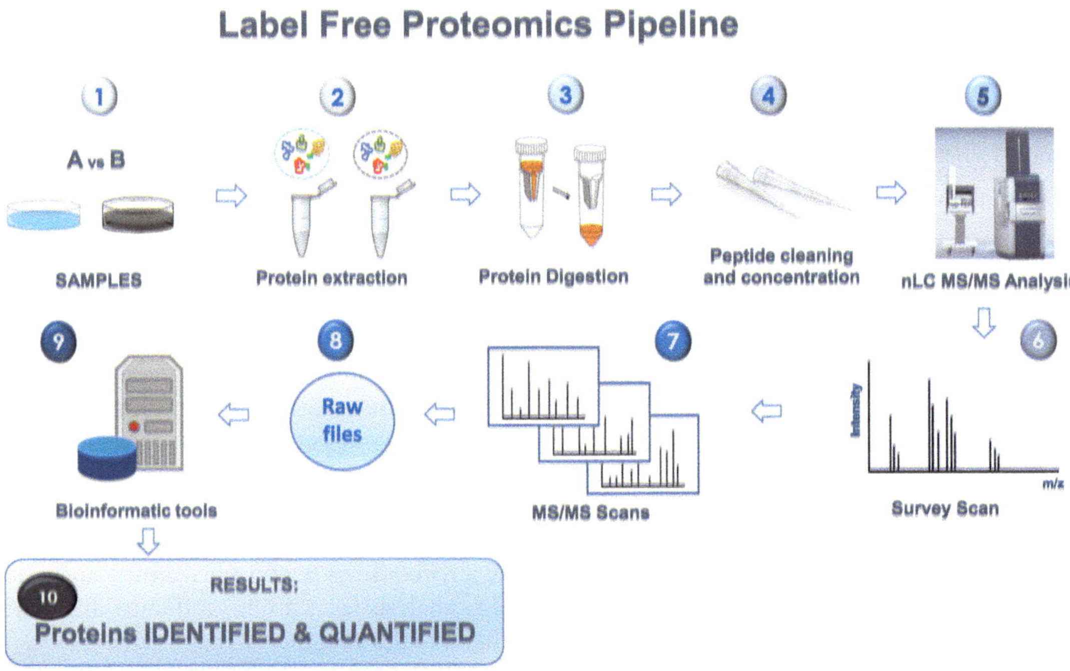

Fig. 1 General scheme of label-free quantification proteomics. (1) Typical differential proteomic study starting from the analysis of two different cell cultures (e.g., treated vs. non-treated); (2) Proteins are extracted from cells; (3) Proteins are digested by trypsin; (4) Resulting peptides are concentrated and desalted by reversed-phase C18 microcolumns; (5, 6, 7) nLC MS/MS analysis is performed where MS (survey scans) and MS/MS scans alternate to give a raw file (8) as an output; (9) Raw data is processed; (10) Proteins are identified and quantified

Fig. 2 Bioinformatic pipeline for label-free data analysis. Once nLC MS/MS is performed, proteins are quantified via Andromeda within MaxQuant (1); later, statistics are applied on Perseus (2), so the list of proteins considered deregulated (up or down) is obtained. Then, different bioinformatic tools can be used to get biological information: DAVID (Database for Annotation, Visualization and Integrated Discovery) provides a set of functional annotation showing the biological meaning behind (3); PANTHER (Protein ANalysis THrough Evolutionary Relationships) classification system allows the classification of proteins according to family (and subfamily), molecular function, biological process and pathways (4); KEGG (Kyoto Encyclopedia of Genes and Genomes) is a database resource which aim is to help in the understanding of high-throughput -omics type of experiments (5). Finally, Ingenuity Pathway Analysis can be used since it may predict downstream effects and identify new targets or candidate biomarkers (6)

Thus, despite the proteome complexity, by using the latest combination of proteomics techniques and methodologies we can nowadays generate near proteome-wide quantitative measurements that together with bioinformatics tools hereby described (e.g., DAVID, Ingenuity Pathway Analysis, etc.) shall help getting deeper insights about the molecular events we want to study (Fig. 2).

2 Materials

Reagents that are routinely used in our lab, and therefore considered as reliable and useful for the protocol provided, are recommended in the following section. Any other source for these materials might be as useful as the products mentioned. However, always check that all the solutions and buffers are prepared using MS or proteomics-grade reagents.

2.1 Sample Preparation

1. PBS: Phosphate Buffer Saline.
2. CLB, Cell Lysis Buffer (CLB): 7 M Urea, 2 M Thiourea, 4% CHAPS (3-[(3-Cholamidopropyl)dimethylammonio]-1-propanesulfonate).

3. UA: Urea solution: 8 M Urea in 100 mM Tris-HCl pH 8.5.
4. AMBIC: Ammonium bicarbonate 99.5% (*see* **Note 1**).
5. DTT: Dithiothreitol (1 M stock): 0.154 g DTT in 1 mL Milli-Q water. Store at −20 °C in 10 μL aliquots.
6. 10 mM DTT: Dilute a 10 μL aliquot of 1 M DTT with 990 μL UA (prepare it fresh).
7. IAA: Iodoacetamide (55 mM): Dissolve 10.17 g of IAA in 1 mL UA (prepare it fresh).
8. FA: Formic acid.
9. FA (0.1%): Dissolve 1 mL FA in 999 mL Milli-Q water (*see* **Note 2**).
10. ACN: Acetonitrile.
11. Speed Vac: Rotational-Vacuum-Concentrator.
12. Reverse-Phase Microcolumns (such as Omix C18 Pipette tips, 2–10 μL, Agilent).
13. Optimal water LC/MS quality.
14. Trypsin Gold-Mass Spectrometry Grade (100 μg).
15. Trypsin Gold 2 μg/μL stock (*see* **Note 3**).
16. Trypsin gold 1 μg/μL: Dilute the 5 μL aliquot of Trypsin gold 2 μg/μL with 5 μL AMBIC 50 mM.
17. Amicon 30 kDa cut-off filter (Merck).
18. BSA: Bovine serum albumin.
19. Bradford reagent: Protein Assay.
20. ACN/FA (0.1%) [70:30] (for 25 mL mix): 17.5 mL CAN, 7.5 mL FA 0.1%.
21. 0.1% FA/3% CAN (for 25 mL mix): 750 μL CAN, 24.25 mL FA 0.1%.

2.2 MS Analysis

Described nLC MS/MS system together with analytical parts, such as columns, can vary from lab to lab. Nevertheless, the hereby described systems allow us to get a very good performance in a cost-efficient manner.

1. FA: Formic Acid.
2. FA (0.1% in water): 1 mL FA in 999 mL Milli-Q water (*see* **Note 2**).
3. FA (0.1% in CAN): 1 mL FA in 999 mL CAN (*see* **Note 2**).
4. ACN: Acetonitrile.
5. Glass vials (LCGC certified clear glass 12 × 32 mm total recovery with cap and Prelist PTFE/Silicone, Waters).
6. Nanoscale Ultra performance liquid chromatography (UPLC) system (nanoElute, Bruker).

7. Reversed-phase UPLC trapping column (Acclaim PepMap 100 C18 (Thermo scientific, PN164564), 100 μm × 2 cm, 5 μm, 100 A).
8. Reversed-phase UPLC column for peptide separation (Bruker Fifteen C18 (Bruker Daltonics, PN1842621), 75 μm × 15 cm, 1.9 μm, 120 A).
9. Mass spectrometer for large-scale proteomics (TIMS Tof Pro, Bruker).
10. Ultrasonic cleaning bath.

3 Methods

Prepare all solutions using Milli-Q water (18 MΩ cm at 25 °C) and MS-grade reagents. Prepare and store all reagents at 4 °C (unless indicated otherwise) and follow all waste disposal regulations when disposing waste material.

3.1 Protein Extraction

1. Wash cells three times with cold PBS to avoid any carryover of fetal bovine serum proteins present in cell culture media. Add PBS at 4 °C enough to cover cells, spin down cells, and discard supernatant (repeat three times).
2. Lyse cells in Cell Lysis Buffer (CLB). Two hundred microliters CLB will be added to pelleted cells. Shake it in thermomixer at room temperature for 60 min (*see* **Note 4**).
3. Sonicate the tube for 3 min in a sonication bath to reduce viscosity by breaking DNA (*see* **Note 5**).
4. Spin down debris at 16,000 × *g* at room temperature for 10 min.
5. Take supernatant and transfer it to a new Eppendorf tube.

3.2 Protein Quantification

1. Use BSA as standard in duplicate in a 96-well plate: Mix 187 μL Milli-Q water and 13 μL BSA [1 μg/μL]; mix and make 8 following serial dilution by mixing 100 μL of previous dilution with 100 μL Milli-Q water; use 100 μL of Milli-Q water as blank in an extra well.
2. Prepare sample by mixing 100 μL Milli-Q and 1 μL sample (*see* **Note 6**).
3. Add 100 μL of Bradford dye reagent to each well (*see* **Note 7**).
4. Read plate at 595 nm in the following 30 min (*see* **Note 8**).

3.3 Protein Digestion by FASP (Filter Aided Sample Preparation)

The method described here is based in the FASP method, described by Wiśniewski et al. [3].

1. Activate the 30 KDa cut-off amicon filter by adding 200 μL UA and centrifuge by applying 16,000 × *g* 15 min at R/T.

2. Take 30 μL of lysate and add 200 μL of UA solution (*see* **Note 9**).
3. Take the mixture and transfer it to 30 kDa cut-off amicon filter.
4. Locate the filter within the corresponding centrifugation tube.
5. Filter by centrifugation applying 16,000 × *g* 15 min at R/T.
6. Add 200 μL more UA to the filter and repeat **step 4**.
7. Discard the filtered/flow-through liquid.
8. Add 200 μL 10 mM DTT in UA to the filter. Place the filter in a thermomixer and apply 600 rpm for 1 min at R/T. Incubate 30 min in thermomixer at 600 rpm, 56 °C.
9. Filter by centrifugation applying 16,000 × *g* 15 min at R/T.
10. Add 100 μL 50 mM IAA in UA to the filter.
11. Place the filter in a thermomixer and apply 600 rpm for 1 min at R/T; then let the sample 20 min at R/T and in the dark.
12. Filter by centrifugation applying 16,000 × *g* 15 min at R/T.
13. Add 200 μL more UA to the filter and filter by centrifugation applying 16,000 × *g* 15 min at R/T.
14. Repeat **step 12** two more times.
15. Add 100 μL AMBIC 100 mM to the filter and filter by centrifugation applying 16,000 × *g* 15 min at R/T.
16. Repeat **step 14** two more times.
17. Add the trypsin (1 g/μL) diluted in 50 mM AMBIC in a 1:10 trypsin/protein ratio.
18. Add 80 μL AMBIC 50 mM to the filter.
19. Place the filter in a thermomixer and apply 600 rpm for 1 min at R/T and incubate for digestion overnight at 37 °C (16 h).
20. Place the filter in a new tube and filter by centrifugation applying 16,000 × *g* 15 min at R/T. Take care: Digested peptides are now in the filtered faction! Do not discard it.
21. Add 50 μL 50 mM AMBIC to the filter and filter by centrifugation applying 16,000 × *g* 15 min at R/T.
22. Take filtered fraction and transfer it to a new Eppendorf tube.
23. Digested peptides are dried by speed vacuum.

3.4 Peptide Concentration and Desalting by Reverse-Phase Microcolumns

1. Resuspend the dried samples adding 10 μL 0.1% FA and sonicate for 5 min in the ultrasonic cleaning bath.
2. C18 reverse-phase microcolumn conditioning: Flush the reverse-phase tip with 10 μL ACN two times.
3. Flush the reverse-phase tip with 10 μL 0.1% FA two times.
4. Load the sample onto reverse-phase tip by taking and discarding 10 μL. Repeat this last **step 10** times (*see* **Note 10**).

5. Wash peptides with 10 μL 0.1% FA (*see* **Note 11**).
6. Elute the peptides using a solution containing [70:30] ACN/0.1% FA. For this purpose, take 10 μL of the eluting solution and release them in a clean Eppendorf tube. Pipette up and down this volume two times in order to release the sample completely from the tip.
7. Dry vacuum the samples in the speed vac. Samples can be stored at −20 °C.

3.5 Sample Load onto the Chromatographic System

1. Resuspend the sample in 20 μL 0.1% FA/3%ACN. Sonicate 5 min in the ultrasonic cleaning bath.
2. Put the resuspended sample in a vial and load the sample into the mass spectrometer (*see* **Note 12**).
3. Peptides are separated on the UPLC system using a combination of reversed-phase trapping and separation columns.
4. The recommended chromatographic gradient may depend on the complexity of the sample. For total cell extracts gradients, and with the columns stated above, between 60 and 90 min can be performed (Table 1). In general, with longer gradients peptides are better resolved and therefore more proteins can be identified and therefore more proteins will be quantified (*see* **Note 13**).

3.6 MS Acquisition

Optimal sample loading is of paramount importance when using the novel TIMS Tof Pro mass spectrometer. The system shows it best when loading 100–200 ng of peptide mixtures. Once the loading is well stablished, default acquisition method routinely can be used for the analysis of tryptic peptides [4].

3.7 MS Data Analysis

Once nLC MS/MS is performed, raw files need to be processed. The processed raw file (e.g., mgf) is then searched against a selected database for protein identification. Typical searching parameters are shown in Table 2. Decoy search is necessary to know the false discovery rate (FDR) at peptide and protein level.

3.8 Label-Free Quantification

Mammary stem cell (or any other cell type) MS results can be analyzed by freely online available MaxQuant software (Max Plank Institute, Germany) in order to obtain accurate label-free relative quantification data. To do so, raw files are uploaded into MaxQuant and the software performs both protein identification (using Andromeda search engine) and quantification, without the need of stable isotope compounds [5, 6]. Instead, a spectral counting or precursor signal intensity approach can be used. Once the MaxQuant program generates the quantification data matrix, other program must be used to perform quantification interpretation and corresponding statistical analysis. Perseus (also online and freely

Table 1
Solvents: A) 0.1% Formic Acid in H2O; B) 0.1 % Formic Acid in Acetonitrile

Time (min)	*A* %	*B* %	Flow
0	97	3	0.3 mL/min
60	60	40	"
61	15	85	"
70	15	85	"
72	97	3	"
90	97	3	"

Table 2
Parameters for database search

Parameter	Default
Precursor mass tol.	20 ppm
Fragment mass tol.	0.05 Da
Dynamic mods.	Oxidation (M) Acetyl (N-term)
Fixed mods.	Carbamidomethyl (C)

available) software can be the tool of choice (Max Plank Institute, Germany) [7]. This software allows high dimensional omics data analysis, covering normalizations, comparisons, and multiple-hypothesis testing.

3.9 Bioinformatic Analysis

Once we obtain the list of proteins that are deregulated, hereby we describe some basic analyses that can be performed in order to extract meaningful biological information.

Using the free online DAVID functional annotation tool ((Database for Annotation, Visualization and Integrated Discovery; https://david.ncifcrf.gov/) can be a good option to start up with the analysis [8, 9]. Usually enrichment in Biological Processes (BP), Cellular Component (CC), and Molecular Function (MF) are analyzed, but the program allows plenty of other interesting analyses, such as KEGG (Kyoto Encyclopedia of Genes and Genomes) Pathways. Processes enriched with a Fisher's exact test p value <0.05 are considered as significantly enriched, although more restrictive cutoffs such as $<5\%$ FDR can also be used for a more reliable analysis. PANTHER (Protein ANalysis THrough Evolutionary Relationships; http://www.pantherdb.org/) classification system is another free tool worth using since it allows the classification of proteins according to family (and subfamily), molecular function, biological process, and pathways [10–12].

In addition, Ingenuity Pathway Analysis (IPA, by QIAGEN) can be performed. This is a web-based functional analysis tool for omics data analysis [13]. It allows visualization of the different biological processes in a graphical way helping to place the obtained experimental results within the context of the studied biological systems. The approach is widely used since it may: provide key regulators and activity to explain expression patterns, predict downstream effects on biological and disease processes, and build interactive models of experimental systems.

4 Notes

1. AMBIC solution does not need pH adjustment (should be around pH 8). 1 M Stock: Dissolve 3.9 g AMBIC in 50 mL water. Store at −20 °C in 500 μL aliquots. AMBIC 100 mM: 500 μL Stock, 4.5 mL Water. AMBIC 50 mM: 500 μL stock, 9.5 mL Water.
2. Do this in glassware instead of plasticware. It is recommended to use dedicated glassware for this purpose so no detergents or any other contaminants interfere with the MS analysis.
3. Dissolve a 100 μg vial of trypsin gold in 50 μL of 50 mM acetic acid. Aliquots (5 μL) can be stored at −20 °C for at least 1 month [2 μg/μL].
4. Do not heat CLB since present urea can cause carbamylation at the N-termini of proteins/peptides and at the side chain amino groups of lysine and arginine residues. This would have a direct and negative impact on the proteomic analysis.
5. If sample gets highly viscose, add extra 200 μL CLB to get rid of DNA by shaking it 60 min in thermomixer.
6. For a 96-well plate prepare Bradford dye solution by mixing 6 mL Milli-Q and 4 mL Bradford in a 10 mL Falcon tube.
7. If sample is too concentrated, dilute it five times or as necessary. Then add 1 μL of diluted sample to 100 μL Milli-Q water and keep on with the procedure.
8. Waiting for longer period could affect the Bradford assay.
9. 1×10^6 cells will typically render 100 μg of protein. In our laboratory, we have observed that the sensitivity of this modified FASP method is valid for 3×10^5 for most cell types. About starting protein amount, we have observed that FASP method works perfectly with as few as 20 μg of total protein as starting material.

10. Sample recovery may increase with the number of uptakes. However, ten times is the recommended number. C18 typical load limit (2 μg for the product referenced in this chapter) should be kept in mind.
11. Flushing 2–3 times is recommended. Washing the samples too much is not recommended since part of the sample might get lost.
12. Sample resuspension volume and amount to be loaded may depend on the starting material and used mass spectrometer. TIMS Tof Pro (Bruker) is a very sensitive equipment when parallel accumulation-serial fragmentation (PASEF) mode is activated; and best performance is achieved with 100–200 ng. Nevertheless, if other type of equipment is used for nLC MS/MS it is possible that higher amounts of digested peptides may be required (typically up to 1 μg).
13. The columns and gradient are illustrative. However, a linear gradient followed by a washing step and an equilibration step are needed as part of the protocol. Length of each phase may depend on sample load, sample complexity and/or column length among others. Adaptation to the system and optimization are therefore needed.

Acknowledgments

CIC bioGUNE is supported by Basque Department of Industry, Tourism and Trade (Etortek and Elkartek programs), the Innovation Technology Department of the Bizkaia County. The Proteomic Platform is supported by The ProteoRed-ISCIII (Grant PRB3 IPT17/0019); CIBERehd Network and Severo Ochoa Grant (SEV-2016-0644).

References

1. Aebersold R, Mann M (2016) Mass-spectrometric exploration of proteome structure and function. Nature 537(7620): 347–355. https://doi.org/10.1038/nature19949
2. Altelaar AF, Munoz J, Heck AJ (2013) Next-generation proteomics: towards an integrative view of proteome dynamics. Nat Rev Genet 14(1):35–48. https://doi.org/10.1038/nrg3356
3. Wisniewski JR, Zougman A, Nagaraj N, Mann M (2009) Universal sample preparation method for proteome analysis. Nat Methods 6(5):359–362. https://doi.org/10.1038/nmeth.1322
4. Meier F, Brunner AD, Koch S, Koch H, Lubeck M, Krause M, Goedecke N, Decker J, Kosinski T, Park MA, Bache N, Hoerning O, Cox J, Rather O, Mann M (2018) Online parallel accumulation-serial fragmentation (PASEF) with a novel trapped ion mobility mass spectrometer. Mol Cell Proteomics 17(12):2534–2545. https://doi.org/10.1074/mcp.TIR118.000900
5. Cox J, Hein MY, Luber CA, Paron I, Nagaraj N, Mann M (2014) Accurate proteome-wide label-free quantification by delayed normalization and maximal peptide ratio extraction, termed MaxLFQ. Mol Cell

Proteomics 13(9):2513–2526. https://doi.org/10.1074/mcp.M113.031591

6. Cox J, Mann M (2008) MaxQuant enables high peptide identification rates, individualized p.p.b.-range mass accuracies and proteome-wide protein quantification. Nat Biotechnol 26(12):1367–1372. https://doi.org/10.1038/nbt.1511
7. Tyanova S, Temu T, Sinitcyn P, Carlson A, Hein MY, Geiger T, Mann M, Cox J (2016) The Perseus computational platform for comprehensive analysis of (prote)omics data. Nat Methods 13(9):731–740. https://doi.org/10.1038/nmeth.3901
8. Huang da W, Sherman BT, Lempicki RA (2009) Systematic and integrative analysis of large gene lists using DAVID bioinformatics resources. Nat Protoc 4(1):44–57. https://doi.org/10.1038/nprot.2008.211
9. Huang da W, Sherman BT, Lempicki RA (2009) Bioinformatics enrichment tools: paths toward the comprehensive functional analysis of large gene lists. Nucleic Acids Res 37(1):1–13. https://doi.org/10.1093/nar/gkn923
10. Mi H, Muruganujan A, Ebert D, Huang X, Thomas PD (2019) PANTHER version 14: more genomes, a new PANTHER GO-slim and improvements in enrichment analysis tools. Nucleic Acids Res 47(D1): D419–D426. https://doi.org/10.1093/nar/gky1038
11. Mi H, Muruganujan A, Huang X, Ebert D, Mills C, Guo X, Thomas PD (2019) Protocol update for large-scale genome and gene function analysis with the PANTHER classification system (v.14.0). Nat Protoc 14(3):703–721. https://doi.org/10.1038/s41596-019-0128-8
12. Mi H, Thomas P (2009) PANTHER pathway: an ontology-based pathway database coupled with data analysis tools. Methods Mol Biol 563:123–140. https://doi.org/10.1007/978-1-60761-175-2_7
13. Kramer A, Green J, Pollard J Jr, Tugendreich S (2014) Causal analysis approaches in ingenuity pathway analysis. Bioinformatics 30(4): 523–530. https://doi.org/10.1093/bioinformatics/btt703

Chapter 6

Orthotopic Transplantation of Mouse Mammary Epithelial Cells

Marisa M. Faraldo, Marina A. Glukhova, and Marie-Ange Deugnier

Abstract

The orthotopic transplantation assay has provided important insights into mammary development, stem cell function, and tumorigenesis. Technically, it consists in grafting mammary tissue fragments, organoids, mammospheres, or isolated cells into the fat pads of prepubertal mice from which the endogenous epithelium has been surgically removed, thereby allowing growth and differentiation of mammary epithelial cells in their physiological environment. Here, we describe how is conducted transplantation of epithelial fragments and cells isolated from mouse mammary glands, report the various approaches currently used to evaluate the regeneration and self-renewal properties of mammary stem cells, and highlight the strengths and limitations of this in vivo grafting assay.

Key words Mammary epithelium, Transplantation, Tissue regeneration, Stem cell, Tumor-initiating cell

1 Introduction

The mammary gland consists in a ramified epithelial tree embedded into a complex stroma, also known as mammary fat pad. The mammary epithelium is a hormonally controlled tissue whose development and remodeling are mainly postnatal. Puberty is marked by the rapid growth of branched epithelial ducts, whereas each pregnancy leads to the formation of a dense network of milk-secreting alveoli that involutes upon weaning.

The postnatal mammary epithelium has remarkable regenerative capacities, as evidenced by the formation of new secretory tissue during each pregnancy and, experimentally, in orthotopic transplantation assays. This in vivo grafting assay, originally described by DeOme et al. in 1959 [1], is still widely used to analyze the cellular and molecular mechanisms of mammary development as well as mammary tumor initiation and progression [2–6].

Maria dM. Vivanco (ed.), *Mammary Stem Cells: Methods and Protocols*,
Methods in Molecular Biology, vol. 2471, https://doi.org/10.1007/978-1-0716-2193-6_6,

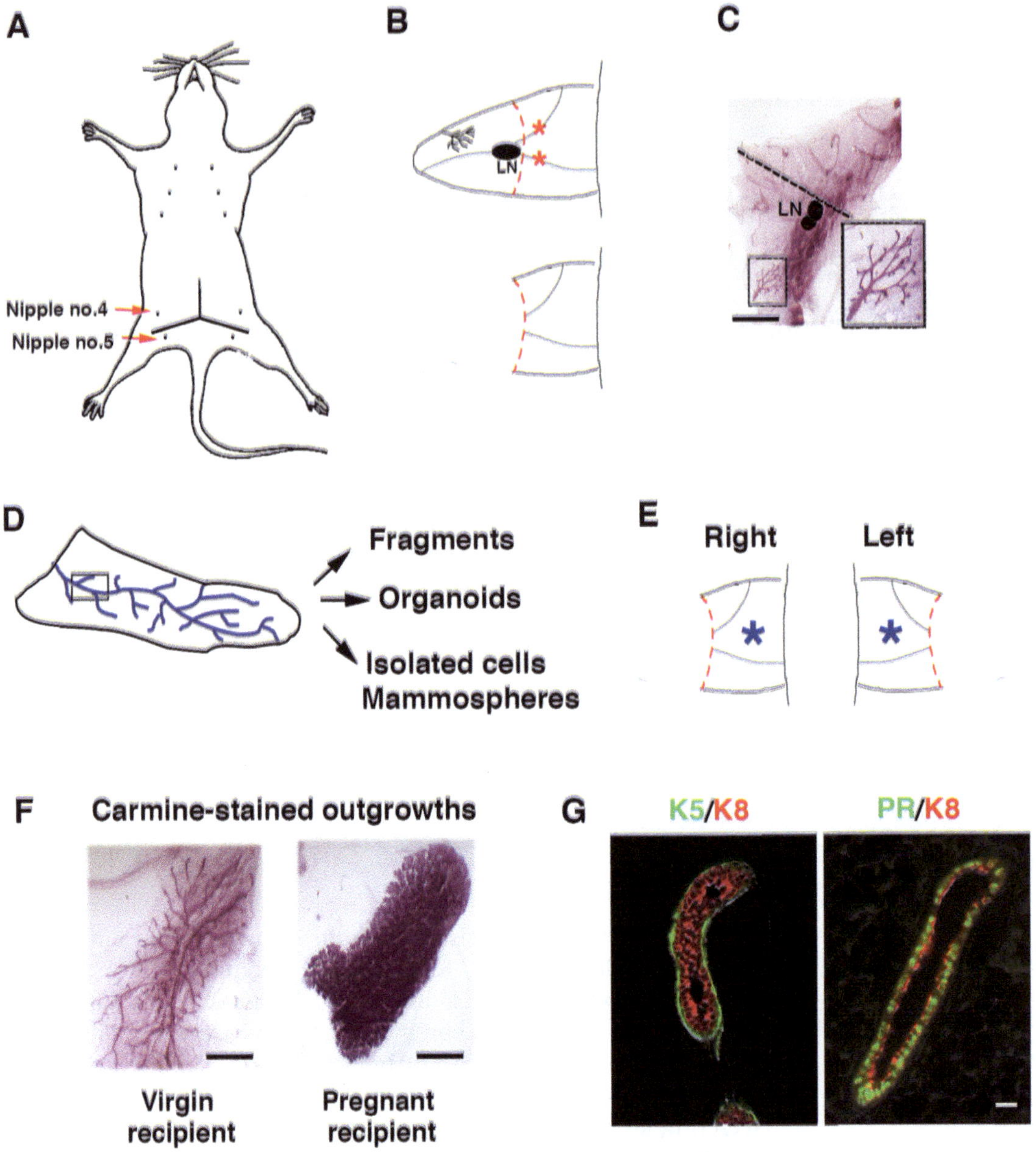

Fig. 1 Transplantation of mouse mammary epithelial cells into the cleared mammary fat pad and outgrowth analysis. (**a**) Preparation of the prepubertal recipient mouse. One ventral midline incision, 1–1.5 cm long, starting a few mm above the pelvis, and two lateral incisions ending between the nipples of the glands #4 (inguinal) and #5 are made. (**b**) Cauterization and excision of the fat pad portion containing the endogenous epithelium from gland #4. *LN* lymph node. Asterisks indicate the cauterization sites of the two arteries leading to the node. (**c**) Carmine-stained whole mount of an inguinal mammary gland from a 3-week-old recipient mouse showing the presence of a rudimentary epithelial tree. Bar, 2.5 mm. (**d**) Preparation of the cellular material from a donor mouse mammary gland. Mammary epithelial fragments, organoids, isolated cells, and mammospheres can be used for transplantation. (**e**) Grafting of the prepared material into the de-epithelialized fat pad of the recipient mouse. Contralateral cleared fat pads can be engrafted with mammary specimens from different donor mice. Asterisks indicate the grafting sites. (**f**) Carmine-stained whole-mount preparations of outgrowths obtained after injecting freshly isolated mammary basal cells. Outgrowths were analyzed

In orthotopic transplantation assays, the fat pad of a prepubertal recipient mouse cleared of the endogenous rudimentary ductal tree serves as a physiological environment for grafting mammary fragments or epithelial cells originating from a donor (Fig. 1). After several weeks, the transplanted tissue or cells expand, repopulating the host fat pad. When normal cells are grafted, they give rise to a ramified epithelial tree that displays normal morphogenetic and differentiation features in response to the hormonal context of the recipient mouse (Fig. 1). Serial transplantations can be carried out to evaluate the repopulating activity of the tested specimens at long term. This strategy is commonly used to reveal the presence of tissue-resident stem cells with self-renewal capacity [2, 3]. Both mouse and human mammary cells have been successfully grafted in the cleared mammary fat pad of recipient mice. For xenografts, immunodeficient host mice such as NOD/SCID are required and their fat pads are humanized by injecting human fibroblasts [7–9].

The mammary epithelium is composed of luminal and basal myoepithelial cells sitting on a basement membrane [2, 3]. During lactation, the secretory luminal cells produce milk components, whereas the contractile myoepithelial cells serve for milk expulsion. Cell separation by flow cytometry has allowed to examine the in vivo regenerative capacities of luminal and basal cell populations freshly isolated from the normal mammary epithelium [10–12]. Numerous studies have now shown that only the basal cell population is capable of regenerating a bilayered epithelial tree in vivo, including at long term upon serial transplantation, allowing to conclude that the basal compartment contains multipotent stem cells [3, 5]. Similarly, the fractionation of breast tumor cells has permitted to identify tumor-initiating cells possessing stem cell-like properties [6, 13]. It is worth noting that tissue fragments or cells isolated from mammary malignant or premalignant lesions can be perpetually regrafted, whereas normal mammary tissue fragments can be successfully retransplanted five or six times only [14, 15].

Most of the cell surface markers used to enrich mammary stem cells or tumor-initiating cells are adhesion molecules, such as integrins, EpCAM, CD24, and CD44 [3]. This indicates that cell-cell and cell-matrix adhesion are important for maintaining a functional stem cell population. Notably, basal rather than luminal cells are enriched in integrins binding to laminins, the major components of

Fig. 1 (continued) 10 weeks post-transplantation in a virgin recipient mouse (left) and a pregnant host previously mated (right). The grafted epithelium gave rise to a typical ductal network in the virgin host and to lobulo-alveolar development during pregnancy. Note that due to surgery, the outgrowths are not connected to the nipple and remain therefore improper for pup feeding. Bar, 4 mm (left), 6 mm (right). (**g**) Immunodetection of basal-specific (*K5* keratin5) and luminal-specific (*K8* keratin 8, *PR* progesterone receptor) markers in ductal structures derived from transplanted basal cells. Bar, 30 μm

the mammary basement membrane [16]. Transplantation of isolated basal cells suspended in 25–50% Matrigel, a laminin-rich basement membrane preparation, significantly increases the take rate for outgrowths, probably by increasing the survival of transplanted cells and favoring early steps in morphogenesis, such as cell polarization [17].

Mammary stem cell activity can be estimated using limiting dilution assays [18]. To this aim, the single-cell suspensions to be tested are serially diluted and transplanted into the cleared fat pads of host mice. The frequency of mammary regenerating cells, i.e., functional stem cells, is calculated from the rates of successful outgrowth formation for each dilution (http://bioinf.wehi.edu.au/software/elda/index.html). The reported frequency of regenerating cells in the mammary basal compartment varies between studies, depending on mouse strain, genetic background, details of tissue digestion, cell sorting, and transplantation protocols, but it does not exceed 5% for wild-type mammary epithelium [3, 10, 11, 19]. The same strategy is used to estimate the frequency of tumor-initiating cells [20].

Importantly, orthotopic transplantation techniques allow to compare mammary development and stem cell activity in specimens originating from two different donors, such as wild-type and mutant mice. For such assays, the specimens to be compared are transplanted into contralateral fat pads of the same mouse, so that the regenerative potential of the tissues or cells examined is evaluated under the same physiological hormonal context. This approach permits to reveal mammary epithelium intrinsic phenotype. It is particularly pertinent for analyzing the mammary phenotype of genetically modified mice that might present hormonal perturbations affecting mammary development or display early mortality [2, 21]. Tissue recombination techniques are also used to assess the ability of a mutant stromal environment to support the growth of wild-type mammary cells [2].

The co-transplantation of two mammary epithelial cell populations can be useful to evaluate their respective repopulation capacities and fate decisions in a competition context as well as their cooperative ability [2, 19, 22]. These experiments require the presence of a genetic marker (GFP, lacZ, etc.) in at least one of the populations compared, for estimating the relative contribution of each population to the chimeric outgrowths. Such type of studies has been particularly fruitful to reveal luminal-to-luminal and luminal-to-basal paracrine signaling [2, 22].

Advances in three-dimensional culture methods have enabled the establishment of mammospheres from isolated cells and mammary organoids preserving tissue architecture [19, 23, 24]. Both systems require to be cultured in the presence of extracellular matrix components, such as Matrigel and/or collagen.

Interestingly, they are amenable to in vitro manipulation before being grafted, offering the possibility of loss- and gain-of-function studies using Adeno-Cre or lentiviral transduction strategies [16, 25, 26]. In mouse, primary tissue-derived mammary organoids maintaining stem cell activity at long term have been established and successfully transplanted into cleared mammary fat pad [27].

Over years, orthotopic transplantation assays have greatly contributed to our understanding of mammary development, stem cell function, and tumorigenesis. Importantly, several lineage-tracing studies have shown that after birth, mammary development and homeostasis are essentially sustained by basal and luminal unipotent stem cells, revealing a discrepancy between the restricted fate of basal cells in situ and their multipotency upon transplantation [5, 22, 28]. It is now considered that under regenerative conditions and upon oncogene expression, lineage-restricted stem cells isolated from the adult mammary epithelium can reactivate a multipotency program reminiscent of that observed during embryogenesis. Interestingly, recent data have shown that luminal cells can restrict the fate of basal cells in situ, by secreting tumor necrosis factor [29].

2 Materials

2.1 Preparation of Mammary Epithelial Cells and Fragments for Transplantation

1. Scissors and forceps for tissue dissection.
2. Scalpels.
3. CO_2-independent medium with 5% FCS.
4. DNAse I: Prepare a 10 mg/mL solution and store aliquots at −20 °C.
5. Trypsin/Versene solution: 0.25% trypsin, 0.1% versene (EDTA) in PBS w/o Ca^{2+}, Mg^{2+}.
6. 0.05% trypsin.
7. Ammonium chloride solution.
8. Sterile cell strainer, 40 μm nylon mesh.
9. Trypan blue solution (0.4%).
10. Growth factor-reduced Matrigel.
11. Dissecting microscope with transmitted illumination.
12. Digestion solution: 3 mg/mL collagenase, 100 units/mL hyaluronidase in CO_2-independent medium, 5% FCS, 2 mM L-glutamine.
13. Dispase solution: 5 mg/mL dispase, 5% FCS in CO_2-independent medium.

2.2 Adenovirus Infection and Mammosphere Culture

1. Adenovirus expressing Cre recombinase (Ad-Cre, Tebu/SignaGen Labs, Cat # 189SL100707).
2. DMEM/F12 Mammosphere medium: DMEM/F12 medium, B27, 20 ng/mL EGF, 20 ng/mL bFGF, 4 μg/mL heparin, 10 μg/mL insulin, and 2% Matrigel [19, 23].
3. Growth factor-reduced Matrigel.
4. Ultra-low adherence 24-well plates.
5. 5 mL polystyrene round-bottom tubes with cell-strainer cap.

2.3 Animal Surgery

1. Dissecting microscope with direct illumination.
2. Fiber-optic light source.
3. Electric thermocautherizer (i.e., Harvard Apparatus Cat # 72-6067).
4. Shaver blade (only required when hairy mice are used as hosts).
5. Iodine solution, sterile gauze.
6. Surgeon's material: clean lab coat, mask, cap, and gloves.
7. Surgery table (about 20 cm square).
8. Rubber or tissue band, 0.5 cm wide.
9. Anesthetic solution: 0.4 mg/mL xylazine, 8 mg/mL ketamine, 30 μg/mL flunitrazepam in PBS. Alternatively, isoflurane gas system.
10. Eye gel.
11. 1 mL syringe and 25-gauge syringe needles.
12. Scissors, 110 mm, S/S.
13. Forceps with teeth (skin forceps).
14. Dissecting forceps (jeweler's type).
15. Microsurgery iris scissors, angle to side, 70–100 mm, S/S.
16. Autoclave or dry oven.
17. Cotton swabs.
18. Hamilton syringe, 25 μL.
19. Sterile PBS and water.
20. 9 mm wound clips, clipping machine, and clip remover. Alternatively, braided absorbable suture Vicryl (6–0) with a multipass curved micro-needle and a needle holder.
21. Heating lamp or heating pad.
22. Three-week-old female mice, weighing 8–10 g (*see* **Notes 1** and **2**).

2.4 Outgrowth Analysis by Carmine-Stained Whole Mounts

1. Glass histology slides.
2. Dissection material (scissors, forceps).
3. Methacarn fixation solution (methanol:chloroform:acetic acid/6:3:1).
4. Carmine Alum solution.
5. 70%, 95%, and 100% ethanol.
6. Xylene or methyl salicylate.

3 Methods

Before planning experiments, check the local recommendations for care and use of laboratory animals. The transplantation procedure must be approved by a relevant ethic committee. We describe below a procedure established in accordance with the current recommendations of the European Community (2010/63/UE) in compliance with the international guidelines.

3.1 Preparing Material for Surgery

1. Sterilize surgery instruments by autoclaving or dry heat sterilization (Poupinel oven: e.g., 4 h at 140 °C, 2.5 h at 160 °C, or 30 min at 180 °C) (*see* **Note 3**).
2. Prepare anesthetic solution. This solution is stable at 4 °C for several weeks (*see* **Note 4**).
3. Prepare the operating table surface: First clean carefully and rinse with 70% ethanol, then cover with a sterile medical sheet. Clean the dissecting microscope and the fiber-optic light source and cover them with plastic films to minimize contaminations.
4. Surgeon's preparation: Wash hands carefully and wear clean labcoat, mask, cap, and gloves during surgery.

3.2 Clearing the Fat Pad

1. Inject 100 μL anesthetic solution intraperitoneally into 3-week-old female recipient mice; the mice will be anesthetized within about 2 min and will remain so for at least 1 h. Alternatively, administrate isoflurane gas using an adapted mask during the whole surgery period (*see* **Notes 5** and **6**).
2. Shave the mice, if hairy mice are used.
3. Disinfect the skin with a sterile gauze embedded of iodine solution.
4. Put the mouse on the surgery table, ventral side up, and fix the hind legs with the rubber band, tightly enough to immobilize the animal in place but not so tightly as to hamper blood circulation.

5. Apply eye gel to prevent ocular dehydration and potential eye dryness, as the blinking reflex is inhibited by anesthesia.
6. Disinfect the skin by swabbing the inguinal area of the mice with a sterile gauze embedded with iodine solution.
7. Grasp the skin above the pelvis with teeth forceps, lifting it up from the abdomen. Make a midline incision starting just above the pelvis and extending over 1–1.5 cm from the pubis to the sternum (Fig. 1a). Take care not to pierce the peritoneum and damage the abdominal musculature.
8. From this middle incision, make two oblique incisions towards the hind legs, ending at about the mid-point between nipples 4 and 5 (Fig. 1a).
9. Lift up the skin with the forceps and pull the body wall with the cotton swab to expose fad pad #4 (Fig. 1b); pin the skin to the board with a 25 G needle.
10. Under the microscope, cauterize the arteries of nipple #4 leading to the node and margins of the fifth gland, to prevent ingrowth (Fig. 1b).
11. With the scissors, carefully excise the proximal part of the fat pad in one piece, including the lymph node, making clean and precise cuts, starting from the nipple (Fig. 1b, c, *see* **Note 7**).
12. Clear the contralateral fat pad (*see* **Note 8**).

3.3 Preparation and Transplantation of Mouse Mammary Epithelial Fragments

1. Dissect gland #4 from a recently euthanized female mouse; spread the gland onto a sterile glass histology slide and cover it with a drop of CO_2-independent medium (*see* **Note 9**).
2. Under a dissecting microscope with transmitted illumination, identify the primary duct and the downstream secondary ducts (Fig. 1d).
3. Using a scalpel, cut tissue fragments of about 1 mm^3 in volume and containing epithelial ducts (*see* **Note 10**). Steep the dissected fragments in 0.5–1 mL of CO_2-independent medium in a 24-well plate (one fragment per well) and keep the plate at 4 °C until transplantation (*see* **Note 11**).
4. Under a dissecting microscope with direct illumination, make a small pocket in the middle of the cleared fat pad with iris scissors with an angled tip (Fig. 1e). The pocket should be deep enough to contain the tissue to be grafted, but care should be taken not to cut through the fat pad completely.
5. Insert a piece of donor epithelium into the pocket with the dissecting forceps. Hold the pocket with the skin forceps to keep it closed as you remove the dissecting forceps, so that the inserted tissue remains in place.

3.4 Preparation of Single Mouse Mammary Cells

1. Weigh the required amount of digestion enzymes and keep at 4 °C until use. We use 15 mL of digestion solution per sample for the pooled glands from 3 to 4 mice.
2. Dissect both #4 glands, remove the lymph nodes and put the glands on ice in a 100 sterile Petri dish with 500 μL CO_2-independent medium, to prevent the tissue from drying out (*see* **Note 12**).
3. Mince the tissue, first with curved scissors and then with two scalpels, one in each hand, for at least 5 min, until a homogenous paste is obtained (*see* **Note 13**).
4. Transfer the minced tissue to the digestion solution. We recommend using 50 mL tubes in this step for optimal shaking.
5. Incubate for 90 min at 37 °C with shaking (150 rpm). Midway through the incubation, shake the tube vigorously to favor dissociation.
6. Transfer the samples to 15 mL tubes and centrifuge for 5 min at 450 × *g*.
7. Discard the upper fat layer and supernatant. Wash the pellet in 15 mL of CO_2-independent medium and centrifuge the tubes for 5 min at 450 × *g*.
8. Resuspend the pellet in 1 mL of the Trypsin/Versene solution prewarmed to 37 °C and pipette up and down with a 1000 μL tip for 1 min.
9. Add 8 mL of CO_2-independent medium and centrifuge for 5 min at 450 × *g*.
10. Carefully remove the supernatant, and resuspend the pellet in dispase solution, 2 mL per sample.
11. Add DNAse I to a final concentration of 0.1 mg/mL. Incubate at 37 °C for 5 min.
12. Dilute dispase solution with 10 mL of CO_2 independent medium and centrifuge 5 min at 450 × *g*.
13. Resuspend in 1 mL of cold ammonium chloride solution and immediately centrifuge for 5 min at 450 × *g* and 4 °C (*see* **Note 14**).
14. Resuspend in 1 mL of serum-free CO_2-independent medium and filter the cell suspension through a nylon mesh cell strainer with 40 μm pores inserted into a 50 mL tube. Wash the filter with 8 mL of serum-free medium.
15. Transfer to a 15 mL tube and centrifuge for 5 min at 450 × *g*.
16. Resuspend the pellet in 1 mL of serum-free CO_2-independent medium and count cells.

17. Immunostain the single-cell suspension and proceed with flow cytometry separation using combination of the lineage-specific markers previously validated [10–12, 19, 30, 31]. Unstained whole mammary cell preparations can be used for transplantation as well [32].

3.5 Injection of Mammary Epithelial Cells into Cleared Fat Pads

1. Check cell viability by staining with trypan blue and correct cell density accordingly.
2. Prepare the required amount of 25% growth factor-free Matrigel, by diluting it with CO_2-independent medium (*see* **Note 15**).
3. Prepare the required cell dilutions by resuspending cells in 25% Matrigel, such that 10 μL cell aliquots can be used for transplantation (*see* **Note 16**). Try to avoid making bubbles during the resuspension of the cells.
4. Using a 25 μL Hamilton syringe, inject 10 μL of cell suspension into the middle of the cleared fat pad (*see* **Note 17**). Place the syringe in an oblique position, such that 1–2 mm of the tip penetrates the fat pad, without passing all the way through it. After inoculation, the injected liquid forms a bolus, visible under the fat pad (*see* **Note 18**).
5. Place the skin in its original position and then inject cells into the contralateral fat pad (*see* **Note 8**).

3.6 Injection of Mammary Cells After Adeno-Cre Infection

Due to early lethality, some mutant models preclude the study of repopulating activity of mammary epithelial cells following transplantation. An option is to isolate mammary cells from floxed mice and to induce gene excision in vitro by infection with an adenovirus expressing the Cre recombinase prior to cell injection in the cleared fat pad [16].

1. Wash the sorted mammary cells with cold PBS or DMEM/F12 medium and centrifugate for 5 min at 450 × *g* and 4 °C.
2. Resuspend the cells in 300–500 μL of DMEM/F12 medium. Add the adeno-Cre virus and mix gently (*see* **Note 19**).
3. Incubate at 37 °C for 1 h with shaking (150 rpm).
4. Wash with fresh DMEM/F12 medium, centrifugate for 5 min at 450 × *g* and 4 °C.
5. Carefully discard the supernatant. Check cell viability by staining with trypan blue and correct cell density accordingly. Resuspend the cells in Matrigel at the adequate concentration and proceed for injection as explained above.

3.7 Injection of Cultured Mammospheres

Mammospheres can be established from isolated basal and luminal cells however under normal culture conditions, only basal cell-derived mammospheres display significant morphogenetic potential upon transplantation [19, 30]. Resuspend the cells in mammosphere medium and plate them in 24-well low-adherent plates at 5000 cells/well (*see* **Note 20**).

1. Culture mammospheres for 12–14 days with addition of fresh medium (50–100 μL) twice a week.
2. Collect mammospheres: Wash with PBS or medium and centrifugate at 150 × *g* (gentle centrifugation is important to preserve mammosphere structures).
3. For injection as whole structures, resuspend the mammospheres in 25% growth factor-reduced Matrigel. The number of transplanted mammospheres required to observe regeneration must be determined for each mouse model [30].
4. Alternatively, mammospheres can be dissociated with 0.05% trypsin at 37 °C for 5 min and cells filtered in a 5 mL tube with a cell-strainer cap before counting and adjusting concentration for transplantation [19].

3.8 Closing/Revival of Mouse

1. Use forceps to bring the opposing margins of the skin together, and close with wound clips. Take care not to attach the abdominal wall to the skin. Two clips are normally sufficient to close the midline incision, with one clip required for each lateral oblique incision (*see* **Note 21**).
2. Ear-tag the mouse if required.
3. Place the mouse on a paper towel on a heating pad (*see* **Note 22**). When awaken, put the mice in a clean and dry cage, give water and food (*see* **Note 23**).
4. Administrate analgesics (e.g., buprenorphine injected subcutaneously at 0.1 mg/kg body weight 12 h after surgery).
5. The wound clips should be removed after 2–3 weeks.

3.9 Outgrowth Analysis by Carmine-Stained Whole Mounts

Typically, outgrowth analysis is performed in virgin hosts 6–10 weeks after transplantation or in pregnant hosts previously mated (*see* **Notes 24** and **25**). Carmine-stained whole mounts allow global analysis of the outgrowth development patterns and can be embedded in paraffin for further immunohistological analyses (Fig. 1f, g). Alternative visualization strategies have recently emerged, such as three-dimensional fluorescence imaging of whole organs [26, 33].

1. Euthanize the mouse and place it, on its back, on a corkboard, fixing the limbs in position with needles.
2. Spray the mouse with 70% ethanol. Make an incision along the longitudinal axis, and a lateral incision between nipples 4 and 5.
3. Using teeth forceps, lift up the skin and gently pull the body wall with the cotton swab until the transplanted fat pad is

exposed. The fat pad generally remains stuck to the skin, but it may, in some cases, be attached to the peritoneum.

4. Carefully dissect the transplanted fat pad. For this purpose, pull up the fat pad with tissue forceps, while cutting the transparent membrane attaching the gland to the skin, with sharp fine scissors.
5. Spread the removed fat pad out on a glass histology slide. Make the gland as flat as possible and wait 5 min, to allow the tissue to stick well to the slide. Fix with Methacarn for 6–18 h.
6. Place slides in 70% ethanol for 20 min. Rinse twice with distilled water.
7. Stain with carmine alum overnight at room temperature.
8. Rinse with distilled water. Dehydrate the tissue by immersion in successive baths of 70%, 95%, and 100% ethanol, for 20 min per bath.
9. Clear in Xylene or methyl salicylate until the tissue becomes translucent. For examination and imaging under the dissecting microscope, place the slide in a glass Petri dish with some xylene or methyl salicylate. Peel the tissue off of the glass slide for embedding in paraffin, sectioning and analysis by standard immunohistological methods.

Unstained outgrowths can be processed as mammary tissue, by dissociating cells for flow cytometry studies as described in Subheading 3.4 (*see* **Note 26**). Sorted cells can be retransplanted, manipulated in in vitro assays, or analyzed for their gene expression patterns using qPCR or microarray strategies.

4 Notes

1. Use prepubertal 3-week-old female mice as recipient. Nude mice are expected to weight 8–10 g at this stage. Avoid tiny nude females of less than 8 g in which poor outgrowth development is frequently observed and females weighing more than 12 g in which the mammary rudiment may be overdeveloped. If incompletely removed, the remaining endogenous epithelium can lead to an important ingrowth.
2. Compound-mutant donor mice often have a mixed genetic background, making the use of syngeneic host mice impossible. Immunodeficient nude mice are routinely used to prevent graft rejection.
3. All consumables should be sterilized by an adapted method (autoclaving, irradiation, dry heat). Working under good aseptic conditions is recommended for experimental reliability and reproducibility. Administration of antibiotics to the mice is an

option for preventing infections. A standard treatment is 15 mg/kg body weight of amoxicillin in drinking water 1 h before surgery and 3 days for postoperative care.

4. A simple xylazine/ketamine solution has anesthetic properties, but we recommend the addition of flunitrazepam, as it significantly improves the efficacy of anesthesia and the tranquility of the mice during surgery.
5. For isoflurane concentration, follow the instructions of your particular system taking into account the weight of the animals. Use small masks, adapted as much as possible to the size of the mice.
6. Previous administration of analgesics (e.g., ibuprofen 0.05 mg/kg body weight) is necessary when using isoflurane anesthesia. This is not required when using injectable drugs since they display analgesic properties.
7. Not all protocols entail removal of the lymph nodes during fat pad clearance. However, whole-mount Carmine staining of the removed gland fragments shows that in some 3-week-old animals, the epithelium can reach the lymph node. We therefore recommend staining the removed fat pad fragments in carmine alum, as described in Subheading 3.9, to check for cleared margins.
8. When fragments (or cell populations) from different donors are compared (for example, wild-type and mutant epithelia), they should be transplanted into contralateral fat pads of the same animal. This allows to monitor their growth in the same hormonal environment.
9. In some mouse strains and in aged animals, the fat pad containing gland # 4 may be thick, making it difficult to dissect out epithelial fragments. Gland #3 can be used instead, in such cases.
10. In serial transplantations of epithelial fragments, it is important to use fragments obtained from the center of the outgrowth. Mammary fragments dissected from any part of the mammary tree can give rise to new mammary tissue, but, when serially transplanted, the peripheral ducts may undergo senescence earlier, probably because they have been through more cell divisions [34].
11. In our experience, keeping fragments in CO_2-independent medium at 4 °C for up to 2 h does not affect their morphogenetic capacity. Some authors report freezing mammary epithelial fragments in 10% DMSO according to a classical cell freezing protocol. Frozen fragments can be stored in liquid nitrogen until transplantation, with a minimal decrease in transplantation efficiency (> 85% efficiency) [35].

12. For the preparation of a single-cell suspension from virgin mouse mammary glands, we recommend pooling the fourth and fifth glands from at least three or four virgin animals. When not enough mice are available, a single animal can be used. In this case, for maximal cell recovery, one can add glands #3, paying attention to avoid muscle contamination during dissection. The volume of digestion solution can be scaled down to 10 mL.
13. Some protocols suggest using a tissue chopper (Mcllwain Tissue Chopper, Mickle Laboratory Engineering Co. Ltd.) for mincing the tissue [32].
14. The volume of ammonium chloride solution used should be adapted to the size of the pellet, please follow the manufacturer's instructions. It is also important to use a cold solution and avoid long incubation times. In our experience, excessive solution volumes, warm temperature, and incubation time may result in decreased cell viability.
15. The use of Matrigel for cell injection is optional. However, Matrigel significantly increases outgrowth formation and is particularly recommended when working with small numbers of cells. For example, it has been reported that the transplantation of 75 cells in the absence of Matrigel yields successful reconstitution in 45% of cases, whereas the addition of 50% Matrigel increases this percentage to 87.5% [17].
16. For evaluation of the frequency of repopulating stem cells in limiting dilution transplantations, it is necessary to transplant cells at a minimum of four to five different dilutions. Increasing the number of dilutions and the number of transplanted fat pads per dilution increases the accuracy of the estimate. Note that, when not all transplanted samples give rise to outgrowths, calculation of stem cell frequency includes the proportion of outgrowth-free fat pads. In our hands, in a typical limiting dilution experiment performed with sorted mammary basal cells from mice of mixed 129SV/C57BL6 genetic background in the presence of 25% Matrigel, between 5 and 150 cells were transplanted [19].
17. We prefer to inject 10 μL aliquots, to prevent cell loss during injection. Always use the same syringe for the injection of a given population and, in limiting dilution experiments, start with the most dilute samples. When two different cell populations are injected, we use two different syringes to avoid the need to rinse the syringes between injections.
18. Formation of a "bolus" in the place of injection is a sign of successful cell inoculation into the fat pad. On the contrary, if some liquid is visible after removal of the syringe, one can doubt of the efficacy of the injection.

19. We recommend to perform a pilot experiment to estimate, in your particular experimental conditions and for each batch of adeno-Cre virus, the optimal MOI for maximal recombination efficacy without compromising cell viability.
20. Mammosphere medium composition [19, 23].
21. Given the fragility of the skin in nude mice, we prefer to stitch up the incision with absorbable sutures (6–7 stitches per mouse) rather than using wound clips. It is always a good idea, in any case, to have some suture available to close the peritoneum if it is accidentally pierced.
22. A heating lamp can be used as an alternative to the heating pad. The lamp should just warm the mouse, and care should be taken to avoid burning the skin.
23. We recommend the use of hydration and nutritional gels during the first days after surgery.
24. When very small numbers of cells are transplanted, complete growth may take up to 10–12 weeks. By contrast, the outgrowths develop more rapidly if tissue fragments, rather than sorted cell populations, are transplanted. For the visualization of terminal end buds, the bulbous structures present at the tips of growing ducts, the outgrowths should be analyzed earlier, before the recipient fat pad is completely filled with the transplanted epithelium. Typically, for grafted wild-type mammary epithelium, terminal end buds can be found in the outgrowths 4–5 weeks after the transplantation of epithelial fragments and 6–7 weeks after the transplantation of sorted mammary basal cells.
25. Grafted tissue fragments or isolated basal cells lead to full alveolar development in pregnant hosts mated 6–10 weeks post transplantation. Alternatively, hormone stimulation of host mice can be achieved by implanting estrogen/progesterone releasing devices subcutaneously [36, 37].
26. Due to their small size, we recommend to pool at least 4–6 outgrowths for optimal cell recovery after dissociation and sorting.

Acknowledgments

The work was supported by grants from the Ligue Contre le Cancer Ile de France (Comité de Paris RS16/75-70 and RS19/75-67). MAG is Directeur de Recherche, MAD and MMF are Chargé de Recherche at the Institut National de la Santé et de la Recherche Médicale (INSERM).

References

1. DeOme KB, Faulkin LJ Jr, Bern HA et al (1959) Development of mammary tumors from hyperplastic alveolar nodules transplanted into gland-free mammary fat pads of female C3H mice. Cancer Res 19(5):515–520
2. Brisken C, Ataca D (2015) Endocrine hormones and local signals during the development of the mouse mammary gland. Wiley Interdiscip Rev Dev Biol 4(3):181–195. https://doi.org/10.1002/wdev.172
3. Fu NY, Nolan E, Lindeman GJ et al (2020) Stem cells and the differentiation hierarchy in mammary gland development. Physiol Rev 100(2):489–523. https://doi.org/10.1152/physrev.00040.2018
4. Lv X, Dobrolecki LE, Ding Y et al (2020) Orthotopic transplantation of breast tumors as preclinical models for breast cancer. J Vis Exp. https://doi.org/10.3791/61173
5. Lloyd-Lewis B, Harris OB, Watson CJ et al (2017) Mammary stem cells: premise, properties, and perspectives. Trends Cell Biol 27(8): 556–567. https://doi.org/10.1016/j.tcb.2017.04.001
6. Zhang M, Lee AV, Rosen JM (2017) The cellular origin and evolution of breast cancer. Cold Spring Harb Perspect Med 7(3): a027128. https://doi.org/10.1101/cshperspect.a027128
7. Kuperwasser C, Chavarria T, Wu M et al (2004) Reconstruction of functionally normal and malignant human breast tissues in mice. Proc Natl Acad Sci U S A 101(14): 4966–4971. https://doi.org/10.1073/pnas.0401064101
8. Ginestier C, Hur MH, Charaffe-Jauffret E et al (2007) ALDH1 is a marker of normal and malignant human mammary stem cells and a predictor of poor clinical outcome. Cell Stem Cell 1(5):555–567. https://doi.org/10.1016/j.stem.2007.08.014
9. Lim E, Vaillant F, Wu D et al (2009) Aberrant luminal progenitors as the candidate target population for basal tumor development in BRCA1 mutation carriers. Nat Med 15(8): 907–913. https://doi.org/10.1038/nm.2000
10. Shackleton M, Vaillant F, Simpson KJ et al (2006) Generation of a functional mammary gland from a single stem cell. Nature 439(7072):84–88. https://doi.org/10.1038/nature04372
11. Stingl J, Eirew P, Ricketson I et al (2006) Purification and unique properties of mammary epithelial stem cells. Nature 439(7079): 993–997. https://doi.org/10.1038/nature04496
12. Sleeman KE, Kendrick H, Robertson D et al (2007) Dissociation of estrogen receptor expression and in vivo stem cell activity in the mammary gland. J Cell Biol 176(1):19–26. https://doi.org/10.1083/jcb.200604065
13. Al-Hajj M, Wicha MS, Benito-Hernandez A et al (2003) Prospective identification of tumorigenic breast cancer cells. Proc Natl Acad Sci U S A 100(7):3983–3988. https://doi.org/10.1073/pnas.0530291100
14. Daniel CW, De Ome KB, Young JT et al (1968) The in vivo life span of normal and preneoplastic mouse mammary glands: a serial transplantation study. Proc Natl Acad Sci U S A 61(1):53–60
15. Medina D (2000) The preneoplastic phenotype in murine mammary tumorigenesis. J Mammary Gland Biol Neoplasia 5(4):393–407
16. Romagnoli M, Cagnet S, Chiche A et al (2019) Deciphering the mammary stem cell niche: a role for laminin-binding integrins. Stem Cell Rep 12(4):831–844. https://doi.org/10.1016/j.stemcr.2019.02.008
17. Vaillant F, Lindeman GJ, Visvader JE et al (2011) Jekyll or hyde: does matrigel provide a more or less physiological environment in mammary repopulating assays? Breast Cancer Res 13(3):108. https://doi.org/10.1186/bcr2851
18. Hu Y, Smyth GK (2009) ELDA: extreme limiting dilution analysis for comparing depleted and enriched populations in stem cell and other assays. J Immunol Methods 347(1–2):70–78. https://doi.org/10.1016/j.jim.2009.06.008
19. Chiche A, Moumen M, Petit V et al (2013) Somatic loss of p53 leads to stem/progenitor cell amplification in both mammary epithelial compartments, basal and luminal. Stem Cells 31(9):1857–1867. https://doi.org/10.1002/stem.1429
20. Chiche A, Moumen M, Romagnoli M et al (2017) p53 deficiency induces cancer stem cell pool expansion in a mouse model of triple-negative breast tumors. Oncogene 36(17):2355–2365. https://doi.org/10.1038/onc.2016.396
21. Taddei I, Deugnier MA, Faraldo MM et al (2008) Beta1 integrin deletion from the basal compartment of the mammary epithelium affects stem cells. Nat Cell Biol 10(6): 716–722. https://doi.org/10.1038/ncb1734
22. Van Keymeulen A, Rocha AS, Ousset M et al (2011) Distinct stem cells contribute to

mammary gland development and maintenance. Nature 479(7372):189–193. https://doi.org/10.1038/nature10573
23. Spike BT, Engle DD, Lin JC et al (2012) A mammary stem cell population identified and characterized in late embryogenesis reveals similarities to human breast cancer. Cell Stem Cell 10(2):183–197. https://doi.org/10.1016/j.stem.2011.12.018
24. Srivastava V, Huycke TR, Phong KT et al (2020) Organoid models for mammary gland dynamics and breast cancer. Curr Opin Cell Biol 66:51–58. https://doi.org/10.1016/j.ceb.2020.05.003
25. Lu Y, Cao J, Napoli M et al (2018) miR-205 regulates basal cell identity and stem cell regenerative potential during mammary reconstitution. Stem Cells 36(12):1875–1889. https://doi.org/10.1002/stem.2914
26. Lagoutte E, Villeneuve C, Fraisier V et al (2020) A new pipeline for pathophysiological analysis of the mammary gland based on organoid transplantation and organ clearing. J Cell Sci 133(12):jcs242495. https://doi.org/10.1242/jcs.242495
27. Jardé T, Llyod-Lewis B, Thomas M et al (2016) Wnt and neuregulin1/ErbB signalling extends 3D culture of hormone responsive mammary organoids. Nat Commun 7:13207. https://doi.org/10.1038/ncomms13207
28. Prater MD, Petit V, Russell IA et al (2014) Mammary stem cells have myoepithelial cell properties. Nat Cell Biol 16(10):942–950. https://doi.org/10.1038/ncb3025
29. Centonze A, Lin S, Tika E et al (2020) Heterotypic cell-cell communication regulates glandular stem cell multipotency. Nature 584(7822): 608–613. https://doi.org/10.1038/s41586-020-2632-y
30. Di-Cicco A, Petit V, Chiche A et al (2015) Paracrine Met signaling triggers epithelial-mesenchymal transition in mammary luminal progenitors, affecting their fate. eLife 4: e06104. https://doi.org/10.7554/eLife.06104
31. Bresson L, Faraldo MM, Di-Cicco A et al (2018) Podoplanin regulates mammary stem cell function and tumorigenesis by potentiating Wnt/beta-catenin signaling. Development 145(4):dev160382. https://doi.org/10.1242/dev.160382
32. Smalley MJ (2010) Isolation, culture and analysis of mouse mammary epithelial cells. Methods Mol Biol 633:139–170
33. Lloyd-Lewis B (2020) Multidimensional imaging of mammary gland development: a window into breast form and function. Front Cell Dev Biol 8:203. https://doi.org/10.3389/fcell.2020.00203
34. Daniel CW, Young LJ (1971) Influence of cell division on an aging process. Life span of mouse mammary epithelium during serial propagation in vivo. Exp Cell Res 65(1):27–32
35. Dunphy KA, Tao L, Jerry DJ et al (2010) Mammary epithelial transplant procedure. J Vis Exp 40:1849. https://doi.org/10.3791/1849
36. Rajkumar L, Kittrell FS, Guzman RC et al (2007) Hormone-induced protection of mammary tumorigenesis in genetically engineered mouse models. Breast Cancer Res 9(1):R12. https://doi.org/10.1186/bcr1645
37. Romagnoli M, Bresson L, Di-Cicco A et al (2020) Laminin-binding integrins are essential for the maintenance of functional mammary secretory epithelium in lactation. Development 147(4):dev181552. https://doi.org/10.1242/dev.181552

Chapter 7

Lineage Tracing Methods to Study Mammary Epithelial Hierarchies In Vivo

Verónica Rodilla and Silvia Fre

Abstract

Lineage tracing is now considered the gold standard approach to study cellular hierarchies and cell fate in vivo (McKenna and Gagnon, Development 146:dev169730, 2019; Kretzschmar and Watt, Cell 148:33–45, 2012). This type of clonal analysis consists of genetically labeling defined cells and following their destiny and progeny in vivo and in situ.

Here we will describe different existing in vivo systems to clonally trace targeted cells and will discuss their respective advantages and inconveniences; we will then provide stepwise instructions for setting up and evaluate lineage tracing experiments, listing the most common downstream analyses and read-out assays.

Key words lneage tracing, Stem cells, Cellular hierarchies, Inducible Cre/Lox system

1 Introduction

Lineage tracing permits to identify stem cells and track their progeny, by genetically labeling cells in their physiological niche, so that their daughters retain marker expression. This strategy has been used to study cell fate and evaluate stem cell potency in vivo in most tissues [1].

Such an approach has been employed to define novel stem cell markers, such as *Lgr5* in the small intestine [2], as well as to discover the existence of unipotent stem cells in the mammary gland, prostate, or lung [3–10].

The most commonly used lineage tracing approach employs an inducible form of the Cre recombinase to trigger the permanent expression of a reporter gene in defined cells. However, other approaches have been designed to trace specific cells in vivo and monitor their progeny at high spatial and temporal resolution, such as *Tet* or *Dre* systems.

In the mammary gland, this strategy has been instrumental to demonstrate that this tissue is maintained by unipotent cell

Maria dM. Vivanco (ed.), *Mammary Stem Cells: Methods and Protocols*,
Methods in Molecular Biology, vol. 2471, https://doi.org/10.1007/978-1-0716-2193-6_7,

populations [3, 11]. Indeed, initial genetic fate mapping studies using lineage-specific promoters, expressed either in the luminal or in the basal compartment of the mammary epithelium, such as luminal (K8) or basal (K5, K14) cytokeratin promoters, indicated that in the adult homeostatic mouse mammary gland, basal cells (BCs) and luminal cells (LCs) could only generate BCs and LCs, respectively, indicating that tissue homeostasis is maintained by unipotent, lineage-restricted progenitors that give rise to only one type of cells in the postnatal gland [3, 12]. Subsequently, other groups have approached this question by genetically marking specific cells within the luminal compartment using different promoters: *Notch* receptors, *Prominin1,* and Estrogen Receptor α (*Esr1).* These experiments showed that some genes label both ER-α-positive and ERα-negative LCs (like Notch3 [9]), whereas some promoters were found to be specific for LCs expressing hormone receptors (*Esr1* and *Prom1* promoters) and others, such as *Notch1*, target uniquely ERα-negative LCs in the postnatal gland [8, 9, 13, 14]. Collectively, all these studies provided strong evidence that in adult mice, BCs and LCs are entirely self-sustained by unipotent progenitors, and this holds true for ERα-positive and negative luminal subsets, representing three independent lineages (Fig. 1). Each of these cell populations maintains their respective lineage even after serial pregnancies, demonstrating long-term self-renewal capacity.

It should be noted that, as lineage tracing approaches are not applicable to humans, some differences in the cellular hierarchy might exist between the mouse mammary gland and the human breast.

2 Materials

2.1 Genetically Modified Mouse Models

The choice of the murine line to use is of utmost importance to properly perform lineage tracing experiments, both when using existing mouse lines and when designing a new transgenic mouse model.

Nowadays, several established in vivo systems to target and monitor specific cells exist and we will summarize them here.

2.1.1 Tamoxifen-Inducible Cre/Lox System

This system requires compound mice carrying at least two transgenes: the first one expresses a temporally inducible form of the Cre recombinase under the control of a promoter expressed in the cells of interest and a reporter mouse line. The inducible Cre recombinase (Cre^{ER}, Cre^{ERT} or Cre^{ERT2}) allows the investigator to decide when to induce cell labeling upon tamoxifen administration and thus the time at which the tracing experiment will start (Fig. 2). To achieve temporal control of the system, Cre is fused with a mutant ligand-binding domain of the human estrogen receptor (CreER)

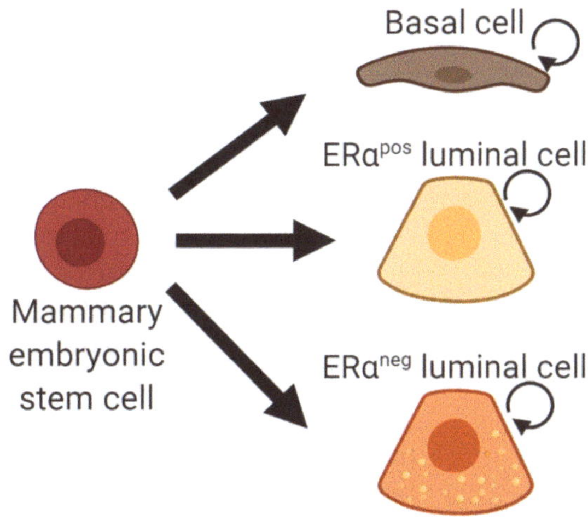

Fig. 1 Mammary epithelial cellular hierarchy resolved by lineage tracing studies. This simplified scheme indicates that multipotent mammary stem cells (SCs) are found exclusively during mouse embryonic development, while after birth distinct unipotent progenitors are responsible for sustaining tissue growth and homeostasis, giving rise to each mammary cell type: basal cells, ERα-positive and ERα-negative luminal cells

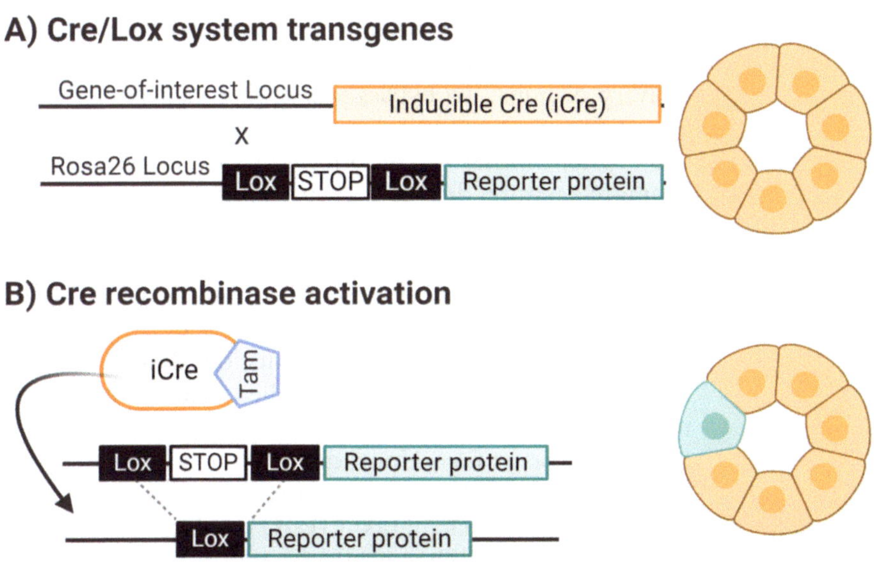

Fig. 2 Tamoxifen-inducible Cre/Lox system. (**a**) Representation of the two transgenes required for this in vivo approach: one mouse carries an inducible Cre recombinase (iCre) under the control of a gene of interest and a reporter mouse brings a Lox-flanked STOP sequence upstream of a reporter gene. (**b**) Cre recombinase activation: upon tamoxifen (Tam) administration, the cells expressing the iCre (determined by the promoter that controls its expression) will be labeled by the reporter protein (i.e., B-galactosidase or a fluorescent protein) due to the Cre-mediated recombination of the DNA at Lox sites, excising the STOP codon in the reporter transgene

[15–17] that does not bind to the endogenous estradiol, but can be activated through experimental exposure of synthetic ligands such as tamoxifen or 4-hydroxytamoxifen (CreERT2). Tamoxifen administration leads to translocation of the CreER fusion protein into the nucleus, where it induces recombination of the DNA at loxP sites. Because this change in DNA is heritable and irreversible, the progeny of these cells can be traced throughout life based on the expression of a reporter. This is the second essential element of the system: the selected Cre line has to be mated with a reporter mouse strain in order to monitor the targeted cells. This system has the advantage of being extensively used by many different groups around the world, not only to trace cells but also to target specific mutations or gene deletions in different tissues, therefore practical issues such as tamoxifen concentrations, experimental timeline, age of induction, etc. have generally been solved. However, the major concern in choosing these lines to study mammary gland biology is the use of tamoxifen. Tamoxifen inhibits the ER signaling pathway by competing with endogenous estrogen for its binding to ER. The effects of tamoxifen delay the growth and expansion of the mammary gland during puberty; however, low concentrations of this drug (up to 1.5 mg/kg of mouse) have negligible consequences on mammary gland development (Supplemental Fig. 9 in [18]). Moreover, for studies of embryonic mammary development, investigators need to consider the fact that tamoxifen impairs natural delivery of pups. To bypass this problem, they can (1) perform a C-section on pregnant induced females at the end of gestation and use a foster mother for adoption [7], or (2) inject progesterone to promote a natural delivery, albeit taking the risk of interfering with the hormonal regulation of mammary growth [19].

2.1.2 *Tetracycline-Inducible* Tet/Lox *System*

Tetracycline-inducible systems exist both as a "tet-OFF" or "tet-ON" method, whereby the administration of tetracycline (or its analog doxycycline) either prevents or allows gene expression, respectively. In tetracycline-controlled systems, a cell-specific promoter drives the expression of a reverse tetracycline-controlled transactivator (rtTA). Upon doxycycline administration, this protein will bind DNA sequences of tetracycline-controlled transcriptional activation elements (Tet-O), which in turn drive Cre expression. Cre will then act on the reporter transgene by recombination of *LoxP* elements (Fig. 3).

Similar to the tamoxifen-inducible Cre/Lox system, this strategy has the advantage that has been widely used. Moreover, since no side effects of doxycycline have been observed in the mammary gland, this system permits to perform lineage tracing at saturation, which was accomplished with the K14-rtTA line to definitively demonstrate lack of multipotent stem cells in the postnatal mammary gland [11]. However, the major problem in choosing this

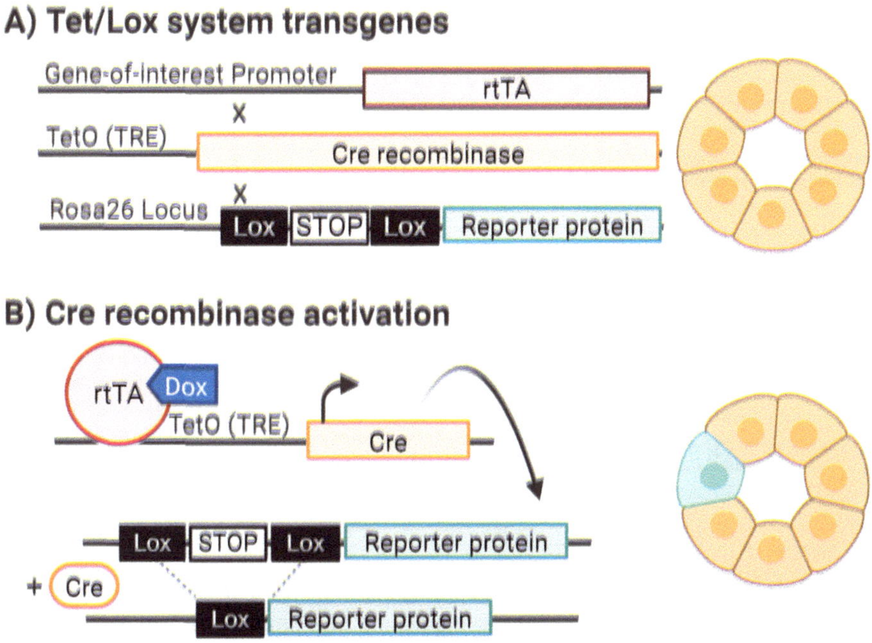

Fig. 3 Tetracycline-inducible *Tet*/Lox system. (**a**) Representation of the three transgenes required for this in vivo approach: the conditional rtTA transgene under the control of a promoter of interest, a doxycycline-inducible rtTA-dependent Tet-O-Cre responder line and a reporter mouse line. (**b**) Cre recombinase activation: upon doxycycline (Dox) administration, the cells expressing rtTA will be labeled by the reporter protein (i.e., B-galactosidase or a fluorescent protein) due to the binding of the rtTA transactivator to TRE sequences, resulting in Cre expression and consequent excision of the lox-flanked STOP codon in the reporter transgene

approach lies in the necessity of generating triple transgenic mice, which implies more animals and an extra generation of crosses in order to obtain the desired genotype.

2.1.3 Dre/Rox System

One drawback of Cre-based approaches for lineage tracing is caused by leakiness or promiscuity of different Cre lines. To circumvent this problem and increase the accuracy of lineage tracing, a recent implementation of this method consists of a dual recombinase system which combines Cre/lox mice with Dre/rox recombination [20], allowing to target different cells with different reporters or the same cells at different times [21]. The dual recombinase system has been successfully used to analyze cardiomyocyte renewal [22, 23] as well as liver [24] and lung regeneration [25]. More recently, this dual system has also been applied to tracing of transient cell fate changes [26]. The major limitation of this approach lies in the current paucity of mouse lines carrying Dre drivers and Rox-flanked reporter lines, which has prevented to date its use for mammary gland studies (Fig. 4).

In Table 1, we have listed the Cre lines used in lineage tracing studies to elucidate epithelial mammary gland hierarchies. Nevertheless, if the investigator is interested in testing a different gene

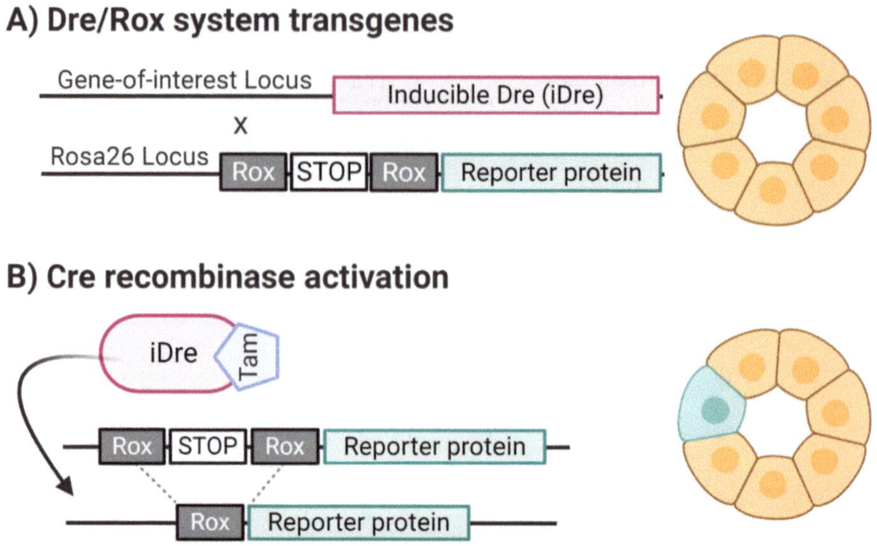

Fig. 4 Dre/Rox system. (**a**) Representation of the two transgenes required in this in vivo approach: a mouse line expressing inducible Dre recombinase (iDre) under the control of a promoter of interest, and a reporter line with a STOP sequence flanked by two loxP sites. (**b**) Dre recombinase activation: upon tamoxifen (Tam) administration, the cells expressing the iDre (defined by the promoter controlling its expression) will be labeled by the reporter protein (i.e., B-galactosidase or a fluorescent protein) due to Dre-mediated recombination of rox sequences and excision of a STOP codon flanked by rox sequences in the reporter transgene

from these lists, Jackson mouse repository offers a large selection of inducible Cre strains (http://cre.jax.org/index.html).

2.2 Reporter Lines

Reporter lines are transgenic mice where a constitutive promoter is engineered upstream a reporter gene, historically encoding the beta-galactosidase enzyme [39] or, more recently, expressing one or several fluorescent proteins. The transgenic construct harbors a loxP-flanked stop signal (loxP-STOP-loxP). The loxP-STOP-loxP sequence is excised upon Cre activation, allowing cells expressing the Cre and the reporter to be visualized and tracked in vivo (Fig. 2).

In Table 2, we have listed the most common murine reporter lines used in lineage tracing studies.

2.3 Inducing Tamoxifen-Mediated Recombination

1. Tamoxifen solution: Dissolve 10 mg/mL Tamoxifen free base in 1 Absolute ethanol: 9 Sunflower oil by vortexing. This solution can be aliquoted in 2 mL Eppendorf tubes and stored at −20 °C, protected from light.
2. 4-Hydroxytamoxifen solution: Dissolve 20 mg/mL 4-Hydroxytamoxifen in 1 Absolute ethanol: 9 Sunflower oil by vortexing. Can be aliquoted in 2 mL Eppendorf tubes and stored at −20 °C, protected from light.

Table 1
List of the driver murine lines used in lineage tracing studies to elucidate epithelial mammary gland hierarchies

Cre line	Recombination system	Targeted cells	Reference
K5-Cre ERT2	Cre/Lox	BC	[3]
K5-rtTA/TetO-Cre	Tet/Lox	BC and LC	[18]
K14-CreER	Cre/Lox	BC	[3, 11]
K14-rtTA/TetO-Cre	Tet/Lox	BC	[3]
Acta2-Cre ERT2	Cre/Lox	BC	[7, 12]
Axin2-Cre ERT2	Cre/Lox	BC	[27]
Lgr5-Cre ERT2	Cre/Lox	BC and LC	[11, 28]
Lgr6	Cre/Lox	BC and LC	[11, 29]
Dll1-Cre ERT2	Cre/Lox	BC and LC	[30]
Procr-CreER	Cre/Lox	BC	[31]
SOX9-Cre ERT2	Cre/Lox	BC and ER-negative LC	[14]
K8-CreERT2	Cre/Lox	LC	[3, 32]
K8-rtTA/TetO-Cre	Tet/Lox	LC	[3]
K19CreERT	Cre/Lox	Mainly LC	[11]
Notch3-Cre ERT2	Cre/Lox	LC	[9]
Esr1-rtTA	Tet/Lox	ER-positive LC	[13]
Prom1-Cre ER	Cre/Lox	ER-positive LC	[14]
Notch1-Cre ERT2	Cre/Lox	ER-negative LC	[8]
Elf5-rtTA	Tet/Lox	LC (ER status not specified)	[18]
Wap-Cre	Cre/Lox	PI-MEC at pregnancy	[33]
Prdm1-CreERT2	Cre/Lox	ER-negative LC	[34]
Notch2CreERT2	Cre/Lox	LC	[35]
Rosa26-Cre ERT2	Cre/Lox	Random cells	[36–38]

3. 1 mL syringe.
4. 26 G needle.

2.4 Inducing Tetracycline-Mediated Recombination

1. 5 mg/mL Doxycycline in PBS. Store at −20 °C, protected from light.
2. 1 mL syringe.
3. 26 G needle.

Table 2
List of the reporter murine lines used in lineage tracing studies to elucidate epithelial mammary gland hierarchies

Reporter line	Official name	Reference
Rosa26-lacZ	B6.129S4-Gt(ROSA)26Sortm1Sor/J	[27, 39]
Rosa26-mTmG	B6.129(Cg)-Gt(ROSA)26Sortm4(ACTB-tdTomato-EGFP)Luo/J	[7–9, 40]
Rosa26-Confetti	B6.129P2-Gt(ROSA)26Sortm1(CAG-Brainbow2.1)Cle/J	[7, 11, 38, 41, 42]
Rosa26-tdT	B6.Cg-Gt(ROSA)26Sortm9(CAG-tdTomato)Hze/J	[3, 43]
Rosa26-eYFP	B6.Cg-Gt(ROSA)26Sortm3(CAG-EYFP)Hze/J	[3, 44]

2.5 Harvesting the Mammary Glands

1. Dissection kit: 1 scissor, 1 forceps, 1 blade.
2. Spray bottle with 70% ethanol.
3. Styrofoam cover.
4. 10–12 needles.

2.6 Immunostaining

1. Whatman paper, 4 cm × 2 cm each.
2. 15 mL Falcon tube.
3. 4% paraformaldehyde.
4. PBS.
5. Ethanol (25%, 50%, 70%, and 100%).
6. Xylene.
7. Paraffin.
8. OCT.
9. Retrieval Buffer: Citrate or TE Solution.
10. Triton X-100.
11. FBS, Fetal Bovine Serum.
12. BSA, Bovine Serum Albumin (5% and 10% solutions).

2.7 Mammary Gland Cell Dissociation

1. Dissociation buffer: 600 U/mL collagenase, 200 U/mL hyaluronidase in DMEM-F12 phenol red free.
2. 5 mg/mL dispase in DMEM-F12 phenol red free.
3. 0,1 mg/mL DNAse I.
4. 0.25% Trypsin-EDTA.
5. Red Blood Lysis Buffer (eBioscience).
6. 40 μm cell strainers.

3 Methods

3.1 Generate the Mouse Colony

As mentioned, the choice of the Cre line is the most critical step of this approach and after reading the advantages and inconveniences of each model exposed in this chapter, we hope that the investigator will select the one that better fits their research questions. Once selected (*see* **Note 1**), the animals have to be crossed with a reporter mouse line.

1. Breed your Cre line of choice with an appropriate reporter line (*see* **Note 2**).
2. Genotype the mice by PCR on genomic DNA extracted from an ear sample (*see* **Note 3**).
3. Separate the double (Cre/Lox system) or the triple (Tet/Lox system) transgenic females and their appropriate controls for lineage tracing experiments.

3.2 Cre Recombinase Activation

1. Select the female mice with the correct genotype (carrying both the Cre driver and the reporter) (*see* **Note 4**).
2. Administer the suitable drug, depending on your lineage tracing system of choice (*see* **Note 5**), through an intraperitoneal injection using 1 mL syringe with a 26 G needle at the desired developmental stage.
   (a) Tamoxifen recommended dose: 0.05 mg/g of mouse body weight [3].
   (b) 4-OHT recommended dose: 0.05 mg/g of mouse body weight [9].
   (c) Doxycycline recommended dose: 1 mg per mouse [3].
3. Discard syringe with 26 G needle in an appropriate residual container.

3.3 Analysis of the Cells of Interest and Their Progeny

At this point it is crucial to distinguish *pulse* versus *chase* experiments (Fig. 5):

- *Pulse experiments* must be performed 18–24 h after induction of Cre recombination and labeled cells represent the cells expressing the promoter driving the Cre. These experiments are particularly important to define the initially targeted cells before they start dividing and generating a progeny that may not express the Cre- driving promoter anymore. Most of the confusion in lineage tracing studies is generated by the lack of proper assessment of the cells expressing the Cre driver in a given tissue (*see* **Note 6**).
- *Chase experiments* can be performed in a wide time schedule, ranging from a few days to several weeks or months after Cre induction. In this case, the labeled cells represent the progeny of

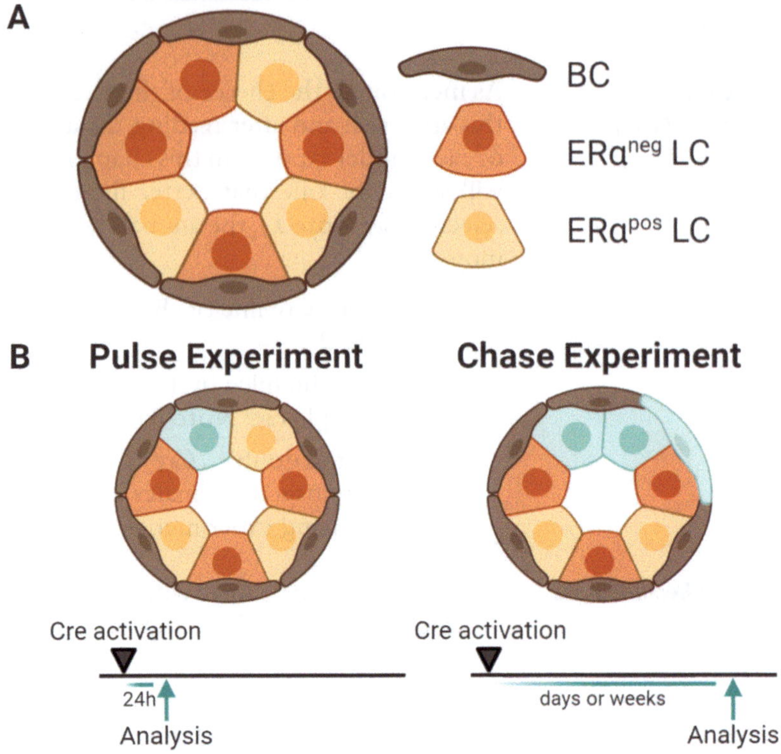

Fig. 5 Lineage tracing experiments. (**a**) Representation of a mammary duct section formed by three epithelial cell types: basal cell (BC), ERα-positive and ERα-negative luminal cells (LC). (**b**) Representation of the two types of lineage tracing experiments: *Pulse experiments*, which permit the analysis and visualization of targeted cells 24 h upon Cre activation; and *Chase experiments*, which consist in the analysis of the clonal progeny of the initially targeted cells (labeled cells in pulse experiments). The cyan-positive cells represent the cells that have activated the recombinase and excised the flox-flanked (or rox) STOP codon in a pulse experiment or the clonal progeny (lineage) derived from initially targeted cells in a chase experiment

the initially targeted cells and they may constitute very heterogeneous cell types, i.e., if multipotent stem cells were initially targeted (as, for example, in the intestine or the early embryonic mammary gland).

Once established if the experiment is designed to analyze "mother" cells (Pulse) or their daughters (Chase):

1. Animals are euthanized and their ten mammary glands dissected using scissors, forceps, and blades (*see* **Note 7**). We recommend opening the females as illustrated in Fig. 6, which will facilitate the dissection of all the glands using needles as pushpins and a styrofoam cover as dissection table.
2. Dissected mammary glands can be further processed for:

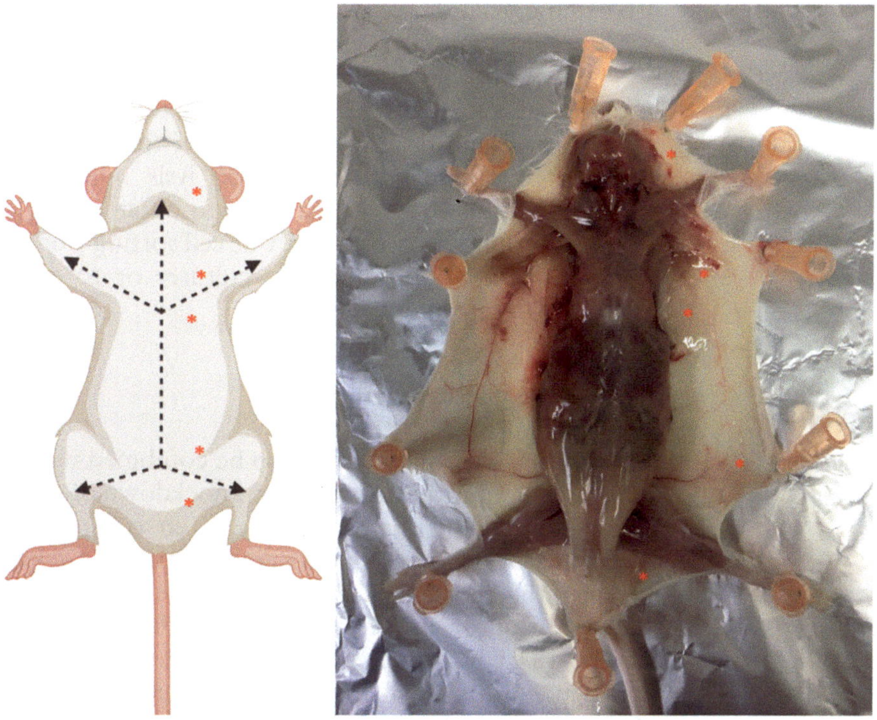

Fig. 6 Mouse preparation for mammary gland dissection. Left, mouse representation in supine position, with black dashed arrows drawing the incisions to be performed with dissection scissors. Right, image of a mouse secured on a dissecting pad with needles and with the skin cut open following the incision lines indicated in the schematic on the left and with an intact peritoneum, ready for mammary glands dissection. Red asterisks indicate the position of the five mammary glands on one side of the mouse

(a) Immunostaining. Extend flatly the mammary gland on a Whatman paper and fix in 4% paraformaldehyde in a 15 mL Falcon tube during 2 h at room temperature. After washing with PBS, the glands can be embedded in paraffin (after an increasing gradient of ethanol and xylene), or in optimal cutting temperature (OCT) medium. The paraffin or OCT blocks are then cut to 4 μm sections; paraffin sections are de-waxed and rehydrated through xylenes and a gradient of ethanol and subjected to antigen retrieval in boiling citrate or TE buffer for 20 min. The tissue is then permeabilized with 0.3% Triton X-100, non-specific antibody binding is blocked with 5% FBS and 10% BSA and then sequentially incubated with primary and secondary antibodies. For further details, check the video-protocols "Immunofluorescence Staining of Paraffin-Embedded Tissue Sections" or "Indirect Immunofluorescence on Frozen Sections of Mouse Mammary Gland" in JoVE (https://www.jove.com).

(b) Single cell dissociation for Flow Cytometry. Freshly dissected mammary glands are incubated in dissociation buffer for 1 h at 37 °C. Cells are then sequentially dissociated in 0.25% trypsin-EDTA for 3 min, 5 mg/mL dispase and 0.1 mg/mL DNase I for 5 min, then incubated 1 min with red blood lysis buffer and filtered through a 40 μm cell strainer to obtain a single cell suspension for flow cytometry or cell sorting (for subsequent use in transplantations or cultures or to extract RNA, DNA, or protein contents).

(c) Whole mounts.

(d) 3D-organotypic cultures.

3.4 Other Lineage Tracing-Related Applications

Lineage tracing experiments can also be combined with functional studies, which imply the activation or knockdown of specific genes, allowing to elucidate their role in a given cell population at a defined time and the consequences of activation or deletion of the said gene in the derived lineages. An example of this type of studies is the expression of an activated form of PI3K in the mammary gland [32, 45], which resulted in the formation of different types of tumors depending on which cells were targeted by the Cre drivers.

Other functional assays include the genetic ablation of a certain cell type using lox-inducible diphtheria toxin A (DTA) or DTA receptor (DTR) mouse models. This approach permits the genetic ablation of a targeted lineage due to the presence of a loxP-flanked STOP cassette preventing expression of DTA or DTR. Upon Cre recombinase activation, the lox-stop fragment is excised resulting in constitutive ablation of the Cre-expressing cells, which can be temporally controlled by injection of DTA in mice expressing the DTA receptor (DTR). This system has been used in the mammary gland to partially eliminate Notch2-expressing cells [35] to assess their role in mammary branching. More recently, the targeted ablation of LCs by a DTA-based approach has illustrated the impressive cellular plasticity of mammary epithelial cells, causing reactivation of multipotency in BCs [46].

The inducible Tet/Lox system allows, as mentioned above, to perform lineage tracing at saturation, enabling targeting all cells in a given population. To achieve quantitative lineage tracing at saturation, triple transgenic female mice, carrying a driver rtTA, a TetO-Cre and a reporter, receive continuous oral administration of doxycycline, through drinking water or by food (*see* **Note 8**), during an extended period of time (i.e., the whole course of pubertal development from 3 to 8 weeks of age or during pregnancy, lactation, and involution). Thus, mice can be analyzed during the DOX treatment (pulse) or after a defined period of time (chase).

Interestingly, a new generation of lineage tracing approaches has been designed. They use unbiased barcoding techniques, allowing to tag and trace every cell individually with a unique and heritable DNA barcode encoding clonal information. Differently from fluorescence-based lineage tracing systems, which are inherently limited by the number of distinct clones that can be simultaneously followed, fate mapping by DNA barcodes theoretically permit to record every single division event in a given tissue. This information is read out retrospectively using high-throughput sequencing and can be combined with other single cell measurements. Targeted barcoding-based methods generally fall into three types:

1. Transgenic integration of exogenous DNA sequences.
2. Recombination of transgenic DNA cassettes.
3. In vivo editing of transgenic DNA targets by CRISPR-Cas9.

In all of these approaches, a DNA-barcoding event permanently alters the genome of an individual cell, the descendants of which inherit the barcode and can be distinguished as a clonal unit. Importantly, DNA barcodes can be analyzed at high throughput, enabling the interrogation of hundreds or thousands of distinct clonal events in parallel by high-throughput sequencing [47].

4 Notes

1. Before using a new Cre driver mouse line (constitutive or inducible Cre), check the characterization of the mouse model on international databases such as MGI website (http://www.informatics.jax.org/).
2. Although most reporter lines can be maintained as homozygous, Cre lines usually have to be kept as heterozygotes, which implies that half of the offspring will be double heterozygous and only half of them will be females. It is of course advisable to optimize your breeding schemes so that litters have a high chance of containing double heterozygous mice, but this is not always possible as not all strains are viable or breed efficiently as homozygous.
3. Jackson Laboratories recommends a quick 2-step DNA extraction protocol, widely used around the world:
   (a) Cut 2 mm of ear and place into an Eppendorf tube or 96-well plate.
   (b) Add 75 μL 25 mM NaOH/0.2 mM EDTA.
   (c) Place in a thermocycler at 98 °C for 1 h, then reduce the temperature to 15 °C.
   (d) Add 75 μL of 40 mM Tris HCl (pH 5.5).
   (e) Centrifuge at 2000 × *g* for 3 min.
   (f) Take 2 μL to perform the PCRs.

4. Cre recombination efficiency should also be carefully assessed, as it can be locus-dependent or vary in different cell types. To examine recombination at the target locus, assess Cre-mediated excision by qPCR in targeted and non-targeted cells. The following experimental and control mice should be analyzed:
   (a) Promoter-Cre^{ER} +/−; $Reporter^{Lox/Lox}$ (experimental model, induced with tamoxifen).
   (b) Promoter-Cre^{ER} −/−; $Reporter^{Lox/Lox}$ (no Cre control, induced with tamoxifen) to check potential reporter leakiness.
   (c) Promoter-Cre^{ER} +/−; $Reporter^{wt/wt}$ (no reporter control, induced with tamoxifen) to check potential recombinase toxicity.
   (d) Promoter-Cre^{ER} +/−; $Reporter^{Lox/Lox}$ (not induced with tamoxifen) to assess potential CreER leakiness.
5. Use tamoxifen or 4-OHT for inducible Cre lines: Cre-ER, Cre-ERT or Cre-ERT2. Use doxycycline for rtTA/Tet-O lines.
6. Cre expression should be checked in different organs and cell types by RT-qPCR, to detect even a small amount of Cre recombinase mRNA and ensure target specificity.
7. Before the first incision, spray the female with 70% Ethanol to avoid contamination of hair in your sample collection.
8. Recommended doxycycline dose: in drinking water 1 mg/mL; in food: 1 g/kg (https://www.bio-serv.com/) [11].

Acknowledgments

We are grateful to the Institut Curie In Vivo Experimental Facility for help in the maintenance and care of mouse colonies. We also acknowledge the Cell and Tissue Imaging Platform (PICT-IBiSA) of the Genetics and Developmental Biology Department (UMR3215/U934) and the Nikon Imaging Centre of Institut Curie, member of the French National Research Infrastructure France-BioImaging (ANR10-INBS-04). SF is supported by Paris Sciences et Lettres (PSL* Research University) (grant # C19-64-2019-228), the French National Research Agency (ANR) grant number ANR-15-CE13-0013-01, the Canceropole Ile-de-France (grant # 2015-2-APD-01-2 ICR-1), the Ligue contre la cancer (grant #RS19/75-101), the Foundation for Medical Research (grant # EQU201903007821 0), the Schlumberger Foundation FSER (grant # FSER20200211117) and by Labex DEEP ANR-Number 11-3 LBX-0044. VR is supported by Josep Carreras Leukaemia Research Institute (IJC) and funded by Ramón y Cajal Programme from Ministry of Science, Innovation and Universities (RYC2018-024099-I).

References

1. Kretzschmar K, Watt FM (2012) Lineage tracing. Cell 148(1–2):33–45. https://doi.org/10.1016/j.cell.2012.01.002
2. Barker N, van Es JH, Kuipers J, Kujala P, van den Born M, Cozijnsen M, Haegebarth A, Korving J, Begthel H, Peters PJ, Clevers H (2007) Identification of stem cells in small intestine and colon by marker gene Lgr5. Nature 449(7165):1003–1007
3. Van Keymeulen A, Rocha AS, Ousset M, Beck B, Bouvencourt G, Rock J, Sharma N, Dekoninck S, Blanpain C (2011) Distinct stem cells contribute to mammary gland development and maintenance. Nature 479(7372): 189–193. https://doi.org/10.1038/nature10573
4. Hong KU, Reynolds SD, Watkins S, Fuchs E, Stripp BR (2004) In vivo differentiation potential of tracheal basal cells: evidence for multipotent and unipotent subpopulations. Am J Physiol Lung Cell Mol Physiol 286(4):L643–L649. https://doi.org/10.1152/ajplung.00155.2003
5. Yang Y, Riccio P, Schotsaert M, Mori M, Lu J, Lee DK, Garcia-Sastre A, Xu J, Cardoso WV (2018) Spatial-temporal lineage restrictions of embryonic p63(+) progenitors establish distinct stem cell pools in adult airways. Dev Cell 44(6):752–761 e754. https://doi.org/10.1016/j.devcel.2018.03.001
6. Tika E, Ousset M, Dannau A, Blanpain C (2019) Spatiotemporal regulation of multipotency during prostate development. Development 146(20):dev180224. https://doi.org/10.1242/dev.180224
7. Lilja AM, Rodilla V, Huyghe M, Hannezo E, Landragin C, Renaud O, Leroy O, Rulands S, Simons BD, Fre S (2018) Clonal analysis of Notch1-expressing cells reveals the existence of unipotent stem cells that retain long-term plasticity in the embryonic mammary gland. Nat Cell Biol 20(6):677–687. https://doi.org/10.1038/s41556-018-0108-1
8. Rodilla V, Dasti A, Huyghe M, Lafkas D, Laurent C, Reyal F, Fre S (2015) Luminal progenitors restrict their lineage potential during mammary gland development. PLoS Biol 13(2):e1002069. https://doi.org/10.1371/journal.pbio.1002069
9. Lafkas D, Rodilla V, Huyghe M, Mourao L, Kiaris H, Fre S (2013) Notch3 marks clonogenic mammary luminal progenitor cells in vivo. J Cell Biol 203(1):47–56. https://doi.org/10.1083/jcb.201307046
10. Ousset M, Van Keymeulen A, Bouvencourt G, Sharma N, Achouri Y, Simons BD, Blanpain C (2012) Multipotent and unipotent progenitors contribute to prostate postnatal development. Nat Cell Biol 14(11):1131–1138. https://doi.org/10.1038/ncb2600
11. Wuidart A, Ousset M, Rulands S, Simons BD, Van Keymeulen A, Blanpain C (2016) Quantitative lineage tracing strategies to resolve multipotency in tissue-specific stem cells. Genes Dev 30(11):1261–1277. https://doi.org/10.1101/gad.280057.116
12. Prater MD, Petit V, Alasdair Russell I, Giraddi RR, Shehata M, Menon S, Schulte R, Kalajzic I, Rath N, Olson MF, Metzger D, Faraldo MM, Deugnier MA, Glukhova MA, Stingl J (2014) Mammary stem cells have myoepithelial cell properties. Nat Cell Biol 16(10): 942–950, 941–947. https://doi.org/10.1038/ncb3025
13. Van Keymeulen A, Fioramonti M, Centonze A, Bouvencourt G, Achouri Y, Blanpain C (2017) Lineage-restricted mammary stem cells sustain the development, homeostasis, and regeneration of the estrogen receptor positive lineage. Cell Rep 20(7):1525–1532. https://doi.org/10.1016/j.celrep.2017.07.066
14. Wang C, Christin JR, Oktay MH, Guo W (2017) Lineage-biased stem cells maintain estrogen-receptor-positive and -negative mouse mammary luminal lineages. Cell Rep 18(12):2825–2835. https://doi.org/10.1016/j.celrep.2017.02.071
15. Danielian PS, White R, Hoare SA, Fawell SE, Parker MG (1993) Identification of residues in the estrogen receptor that confer differential sensitivity to estrogen and hydroxytamoxifen. Mol Endocrinol 7(2):232–240. https://doi.org/10.1210/mend.7.2.8469236
16. Feil R, Brocard J, Mascrez B, LeMeur M, Metzger D, Chambon P (1996) Ligand-activated site-specific recombination in mice. Proc Natl Acad Sci U S A 93(20): 10887–10890
17. Metzger D, Clifford J, Chiba H, Chambon P (1995) Conditional site-specific recombination in mammalian cells using a ligand-dependent chimeric Cre recombinase. Proc Natl Acad Sci U S A 92(15):6991–6995. https://doi.org/10.1073/pnas.92.15.6991
18. Rios AC, Fu NY, Lindeman GJ, Visvader JE (2014) In situ identification of bipotent stem cells in the mammary gland. Nature

506(7488):322–327. https://doi.org/10.1038/nature12948

19. Van der Flier LG, Sabates-Bellver J, Oving I, Haegebarth A, De Palo M, Anti M, Van Gijn ME, Suijkerbuijk S, Van de Wetering M, Marra G, Clevers H (2007) The intestinal Wnt/TCF signature. Gastroenterology 132(2):628–632
20. Sauer B, McDermott J (2004) DNA recombination with a heterospecific Cre homolog identified from comparison of the pac-c1 regions of P1-related phages. Nucleic Acids Res 32(20): 6086–6095. https://doi.org/10.1093/nar/gkh941
21. He L, Li Y, Li Y, Pu W, Huang X, Tian X, Wang Y, Zhang H, Liu Q, Zhang L, Zhao H, Tang J, Ji H, Cai D, Han Z, Han Z, Nie Y, Hu S, Wang QD, Sun R, Fei J, Wang F, Chen T, Yan Y, Huang H, Pu WT, Zhou B (2017) Enhancing the precision of genetic lineage tracing using dual recombinases. Nat Med 23(12):1488–1498. https://doi.org/10.1038/nm.4437
22. Li Y, He L, Huang X, Bhaloo SI, Zhao H, Zhang S, Pu W, Tian X, Li Y, Liu Q, Yu W, Zhang L, Liu X, Liu K, Tang J, Zhang H, Cai D, Ralf AH, Xu Q, Lui KO, Zhou B (2018) Genetic lineage tracing of nonmyocyte population by dual recombinases. Circulation 138(8):793–805. https://doi.org/10.1161/CIRCULATIONAHA.118.034250
23. Liu K, Yu W, Tang M, Tang J, Liu X, Liu Q, Li Y, He L, Zhang L, Evans SM, Tian X, Lui KO, Zhou B (2018) A dual genetic tracing system identifies diverse and dynamic origins of cardiac valve mesenchyme. Development 145(18):dev167775. https://doi.org/10.1242/dev.167775
24. Han X, Wang Y, Pu W, Huang X, Qiu L, Li Y, Yu W, Zhao H, Liu X, He L, Zhang L, Ji Y, Lu J, Lui KO, Zhou B (2019) Lineage tracing reveals the bipotency of SOX9(+) hepatocytes during liver regeneration. Stem Cell Rep 12(3):624–638. https://doi.org/10.1016/j.stemcr.2019.01.010
25. Liu Q, Liu K, Cui G, Huang X, Yao S, Guo W, Qin Z, Li Y, Yang R, Pu W, Zhang L, He L, Zhao H, Yu W, Tang M, Tian X, Cai D, Nie Y, Hu S, Ren T, Qiao Z, Huang H, Zeng YA, Jing N, Peng G, Ji H, Zhou B (2019) Lung regeneration by multipotent stem cells residing at the bronchioalveolar-duct junction. Nat Genet 51(4):728–738. https://doi.org/10.1038/s41588-019-0346-6
26. Li Y, Lv Z, Zhang S, Wang Z, He L, Tang M, Pu W, Zhao H, Zhang Z, Shi Q, Cai D, Wu M, Hu G, Lui KO, Feng J, Nieto MA, Zhou B (2020) Genetic fate mapping of transient cell fate reveals N-cadherin activity and function in tumor metastasis. Dev Cell 54(5):593–607 e595. https://doi.org/10.1016/j.devcel.2020.06.021
27. van Amerongen R, Bowman AN, Nusse R (2012) Developmental stage and time dictate the fate of Wnt/beta-catenin-responsive stem cells in the mammary gland. Cell Stem Cell 11(3):387–400
28. de Visser KE, Ciampricotti M, Michalak EM, Tan DW, Speksnijder EN, Hau CS, Clevers H, Barker N, Jonkers J (2012) Developmental stage-specific contribution of LGR5(+) cells to basal and luminal epithelial lineages in the postnatal mammary gland. J Pathol 228(3): 300–309. https://doi.org/10.1002/path.4096
29. Blaas L, Pucci F, Messal HA, Andersson AB, Josue Ruiz E, Gerling M, Douagi I, Spencer-Dene B, Musch A, Mitter R, Bhaw L, Stone R, Bornhorst D, Sesay AK, Jonkers J, Stamp G, Malanchi I, Toftgard R, Behrens A (2016) Lgr6 labels a rare population of mammary gland progenitor cells that are able to originate luminal mammary tumours. Nat Cell Biol 18(12):1346–1356. https://doi.org/10.1038/ncb3434
30. Chakrabarti R, Celia-Terrassa T, Kumar S, Hang X, Wei Y, Choudhury A, Hwang J, Peng J, Nixon B, Grady JJ, DeCoste C, Gao J, van Es JH, Li MO, Aifantis I, Clevers H, Kang Y (2018) Notch ligand Dll1 mediates cross-talk between mammary stem cells and the macrophageal niche. Science 360(6396):eaan4153. https://doi.org/10.1126/science.aan4153
31. Wang D, Cai C, Dong X, Yu QC, Zhang XO, Yang L, Zeng YA (2015) Identification of multipotent mammary stem cells by protein C receptor expression. Nature 517(7532): 81–84. https://doi.org/10.1038/nature13851
32. Koren S, Reavie L, Couto JP, De Silva D, Stadler MB, Roloff T, Britschgi A, Eichlisberger T, Kohler H, Aina O, Cardiff RD, Bentires-Alj M (2015) PIK3CA(H1047R) induces multipotency and multi-lineage mammary tumours. Nature 525(7567):114–118. https://doi.org/10.1038/nature14669
33. Chang TH, Kunasegaran K, Tarulli GA, De Silva D, Voorhoeve PM, Pietersen AM (2014) New insights into lineage restriction of mammary gland epithelium using parity-identified mammary epithelial cells. Breast Cancer Res 16(1):R1. https://doi.org/10.1186/bcr3593
34. Elias S, Morgan MA, Bikoff EK, Robertson EJ (2017) Long-lived unipotent Blimp1-positive luminal stem cells drive mammary gland

organogenesis throughout adult life. Nat Commun 8(1):1714. https://doi.org/10.1038/s41467-017-01971-w
35. Sale S, Lafkas D, Artavanis-Tsakonas S (2013) Notch2 genetic fate mapping reveals two previously unrecognized mammary epithelial lineages. Nat Cell Biol 15(5):451–460. https://doi.org/10.1038/ncb2725
36. Davis FM, Lloyd-Lewis B, Harris OB, Kozar S, Winton DJ, Muresan L, Watson CJ (2016) Single-cell lineage tracing in the mammary gland reveals stochastic clonal dispersion of stem/progenitor cell progeny. Nat Commun 7:13053
37. Lloyd-Lewis B, Harris OB, Watson CJ, Davis FM (2017) Mammary stem cells: premise, properties, and perspectives. Trends Cell Biol 27(8):556–567. https://doi.org/10.1016/j.tcb.2017.04.001
38. Scheele CL, Hannezo E, Muraro MJ, Zomer A, Langedijk NS, van Oudenaarden A, Simons BD, van Rheenen J (2017) Identity and dynamics of mammary stem cells during branching morphogenesis. Nature 542(7641): 313–317. https://doi.org/10.1038/nature21046
39. Soriano P (1999) Generalized lacZ expression with the ROSA26 Cre reporter strain. Nat Genet 21(1):70–71
40. Muzumdar MD, Tasic B, Miyamichi K, Li L, Luo L (2007) A global double-fluorescent Cre reporter mouse. Genesis 45(9):593–605. https://doi.org/10.1002/dvg.20335
41. Snippert HJ, van der Flier LG, Sato T, van Es JH, van den Born M, Kroon-Veenboer C, Barker N, Klein AM, van Rheenen J, Simons BD, Clevers H (2010) Intestinal crypt homeostasis results from neutral competition between symmetrically dividing Lgr5 stem cells. Cell 143(1):134–144
42. Wuidart A, Sifrim A, Fioramonti M, Matsumura S, Brisebarre A, Brown D, Centonze A, Dannau A, Dubois C, Van Keymeulen A, Voet T, Blanpain C (2018) Early lineage segregation of multipotent embryonic mammary gland progenitors. Nat Cell Biol 20(6):666–676. https://doi.org/10.1038/s41556-018-0095-2
43. Madisen L, Zwingman TA, Sunkin SM, Oh SW, Zariwala HA, Gu H, Ng LL, Palmiter RD, Hawrylycz MJ, Jones AR, Lein ES, Zeng H (2010) A robust and high-throughput Cre reporting and characterization system for the whole mouse brain. Nat Neurosci 13(1): 133–140. https://doi.org/10.1038/nn.2467
44. Srinivas S, Watanabe T, Lin CS, William CM, Tanabe Y, Jessell TM, Costantini F (2001) Cre reporter strains produced by targeted insertion of EYFP and ECFP into the ROSA26 locus. BMC Dev Biol 1:4. https://doi.org/10.1186/1471-213x-1-4
45. Van Keymeulen A, Lee MY, Ousset M, Brohee S, Rorive S, Giraddi RR, Wuidart A, Bouvencourt G, Dubois C, Salmon I, Sotiriou C, Phillips WA, Blanpain C (2015) Reactivation of multipotency by oncogenic PIK3CA induces breast tumour heterogeneity. Nature 525(7567):119–123. https://doi.org/10.1038/nature14665
46. Centonze A, Lin S, Tika E, Sifrim A, Fioramonti M, Malfait M, Song Y, Wuidart A, Van Herck J, Dannau A, Bouvencourt G, Dubois C, Dedoncker N, Sahay A, de Maertelaer V, Siebel CW, Van Keymeulen A, Voet T, Blanpain C (2020) Heterotypic cell-cell communication regulates glandular stem cell multipotency. Nature 584(7822): 608–613. https://doi.org/10.1038/s41586-020-2632-y
47. McKenna A, Gagnon JA (2019) Recording development with single cell dynamic lineage tracing. Development 146(12):dev169730. https://doi.org/10.1242/dev.169730

Chapter 8

Lentiviral Transduction of Mammary Epithelial Cells

Richard Iggo

Abstract

Lentiviral vectors are the workhorses of modern cell biology. They can infect a wide variety of cells including non-dividing cells and stem cells. They integrate into the genome of infected cells leading to stable expression. It is easy to transduce 100% of the cells in a culture and possible to infect cells simultaneously with multiple vectors, greatly facilitating studies on malignant transformation. We present simple protocols to produce and titrate lentiviral vectors, infect mammary epithelial cells, and check for contamination with replication competent viruses.

Key words Lentivirus, Transfection, Infection

1 Introduction

1.1 Lentiviral Vector Production

Lentiviral vectors are convenient tools to express transgenes in a wide variety of cultured cells. At first glance it might seem risky to use vectors derived from HIV as basic research tools, but the labs that conceived them went to great lengths to make them safe [1–4] (*see* **Notes 1** and **2**). Minimally, you should familiarize yourself with level 2 biosafety precautions and make the requisite declarations to the authorities before starting to work with lentiviral vectors.

Unlike transgenes in transfected DNA, transgenes in lentiviral vectors are immediately integrated stably into the genome of the target cell, greatly increasing the chance that they will be expressed stably. Unlike simple oncoretroviral or gamma retroviral vectors, lentiviral vectors can infect non-dividing cells. The critical difference is that the cDNA produced by reverse transcription of lentiviral RNA is professionally imported into the nucleus, whereas cDNA from simple retroviruses must wait for the nuclear envelope to break down at mitosis to gain access to the chromosomes. This simplifies the protocol because there is no need to passage the target cells immediately after infection to make them divide. The vectors are normally pseudotyped with vesicular stomatitis

Maria dM. Vivanco (ed.), *Mammary Stem Cells: Methods and Protocols*,
Methods in Molecular Biology, vol. 2471, https://doi.org/10.1007/978-1-0716-2193-6_8,

glycoprotein (VSVG), which induces fusion of the viral and plasma membranes. Mammary epithelial cells are normally easy to infect with VSVG pseudotyped lentiviral vectors.

1.2 Lentiviral Vector Production

The basic approach is to transfect 293T cells with a vector plasmid and two or three helper plasmids. VSVG induces fusion of the 293T cells to form large syncytia that provide a simple visual clue that the transfection has worked. 293T cells are unique among human cell lines in being easy to transfect with $CaPO_4$. That said, $CaPO_4$ is notorious for giving results that can vary between, and even within, labs. If you get high transfection efficiency with $CaPO_4$, it is the perfect solution. Polyethylenimine (PEI) gives lower transfection efficiency but forgives large departures from the protocol. Commercial reagents like Fugene can also be used but they become very expensive if you need to scale up, for example to produce vectors with large inserts.

1.3 Packaging Vectors

Most labs use so-called second-generation plasmids, in which one helper plasmid contains all the lentiviral helper genes and a second helper plasmid contains VSVG. Other than tat and rev, the lentiviral accessory genes (vif, vpr, vpu, and nef) are all deleted from the helper plasmids. In third-generation systems, the lentiviral genes are split between two plasmids, with the gag, polymerase, protease, and integrase genes on one plasmid and rev on another. In this configuration the tat gene is deleted, which means the vector plasmid requires a tat-independent promoter. A plasmid expressing VSVG is still required, meaning a total of four plasmids must be transfected. The third-generation system is considered safer but the need to transfect an additional plasmid may reduce the titre and increase the risk of interferon induction (*see* **Note 3**). In practice, most users choose the second-generation psPAX2 and pMD2.G packaging vectors developed by the Naldini and Trono labs (Fig. 1a, b, Addgene 12259 and 12260).

1.4 Lentiviral Expression Vectors

There is a wide choice of vector plasmids. In addition to deletion of all complete HIV genes, current vectors have partial replacement of the 5′-LTR by a heterologous promoter to make transcription in the packaging cell independent of tat; a deletion in the 3′-LTR to make both LTR promoters inactive after integration; and the presence of an RNase H-resistant central polypurine tract that primes reverse transcription and promotes nuclear import of the preintegration complex. Many companies now sell lentiviral vectors but a cheaper option for academic labs is to obtain vectors from Addgene. Many commonly used vectors contain two expression cassettes, one for the transgene, the other for a selectable marker (Fig. 2a [5]). It is tempting to assume that the transgene will be expressed as long as the marker is expressed but this is not guaranteed: the promoter for the transgene can be silenced while the

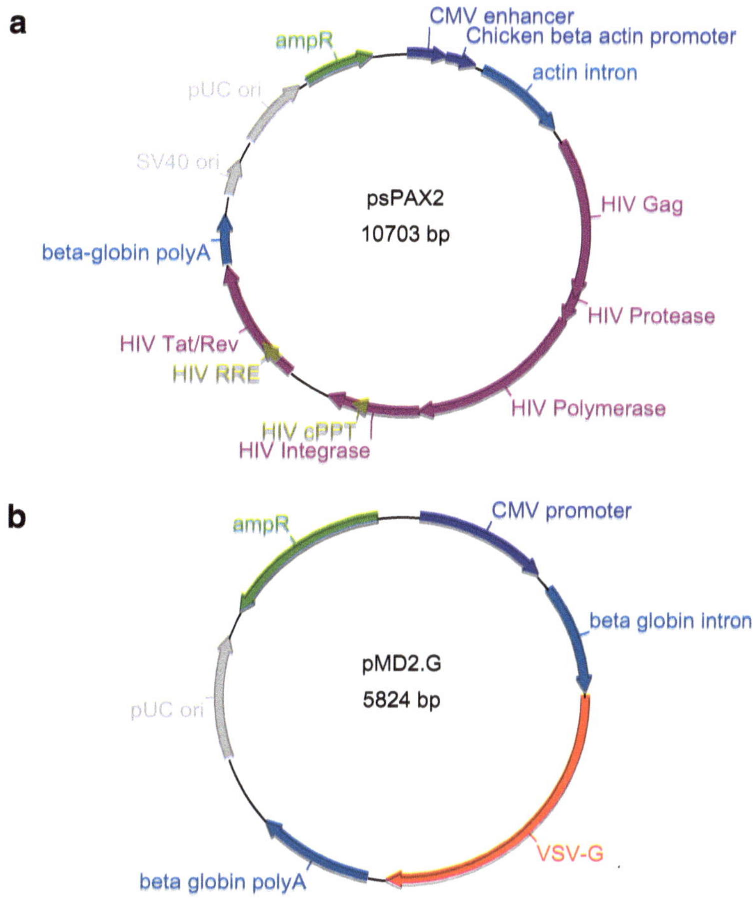

Fig. 1 Lentiviral plasmids. (**a**) psPAX2 contains the main lentiviral genes required for packaging: Gag, Protease, Polymerase, Integrase, Tat, and Rev. (**b**) pMD2.G expresses VSV Glycoprotein, which replaces the normal envelope protein. VSV G induces fusion of the viral lipid envelope with the plasma membrane, allowing the viral particles to infect cells indiscriminately

promoter for the selectable marker remains active. To get around this problem it is now common to express both genes from a single promoter with a picornaviral internal ribosome entry site (IRES) or 2A sequence between the two genes (Fig. 2b). Picornaviruses express multiple proteins from a single transcript, and they do so when cap-dependent translation is blocked. As the name suggests, the IRES allows the ribosome to initiate translation from an internal site in the transcript. The drawback is that this is not terribly efficient, leading to lower expression of the second open reading frame. IRES's form complex stem-loop structures that must fold correctly. If the RNA encoding the transgene interferes with this folding, the IRES will give poor expression, so they can be quite fickle. The second disadvantage is that IRES's are about 1 kb in size. While this may not seem much, the titre of lentiviral vectors

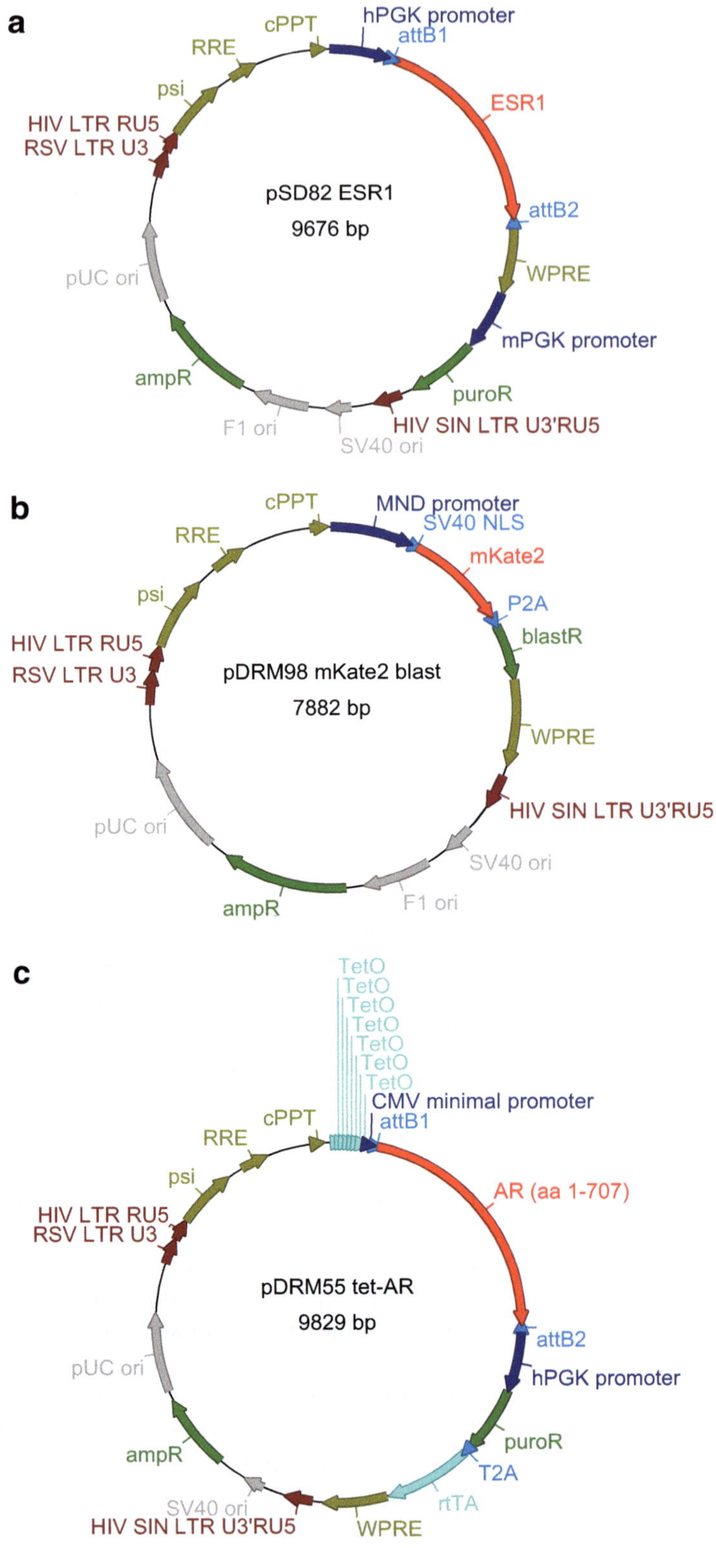

Fig. 2 Vector plasmids. The genomic transcripts made in the packaging cells initiate at a hybrid RSV/HIV promoter. The RSV enhancer is deleted in integrated

falls precipitously as the size increases. These drawbacks explain the popularity of 2A sequences. Picornaviruses generate most of their proteins by protease cleavage of a large polyprotein. The 2A "cleavage site" is an exception: a 20 amino acid sequence ending PGP fools the ribosome into not making a peptide bond [7, 8]. The ribosome carries on regardless, leading to equal expression of the peptides upstream and downstream of the 2A site. The 2A approach is attractive because the 2A sequence adds only 60 bp to the vector but it is not 100% efficient, so a small amount of fusion protein will be produced, and it leaves a 19 amino acid tag on the C-terminus of the upstream protein. It is possible to daisy chain transgenes separated by 2A sequences to express multiple proteins from a single transcript. The choice of promoter is dictated by the propensity of some promoters to become silenced more easily than others (cellular promoters are considered better than viral promoters), and by the characteristics of the cells transduced (PGK works well in mammary progenitors and stem cells). Inducible vectors let you dissociate infection from expression, leading to much better control of the conditions during your experiment. Single vector systems are now available that contain both the reverse tet transactivator (rtTA) and the transgene. We have obtained excellent results with pCW57.1 (Addgene 41393) (Fig. 2c) [6]. It gives very low background expression in the uninduced state and can be titrated with doxycycline from 10 ng/ml to 1 μg/ml to give a wide range of transgene expression. It has the added advantage that Gateway cloning is straightforward if you have an ENTRY vector that already contains your insert; if not, or for more complex designs, you can clone into the attR BsrGI sites by Gibson assembly.

Fig. 2 (continued) proviruses because the 3′-LTR is copied to the 5′-LTR during reverse transcription. There is a deletion removing the enhancer in the U3 region in the 3′-LTR. This deletion is copied to the 5′-LTR after reverse transcription to create integrated proviruses with defective promoters in both LTRs (a so-called self-inactivating or SIN vector). This means transcripts from the integrated proviruses lack the packaging signal because they initiate at the first exogenous promoter. The WPRE stabilises the RNA. The viral RNAs all terminate in the 3′-LTR. It is important delete polyA signals from the transgene to avoid cleaving off the 3′-end of the genomic RNA that is packaged in the viral particles. The vector plasmid contains the SV40 origin so that the copy number in the transfected 293T cells is amplified by T antigen-dependent DNA replication (the "T" in 293T is SV40 T antigen). (**a**) pSD-82 is a typical lentiviral expression vector [5]. The puromycin resistance gene is expressed from a second promoter. (**b**) pDRM98 (Addgene 174720) expresses nuclear mKate2 followed by a 2A site then the blasticidin resistance gene. (**c**) pDRM55 expresses AR from a tet regulated promoter [6]. The reverse tet transactivator (rtTA) and puromycin resistance genes are separated by a 2A site. This ensures that virtually all puromycin resistant cells express rtTA

2 Materials

2.1 Components for Lentiviral Production

1. 293T/17 cells (i.e., ATCC CRL-11268).
2. DMEM medium: DMEM medium, 10% Fetal bovine serum (FBS).
3. Phosphate-buffered saline.
4. 0.05% trypsin/EDTA.
5. TE buffer, pH 8.0.
6. DNAse/RNase-free water.
7. Poly-lysine.
8. OptiMEM 20 mM HEPES medium: Add 1 M HEPES to make OptiMEM 20 mM HEPES.
9. 10 cm tissue culture dishes—for scale-up use 15 cm dishes.
10. 15 and 50 ml conical base tubes.
11. 0.45 and 0.22 μm syringe filters.
12. Vivaspin 20 MWCO 100,000 (i.e., GE Healthcare #28-9323-63).
13. 5% CO_2, 37 °C incubator.
14. 1% Virkon.
15. 70% (v/v) ethanol.
16. Sterile screw cap 1.5 ml plastic vials with O-rings.

2.1.1 For Calcium Phosphate Transfection

1. $CaCl_2$ solution: 18.4 g of 2.5 M $CaCl_2$, for 50 ml; filter sterilize (0.22 μm), store aliquots at −20 °C.
2. HEPES-buffered saline: 2× HeBS.

2.1.2 For PEI Transfection

PEI MAX: 1 g PEI MAX Transfection Grade Linear Polyethylenimine Hydrochloride (MW 40,000): 10 mg/ml in tissue culture grade water, adjust pH to 7.4 with NaOH, filter sterilize 0.22 μm, store aliquots at −20 °C.

2.1.3 Plasmids to Make a Simple GFP Expression Vector

1. pRRLSIN.cPPT.PGK-GFP.WPRE (classic GFP expression vector, Addgene 12252).
2. pMD2.G (for pseudotyping with VSV G protein, Addgene 12259).
3. psPAX2 (contains the main HIV-1 helper genes: Gag, Pol, Tat, and Rev., Addgene 12260) (*see* **Notes 3** and **4**). Many other useful plasmids can be obtained from Addgene. All plasmid maxipreps should be "endotoxin-free" (Qiagen EndoFree or Nucleobond Xtra EF). For convenience, adjust the concentration of plasmids to 1 mg/ml.

2.2 Components for Titration of Viruses

1. Target cells (e.g., MCF7 cells).
2. Cell culture medium: RPMI or DMEM, 10% FCS (for MCF7 cells).
3. Viral suspension.
4. KOVA Glasstic Slide with ten Counting Grids (#87144, www.kovaintl.com).
5. Phosphate-buffered saline.
6. 0.05% trypsin/EDTA.
7. 6-well plates.
8. FACS tubes (polypropylene with lids).

2.3 Components for RCL Test

1. HIV genomic RNA (if available).
2. 293T/17 cells.
3. DMEM medium: DMEM medium, 10% FBS.
4. Phosphate-buffered saline.
5. 0.05% trypsin/EDTA.
6. Trizol LS.
7. RNeasy kit.
8. 6-well plates.
9. 5% CO_2, 37 °C incubator.

2.3.1 Reagents for Routine Molecular Biology

1. Superase inhibitor 20 U/µl.
2. DNase I (RNase-free) 10 U/µl.
3. DNAse/RNase-free water.
4. 10% SDS.
5. Phenol/chloroform/isoamylalcohol.
6. 3 M NaOAc pH 5.5.
7. Glycogen 20 mg/ml.
8. Ethanol 70% and 100% (molecular biology grade).
9. Superscript II First strand kit.
10. Sybr Green qPCR master mix 2×.
11. Forward Primer PAXpolF: tggcagaaaacagggagatt.
12. Reverse Primer PAXpolR: tgcccctgcttctgtatttc.
13. qPCR machine.

3 Methods

Follow national biosafety guidelines when working with lentiviral vectors. Some countries require level 3 containment during production of lentiviral vectors expressing oncogenes but most work with lentiviral vectors can now be performed under level 2 containment. If you are unsure what to do you are almost certainly not legal, so seek advice on local regulations from more experienced colleagues before starting the project. *See* **Note 1** for additional information on biosafety of lentiviral vectors.

3.1 Production of Lentiviral Particles

293T are undemanding cells that grow rapidly in DMEM medium. It is important to split them before they overgrow or exhaust the medium, for example 1:10 twice a week. Coating the dishes with poly-lysine before transfection is not essential but stops the cells detaching during medium changes. Replace cultures every 3 months from frozen stocks that have been checked for mycoplasma. If none of your vectors contains GFP, it is helpful to set up an additional transfection with GFP to check the transfection efficiency.

1. Coat tissue culture plates with poly-lysine by briefly pipetting poly-lysine onto 15 cm plates (allow about 1 ml per plate) and rinsing with PBS.
2. Plate 3×10^7 293T cells per 15 cm dish 24 h before transfection. Make two 15 cm dishes of cells per virus prep for vectors with large inserts.
3. Cells should be about 80% confluent on the day of transfection. It is better to err on the high side—the one absolute certainty is that precipitate landing on bare plastic will be wasted.
4. Replace the medium on the cells with fresh medium immediately before making precipitates. Optional: Add 10 μM chloroquine to this medium (*see* **Note 5**).

3.1.1 Calcium Phosphate Protocol

1. $CaPO_4$. For each transfection of two 15 cm plates, prepare the following plasmid cocktail in a 2 ml Eppendorf tube: 20 μg pMD2.G (VSV-G viral envelope), 50 μg psPAX2 (packaging construct), 50 μg of your expression vector (for a 7 kb vector plasmid). Total DNA = 120 μg in 2250 μl TC grade water. Add 250 μl $CaCl_2$ and mix thoroughly.

   The amount of vector DNA should be adjusted according to the size of the plasmid in the following range: 50 μg <7000 bp, 75 μg 7000 to 10,000 bp, 100 μg >10,000 bp. If you increase the amount of DNA, you should increase the volumes proportionately to keep the DNA concentration below 25 μg/ml [9] (*see* **Note 6**).
2. $CaPO_4$. Prepare 15 ml Falcon tubes containing 2.5 ml 2× HeBS (1 volume).
3. $CaPO_4$. Add the plasmid cocktails drop by drop to the HeBS tubes while vortexing vigorously. If scaling up, use bigger tubes to ensure the solutions mix instantly.

4. $CaPO_4$. Wait 1 min to allow the precipitate to form (longer is not better [9]). You want a very fine (cloudy) precipitate, not big lumps.

3.1.2 PEI Protocol

1. PEI. For two 15 cm plates, prepare the following plasmid cocktails in 15 ml Falcon tubes: 20 μg pMD2.G (VSV-G viral envelope), 50 μg psPAX2 (packaging construct), 50 μg of your expression vector (for a 7 kb vector plasmid). Total DNA = 120 μg in 2.5 ml 0.9% NaCl.

   The amount of vector DNA should be adjusted according to the size of the plasmid in the following range: 50 μg <7000 bp, 75 μg 7000 to 10,000 bp, 100 μg >10,000 bp.
2. PEI. Prepare 15 ml Falcon tubes containing 372 μg PEI in 2.5 ml 0.9% NaCl (=3.1 times the amount of DNA).
3. PEI. Add the DNA cocktails to the PEI tubes. They should be mixed well but vortexing is not required.
4. PEI. Wait 15 min to allow the precipitate to form. The solution should be slightly cloudy without big lumps.

3.1.3 Continue Both Protocols

1. Add the precipitate to the cells by pipetting it drop by drop, trying to homogeneously cover the whole plate. Ensure even spreading of the precipitate by gently swirling the dish. Treat the cells as viral from this point on (dispose of waste accordingly).
2. Incubate overnight at 37 °C 5% CO_2. Replace the medium in the morning with 20 ml OptiMEM 20 mM HEPES.
3. 48 h after adding OptiMEM, harvest the medium, which now contains lentiviral particles. Centrifuge at 1200 × *g* for 3 min if there is abundant debris, then filter sequentially with 0.45 μm and 0.22 μm syringe filters.
4. Rinse a Vivaspin column with 70% ethanol to sterilize it, then rinse with OptiMEM to remove traces of ethanol.
5. Load 20 ml of viral suspension on the Vivaspin and spin at 3000 × *g* at 10 °C in a centrifuge bucket with an aerosol-tight lid. Top up with the remaining viral suspension and spin again until the volume is under 1 ml (50- to 100-fold concentration) (*see* **Note 7**).
6. Transfer the concentrated virus to a 1.5 ml Eppendorf and spin at 5000 × *g* for 5 min at 4 °C in a rotor with an aerosol-tight lid to remove precipitates and debris.
7. Aliquot virus and store at −80 °C. Use vials with screw caps and O-ring seals.

3.2 Infection of Target Cells and Titration of Lentiviral Vectors

Cells differ in their infectability, so viral titres are only strictly valid on the cell line used to calculate the titre. There is nothing to stop you using the cells from the titration for your actual experiment, but it is often not possible to perform a titration on the cells you are most interested in. If the latter, you need to bear in mind that there may be a systematic error when you do the experiment (the effective multiplicity of infection [moi] may be higher or lower than you intended). Infection obeys physical law: the smaller the volume, the higher the concentration, the better the infection. If you do not want to disturb the cells after infection, you can do the infections in 2 ml and leave the virus on the cells. If you have such delicate cells that they do not survive when plated with virus, you can wait for the cells to adhere and perform the infection a few days later. In that case, add an extra well so you can count the number of cells at the time of infection. *See* **Note 8** for more information on calculating viral titres.

If you titred the virus on the cells you are interested in, and you did not let them get overgrown, you can passage one of the wells from the titration and use it for your subsequent experiments. Alternatively, perform another infection following the same protocol using the amount of virus needed for your experiment (very often if you can perform antibiotic selection you will aim for a moi of 1).

3.2.1 Titration by Antibiotic Selection

This is the simplest approach. It only gives a rough idea of the titre but this is good enough for most experiments. A dilution series is performed and the titre is based on finding the first well with no dead cells after antibiotic treatment. This well is conventionally taken to have been infected at an moi of 1. If the cells become confluent before the antibiotic kills them you will be forced to passage a lot of wells, so it is better to titrate the virus on cells you can plate at low density so you can complete the titration without passaging.

1. For tough cells like MCF7 or T-47D, plate 1×10^5 cells per well in 1 ml normal medium in 6-well plates, one plate per virus to be titred. For more delicate cells, plate enough for them to be happy.
2. Infect the cells by adding virus to four wells: 2 μl, 10 μl, 50 μl, and 50 μl virus (*see* **Note 9**). The remaining wells without virus will be used to test antibiotic efficacy. Optional: Add protamine or other compounds that have been reported to increase infection efficiency (*see* **Note 10**).
3. Incubate overnight then replace the medium with 2 ml of fresh medium.
4. 48 h after infection, remove medium and add 2 ml of fresh medium containing antibiotic to one uninfected well and three

of the virus wells (2 μl, 10 μl, and one 50 μl well in the example above). Add 2 ml of fresh medium without antibiotic to one uninfected well and one high dose virus well (the second 50 μl virus well in the example above) (*see* **Note 11**).

Calculate Viral Titre

Observe the cells and score them when there is obvious death in the uninfected well with antibiotic. The goal is to find the well with the lowest amount of virus giving no death. Compare the two wells with the highest amount of virus (50 μl in the example above) to make sure that death is caused by antibiotic rather than a direct toxic effect of the virus (i.e., there should be no death in the well without antibiotic).

Titre (infectious particles/ml)
= number of cells at infection/volume of virus
in ml in the first well with no death.

If you used the amounts above, the titres will be:

1×10^5 cells/0.002 ml = 5×10^7 infectious particles/ml if there is no death with 2 μl virus.

1×10^5 cells/0.01 ml = 1×10^7 infectious particles/ml if there is no death with 10 μl virus.

1×10^5 cells/0.05 ml = 2×10^6 infectious particles/ml if there is no death with 50 μl virus.

Convenient titres for downstream experiments are 10^7–10^8 infectious particles/ml. The antibiotic approach will slightly underestimate the titre (*see* **Note 8** for the reason why).

3.2.2 Titration by Flow Cytometry

It is possible to titre viruses accurately by flow cytometry provided they express a fluorescent protein such as GFP. It is the percentage of cells positive not the intensity of expression that matters.

1. Plate and infect as for antibiotic selection, extending the titration to include an extra virus dilution because you do not need to add antibiotic to an uninfected well. Make sure you count the cells at plating because you will need this number to calculate the titre.
2. Harvest cells for flow cytometry. Three days after infection, aspirate the medium, add 0.5 ml trypsin/EDTA to each well, including the control uninfected well, and incubate at 37 °C until the cells detach. Add 1 ml medium with FBS to neutralize the trypsin, pipette up and down to ensure the cells are dissociated, then transfer the suspension to 5 ml polypropylene FACS tubes.
3. Pellet the cells by centrifuging for 5 min at 1500 rpm and discard the supernatant.

4. If your flow cytometry facility does not accept cells recently transduced with level 2 viruses, kill residual virus by resuspending the pellet in 0.5% formaldehyde in PBS and incubating at room temperature for 5 min. Then pellet the cells again by centrifuging for 5 min at 1500 rpm and discard the supernatant. Note that formaldehyde is compatible with GFP but may abolish the fluorescence of other markers.
5. Resuspend the cells in 200 μl of PBS, put them at 4 °C, and perform flow cytometry within a few hours. Use the uninfected control well to set the gate for background fluorescence. Record the percentage of infected cells in each well.

Calculate Viral Titre

In Excel, make a table with the volume of virus *in ml* in one column and the percentage of fluorescent cells in another column. Add a column called "moi" in which you enter a formula for the "negative log of the fraction of cells not infected." The formula should look like this:

$$= -\mathrm{LN}((100 - \mathrm{POS})/100)$$

where *POS* refers back to the column with the percentage of fluorescent cells, and LN() takes the natural log.

The "moi" column now contains the moi achieved with the specified amount of virus.

Make a graph with volume of virus *in ml* on the *x*-axis and "moi" on the *y*-axis. Exclude any obvious outliers (they may occur at the most extreme virus concentrations), then invite Excel to draw a straight line through the points and ask it to display the slope and intercept of the line. The viral titre in infectious units per ml is given by:

$$\text{titre} = (\text{slope} \times \text{cell number})/(1 - \text{intercept})$$

where cell number is the number of cells present in the well at the time of infection. *See* **Note 8** for additional information on calculation of titres.

3.2.3 Titration by p24 ELISA

It is possible to titre lentiviral vectors by ELISA for p24, a capsid protein derived from the gag polyprotein. This approach measures the amount of p24 in pg, which depends on the number of physical virus particles. The ratio of infectious units to physical particles (typically 1 in 100 to 1 in 1000) is used to assess the quality of viruses, for example to detect batch to batch variation in the quality of preps destined for clinical use. Since the quality of typical academic lab virus preps is highly questionable, it is necessary to make some very large assumptions when extrapolating from physical titres to infectious titres. The approach is only useful if you have a burning desire to know the titre and there is no other way to do it (no antibiotic marker, no fluorescent marker, no transgene that can be detected in transduced cells by IF or IHC). When forced to measure physical titres, we use the Lenti-X p24 Rapid Titer Kit (#632200, Takarabio.com) according to the manufacturer's instructions. This kit is sometimes used to declare cells free of virus but this should only be done at the behest of regulators (it takes 3–4 passages to dilute out the incoming virus). *See* **Note 12** for unwelcome surprises about resistance to antibiotics.

3.3 Replication Competent Lentivirus (RCL) Test

See **Note 2** for a discussion of possible mechanisms leading to the formation of RCL. The protocol below is used to test concentrated virus stocks but the supernatant of any cells could be tested if required. The goal is to show that infected cells do not release replication competent viruses. In some jurisdictions, you may be asked to test for RCL before removing cells from the level 2 lab. It is rarely done now because RCL has not emerged as a significant problem after decades of widespread use of lentiviral vectors. It is also extremely difficult to prove a negative (not a single replication competent virus is present in a prep containing 10^9 replication-incompetent viruses). Furthermore, to prove that an RCL test is truly effective requires the use of live virus as a positive control. In principle, this virus should contain VSV G instead of the normal envelope gene but this would be a Frankenvirus that no-one should deliberately create, least of all in the pursuit of biosafety. It would require level 3 containment and would expose large numbers of non-virologists to a wholly unjustifiable risk of contracting AIDS or worse. If you have access to it, you can use pure HIV RNA as a positive control. Failing that you can use psPAX2 plasmid, which will at least allow you to confirm that the qPCR works.

3.3.1 Infect 293T Cells

This approach will detect recombinant viruses that can infect 293T cells (we assume that a recombinant would use VSV G instead of the normal envelope gene, which is not present in the vectors). To detect viruses using the normal envelope, Escarpe et al. (ref [10]) recommend using the C8166-45 cell line.

1. Coat the plastic with poly-lysine, then plate 5×10^4 293T cells per well in a 12-well tissue plate in 400 μl final volume of DMEM 10% FCS medium. Immediately add 100 μl of viral supernatant to infect the cells as they adhere.
2. Remove virus from cells after 6 h. Wash the cells twice with fresh medium.
3. Incubate at 37 °C 5% CO_2 for 6 days, replacing the medium after 3 days.
4. Coat the plastic with poly-lysine, then plate 1×10^5 293T cells per well in a 6-well tissue plate in 1.5 ml medium per well and add the supernatant from the cells in the 12-well plate.

5. Incubate at 37 °C 5% CO_2 for 6 days, changing the medium after 3 days.
6. Harvest the medium (supernatant) containing putative RCL.

3.3.2 Extract RNA

(Where not specified, spin at maximum speed in a microfuge, rcf >12,000 × *g*)

1. Mix 1 ml of supernatant with 500 μl Trizol LS.
2. Incubate 5 min at room temperature.
3. Add 100 μl chloroform, vortex 15 s, then incubate 3 min at room temperature.
4. Centrifuge for 15 min at 4 °C.
5. Transfer the upper phase to a new tube, add 600 μl of ethanol 70%, and vortex to mix.
6. Load the mixture on an RNeasy Mini spin column.
7. Centrifuge for 15 s. Discard the flow-through.
8. Add 500 μl Buffer RPE and centrifuge for 15 s. Discard the flow-through.
9. Add 500 μl Buffer RPE and centrifuge for 2 min.
10. Place the spin column in a new 2 ml collection tube and centrifuge for 1 min.
11. Place the spin column in a new 1.5 ml collection tube and add 30 μl RNase-free water.
12. Wait 1 min then centrifuge for 1 min to elute the RNA.

3.3.3 Treat with DNase I

1. To each 30 μl RNA prep add: 1 μl Superase inhibitor, 5 μl 10× DNase I buffer, 13 μl RNase-free water, 1 μl DNase I (RNase-free) 10 U/μl.
2. Incubate at 37 °C for 30 min to digest DNA.
3. Add 1 μl 10% SDS.
4. Add 50 μl phenol/chloroform/isoamylalcohol.
5. Vortex, spin 10 min.
6. Take the upper phase (~40 μl) and add: 4 μl NaOAc 3 M pH 5.5, 1 μl glycogen, 120 μl ethanol 100%.
7. Mix, put on dry ice 5 min, spin 20 min, discard the supernatant.
8. Add 100 μl ethanol 70%, spin 5 min, discard the supernatant.
9. Dry the pellet 3 min, add 10 μl RNase-free water.
10. Put on ice for 5 min to resuspend the RNA, then store at −80 °C.

3.3.4 Reverse Transcribe

Using reagents from the Superscript kit:

1. Mix together 1 μl oligo-dT (0.5 μg/ml), 1 μl dNTP (10 mM), and 8 μl RNA.
2. Heat to 65 °C for 10 min, then cool to 4 °C and spin briefly.
3. Prepare 2× reaction mix: 2 μl 10× first strand buffer, 4 μl 25 mM $MgCl_2$, 2 μl 100 mM DTT, 1 μl RNAseOUT (40 U/μl).
4. Mix the 2× reaction mix with the RNA/oligo-dT mix and place at 42 °C for 2 min.
5. Add 1 μl Superscript II reverse transcriptase (50 U/μl).
6. Incubate for 1 h at 42 °C then store at −80 °C.

3.3.5 Perform Quantitative PCR

Set up standard Sybr Green qPCR reactions using PAXpolF and PAXpolR primers according to your normal protocol with 40 cycles and annealing at 60 °C. The product length is 94 bp. You should get no signal after 40 cycles in the absence of the positive control template.

4 Notes

1. General safety issues. Modern lentiviral vectors are classified in most jurisdictions as level 1 or level 2. Importantly, the viral particles used to transduce target cells contain no complete genes coding for HIV proteins. Furthermore, in common with all retroviruses, HIV vector particles have a lipid envelope which renders them sensitive to inactivation by simple detergents and alcohol. Thanks to their fragility, lentiviruses can only be transmitted by direct contact with penetration, for example by needle stick injury. In the lab, this means avoiding the use of glass or needles. Otherwise, good microbiological practice, as commonly applied in level 2 tissue culture labs, provides adequate protection. The conditions where retroviral vectors have been demonstrated to produce malignant transformation in humans are known thanks to gene therapy experiments on children with severe combined immunodeficiency (SCID). It is inconceivable that these conditions could ever be met in a normal laboratory, even as a deliberate criminal act. Specifically, patients were treated with drugs to depress their already deficient immune response, their hematological stem cells were infected in massive numbers ex vivo, and their immune system was regenerated entirely from virally transduced cells. The goal of the protocol was to give the transduced cells the maximum encouragement to engraft and expand clonally. Crucially, the patients were defenseless against foreign antigens because the only cells that could mount an immune

response were the transduced cells themselves. One quarter of the patients developed leukemia [11]. Gene therapy has also taught that a normally functioning immune system is extremely effective at eliminating transgenes, even if they differ by only a single amino acid from a patient's own version of the gene. In a laboratory accident, all integrated vectors are likely to express antibiotic resistance genes or fluorescent protein genes in addition to the experimental transgene, virtually guaranteeing the rapid immune killing of infected cells.

The rules for working in level 2 labs are specified by national authorities. Two sensible precautions when hiring new staff are to exclude people who are either immune deficient or suffer from skin diseases that impair the barrier function of the skin, the former because they are defenseless against foreign antigens and could, if HIV-positive, propagate defective viruses, the latter because their underlying condition makes them easier to infect (and is incidentally likely to be made worse by wearing gloves all day so they will probably spend long periods banned from the level 2 lab).

It is good practice to inactivate viral suspensions under the laminar flow hood by collecting them after use in a beaker containing disinfectant (for example, 1% Virkon), then waiting 15 min before aspirating them to the waste bottle. Avoid centrifuging cells or medium containing concentrated virus; if you use a centrifuge, make sure the rotor has an aerosol-tight lid and never use polystyrene tubes that can shatter easily. An underappreciated route of infection is contact with the eyepieces of microscopes; this happens because the first act after putting petri dishes on the microscope stage is to adjust the separation of the eyepieces—hence if medium has slopped over the side of the dish, it will promptly find its way onto your gloves and then the eyepieces. The solution is to wear safety glasses.

We use Kova slides (disposable plastic hemocytometers) to count cells in the virus lab, so you do not have to worry about cleaning them or breaking glass haemocytometers.

Three situations stand out as requiring careful thought. The first is injection of concentrated virus into live animals. Sharp needles and sudden movement are the perfect recipe for disaster. The second is inclusion of multiple oncogenes in a single vector. They should only be used by experienced staff who understand the risks. The third is cloning genes into the vector that can supply the functions transferred to the packaging plasmids. Additional biosafety precautions and routine testing for RCL should be performed when making vectors modified in this way.

2. Replication competent lentiviruses (RCL). Recombination between vector and packaging plasmids leading to the creation of replication competent lentiviruses (RCL) is the major theoretical safety concern when working with lentiviral vectors. To reduce this risk, the helper genes are split between two ("second generation") or three ("third generation") plasmids. In normal academic use, the quantities of virus produced are too small for RCL to be a problem. Attempts to produce RCL have failed in tests scaled up to 1.4×10^{10} transducing units [10].

   Detecting RCL in the context of lentiviral vector production is not as easy as it sounds. Simple arithmetic shows that the number of molecules of plasmid DNA transfected into lentiviral packaging cells is around 10^{12}. This makes it pointless to test the viral supernatant itself. There is no way so many copies of contaminating plasmid DNA could be brought down to <10 by any simple process. The protocol we use tests for viral RNA in the supernatant of infected cells. It is based on qPCR, since it is relatively easy to assess the sensitivity of qPCR assays on control DNA or purified HIV RNA templates, and qPCR machines are readily available in modern research labs. It is not perfect because it does not use replication competent virus as a positive control, but this is a price most non-virologists would be willing to pay. In our experience it is better than some commonly used gel-based PCR assays, whose sole aim seems to be to get a negative result (data not shown). A full RCL assay based on ELISA for p24 and intended for testing of clinical grade virus preps is described in ref. [10].

   A more plausible situation where RCL could arise in normal use would be transduction of a cell line that was already contaminated with wild-type HIV. Care should therefore be taken when using primary tissue samples that have not been screened for HIV. If you use biopsies from patients who are not routinely screened for HIV, you should consider testing the biopsies for HIV before using them (anonymously, of course). Outside professional virology labs, cell lines contaminated with wild-type HIV should be destroyed immediately (they are automatically classified as level 3). Since the packaging signal in the vector is placed upstream of the active promoter in SIN vectors, the packaging signal is not present in transcripts produced from integrated vector genomes. Hence, the simple presence of wild-type virus is not sufficient to package vector genomes; to transcribe the packaging signal the proviral vector DNA must either be integrated in a site where it can be transcribed from an upstream cellular promoter or it must recombine with integrated wild-type proviral DNA through the region between U5 and the packaging signal. A hybrid virus arising in this situation would most likely resemble a classic retroviral oncogene-encoding virus. Propagation of such defective viruses typically requires the presence of a helper wild-type virus. This in turn requires the use of high titres to prevent loss of the recombinant through competition with the wild-type virus. Trying to assess the risk posed by such a recombinant virus is an interesting academic exercise that pales to insignificance compared to the very real risk posed by the wild-type virus in the mixture.

3. Interferon induction. The interferon system blocks translation, degrades mRNA, and arrests the cell cycle, phenomena that can easily dwarf whatever effects you may be trying to study. We previously reported that lentiviral shRNA vectors can trigger an interferon response [12]. This is a transient phenomenon which occurs soon after infection, probably from a convergence of multiple factors. We originally reported it with shRNA vectors but we have also seen it with ordinary polII expression vectors. It is not a subtle effect. Gene set enrichment analysis (GSEA) of microarray data using the Broad GSEA server (http://www.broadinstitute.org/gsea/) easily detects it, but simple qPCR for an interferon-stimulated gene (ISG) like OAS1 also works well and is less expensive (primers: aggtggtaaagggtggctcc and acaaccaggtcagcgtcagat).

   It is normal for cells to fight viruses by activating the interferon system, and lentiviruses are no exception. A growing number of steps in the lentiviral life cycle are known to be recognized as alien by the host cell: the viral capsid is recognized by TRIM5 immediately following cell entry [13]; SAMHD1 destroys nucleotides to prevent reverse transcription of viral RNA [14]; the reverse transcribed DNA is mutated by APOBEC3G to scramble the viral sequence [15]; and MX2 blocks nuclear import [16]. The genes used by lentiviruses to block these responses are the very genes that are deleted in lentiviral vectors, so it should come as no surprise that interferon responses are occasionally seen following lentiviral vector infections.

   Depending on the type of transgene being expressed it is also possible for the foreign transcript to be mistaken for a viral RNA, leading to activation of the interferon system through the mechanisms, such as PKR (EIF2AK2), that defend the cell against RNA viruses. This probably explains the interferon induction we reported with the U6 promoter [17]. An important theoretical advantage of polII-driven miRNA vectors over polIII-driven shRNA vectors is that saturation of Drosha should just lead to the accumulation of superfluous non-coding polII transcripts; they are unlikely to be mistaken for intruders or to otherwise offend the cell. In contrast, shRNAs produced by polIII promoters skip the Drosha step.

Saturation of Dicer then leads to the accumulation of novel small RNAs that could very easily be mistaken for small viruses.

Besides these biological explanations, there is a technical explanation which probably accounts for the sporadic nature of the problem. Production of lentiviral vectors requires transfection of packaging cells with at least three plasmids. If any one of them is contaminated with bacterial endotoxin, the packaging cells themselves will release interferon, leading to the production of a viral suspension contaminated with both endotoxin and interferon. This is why it is important to use "endotoxin-free" maxiprep kits to purify lentiviral plasmids.

Once the proviral DNA has been integrated into the host genome, there are few clues to inform the cell that it harbors a new virus. Hence interferon responses are a nuisance if cells must be used immediately after infection, but of little or no importance once cells have been passaged. For this reason, it is infinitely preferable to design experiments such that biological measurements can be delayed until the cells have been passaged and any antibiotic selection is complete. The easy way to do this is to use inducible vectors. Otherwise, the effect one is trying to measure can easily be confounded by a mixture of interferon effects and antibiotic-induced death.

4. Plasmid problems. Recombination between the LTRs leading to deletion of the intervening vector sequences is occasionally a problem during production of lentiviral plasmids. One situation where it is almost guaranteed is when there is a toxic gene in the vector. The classic example is a Gateway Destination vector: the *ccdB* gene is used to kill bacterial cells that have taken up the parental vector after a clonase reaction. Special strains containing a *gyrA* mutant that confers resistance to *ccdB* are used to produce Gateway Destination vectors, but it is still necessary to select for a *cat* (chloramphenicol resistance) gene that is adjacent to *ccdB* in the plasmid. Gateway *ccdB* cassettes are deliberately toxic, but even banal inserts may be mildly toxic in lentiviral plasmids. When this occurs, the transformed bacteria may be noticeably temperature sensitive. The solution is to grow bacteria at 30° or 32° at all steps from cloning to maxiprep. Another precaution commonly taken to prevent recombination is to use bacteria deficient in recombination enzymes, for example stbl3. Toxicity seems to be strain-dependent, so if you have problems it is worth simply trying other commonly used bacterial strains (DH5a, DH10B, stbl2, stbl3, stbl4, etc.).

   Recombination should not be a problem with existing plasmids. It is more likely to occur when constructing new plasmids. One solution is to use Gateway cloning, which eliminates the gel purification and ligation steps in conventional cloning. If you do use conventional cloning, it is worth checking the wavelength of your UV light box. Most labs now use 312 nm UVB to purify DNA fragments, despite Maniatis' injunction to use 365 nm UVA. UVB specifically damages DNA, which reduces the number of colonies ~100-fold and triggers a DNA damage response that increases the risk of recombination. Finally, to avoid caramelizing the medium it may help to reduce the duration or temperature of the autoclave cycle used to prepare the media (we use 110 °C for 2 min).

When constructing lentiviral vectors, it is important to delete polyA signals from transgenes. To produce the virus, the packaging cell has to transcribe all the way to the end of the 3′-LTR. If the RNA is cleaved at an internal polyA site, the 3′-LTR sequences will be deleted, preventing reverse transcription of the complete viral RNA. This will give a dramatically reduced viral titre. Apart from that, the commonest cause for low titres is simply that the insert is too big. It is easy to get titres of 10^8 infectious units/ml in unconcentrated supernatant when the total size of the vector is 6.5 kb, but impossible when the size exceeds 10 kb. Hence it is a good idea to delete 5′ and 3′-untranslated regions from transgenes. For large genes, this can be done cheaply by Gateway cloning of orf clones from dnasu.org. For open reading frames under 1 kb, it is better just to order double strand DNA fragments, for example gBlocks from iDT (idtdna.com), and then clone them yourself by Gateway or Gibson. An incidental advantage of gBlocks is that it is trivially easy to add cloning sites and epitope tags or to codon switch the sequence to remove shRNA target sites. The main advantage of gBlocks is that they are MTA-free, allowing you to produce any vector in a matter of days, rather than waiting months for bureaucrats to sign an MTA.

5. Chloroquine intercalates into DNA to prevent recognition by innate immune DNA receptors in endosomes (TLR9 in dendritic cells) and the cytosol (cGAS in all cells) [18]. As a weak base, it is concentrated in endosomes, where it transiently raises the pH. Traditionally, it is used at 25 μM but any early benefit is probably more than offset by late toxicity. Hence it should not be added to the OptiMEM when the medium is changed. Even so, it persists and contaminates the virus prep. This is easily confirmed with quinacrine, a close analog that is visible with GFP filters.

6. 0.22 μm filter sterilization of the plasmid solution before adding $CaCl_2$ may give more reproducible precipitates and relieve any worries you may have about the sterility of your plasmid preps. Increase the amounts by 20% if filter sterilized, to allow for losses in the syringe filter, so you can add exactly 1 volume to the HeBS.

7. Concentration of virus. You can only concentrate virus with a Vivaspin if you harvest it in serum-free medium (e.g., Opti-MEM). If you were to use DMEM 10% FBS, the filter would concentrate unwanted proteins from the serum and become clogged. 293T cells will grow happily in almost anything, so, if your target cells need a particular medium, you can use it instead of OptiMEM. Alternatively, use the Vivaspin to exchange the medium once you have reduced the volume to 200 μl. After PEI transfection, Vivaspins seem to concentrate debris that is presumable just PEI precipitate. This can be removed by a brief spin in an Eppendorf at the end—if it troubles you, just use the $CaPO_4$ protocol. Before you embark on extensive centrifugation of virus, check that you have taken reasonable steps to contain aerosols that might be released by leakage of viral supernatant. If the first 20 ml goes straight through the Vivaspin in the first 5 min, there is a probably a hole in Vivaspin; in that case just filter sterilize the flow-through and apply it to a new Vivaspin. A single Vivaspin can be used to concentrate up to 60 ml. For larger volumes, use two Vivaspins or the flow rate at the end will be unbearably slow. Concentrating the virus is highly recommended if you want to infect delicate cells. It is also much more practical to store 4 tubes of concentrated virus than 40 tubes of raw supernatant. Faced with the need to label so many tubes and find so much space in the freezer, people cram as much virus as possible into each tube, forgetting that water expands when frozen. It then leaks all over the freezer box. There is only a small loss of titre when aliquots are thawed so you do not need to bin the tube when you thaw a concentrated aliquot of virus.
8. Viral titres. One advantage of lentiviral vectors compared to simple oncoretroviral vectors is that titres are higher and virus can be freeze-thawed with minimal loss of titre. The precise viral titre is rarely important in routine lab experiments. A more important question is whether you want the cells to contain only a single copy of the provirus, or you want to have multiple copies. If the former, it is worth erring on the low side by using an moi of 0.2. This presupposes that you will be able to select transduced cells, either by antibiotic treatment or FACS for fluorescent proteins. Reasons to use multiple copies include needing to achieve high level expression of a transgene, or wanting to avoid big differences between individual cells if some proviruses are silenced. A classic situation where multiple copies may be required is the use of shRNA expressed from polIII promoters (H1 or U6) because a single copy frequently gives very poor silencing.

   Viral infection obeys Poisson's law. In its simplest form, this states that the moi is 1 virus per cell when one-third of the cells are not infected. The critical parameter is the fraction of cells that are not infected, which is equal to $1/e$ or 37% when cells are infected at an moi of 1. This is because Poisson's law states that $P(k) = e^{-m} \times m^k/k!$, where m is the multiplicity of infection, k is the number of integrated proviruses per cell after infection, and $P(k)$ means the probability of having k viruses in the cell. When $m = 1$ and $k = 0$, the probability that a cell contains no virus $P(0) = e^{-1} \times 1^0/0! = e^{-1}$. It is easy to estimate the fraction of cells with $k = 0$: they are the unstained cells in flow cytometry or the cells that die in antibiotic. Poisson's law can also be used to calculate a few other (mildly) interesting numbers. At an moi of 3, 5% of cells will contain no virus, and 22% of cells will contain three viruses. This probably corresponds to the commonly used yardstick of "no death with antibiotic"—from which we can surmise that people using the approach in Subheading 3.2.1 will mildly underestimate the titre. If it is not possible to apply any selection, it is worth knowing that at an moi of 5 there will be 0.67% of cells uninfected, whereas at an moi of 10 there will only be 0.0045% of cells uninfected. If cells divide daily, it will take about a week for uninfected cells to overgrow the culture in the former case but over 2 weeks in the latter case. What this tells you is that infection without selection gives you a very narrow window of time to test tumor suppressor genes before the uninfected cells overgrow the culture, even if you use an moi of 10. Much higher multiplicities can be used to get around this problem, but they commonly lead to non-specific toxicity or wildly unphysiological levels of transgene expression.

9. The volume of virus should be adapted to the size of the virus. For a small virus (for example, made with a plasmid like that in Fig. 2b), the titre will be $>10^8$ infectious units/ml so you should probably use smaller volumes of virus covering a wider range (for example, 10 μl, 1 μl, 0.1 μl of virus). Make serial dilutions rather than trying to pipette volumes below 2 μl. For large viruses (for example, made from the plasmid in Fig. 2c), you are unlikely to get $>10^7$ infectious units/ml, so the amounts given in the protocol should be about right. If you have produced a vector before and know it is not toxic you may choose to omit the second well with 50 μl virus, to avoid wasting virus.

10. Addition of polybrene 8 μg/ml to medium during infection of target cells was always done with retroviral vectors. It neutralizes charge leading to better receptor binding, but has to be removed promptly to prevent toxicity. It is not necessary with VSV G pseudotyped lentiviral vectors. Protamine sulfate 4 μg/mL has been reported to increase transduction by lentiviral

vectors [19]. Neuraminidase may improve transduction of luminal mammary epithelial cells [20].

11. The concentration of antibiotic required depends on the cells so before testing the virus you should titrate the antibiotic to find the concentration that kills uninfected cells in a reasonable time—for MCF7 and T-47D it is likely to be around 2 μg/ml puromycin, 200 μg/ml hygromycin, 10 μg/ml blasticidin, and 750 μg/ml G418. The time to kill 100% of the cells should be 2–3 days for puromycin, and 5–7 days for blasticidin, hygromycin and G418.

12. Unwelcome surprises. Students often imagine that resistance to antibiotic guarantees successful expression of a transgene. Sadly, this is not the case when the transgene and marker are expressed from different promoters. Antibiotic resistance means the DNA has been successfully integrated, but for a transgene that is even mildly toxic, there is always a good chance that its promoter will be silenced. To work out whether the transgene is expressed people normally perform a western blot, but this can hide a multitude of sins. In particular, a transgene expressed in only 10% of the cells will give a nice band on a Western. To find out how many cells express the transgene, it is possible to do immunofluorescence (IF) on coverslips, immunohistochemistry (IHC) on plastic dishes or flow cytometry with antibodies against your protein.

    Lentiviruses use complex alternative splicing to express their own genes efficiently. Their predilection for arcane splicing carries over into lentiviral vectors [21]. It can generate perplexing results, for example generating fusion proteins between a transgene and GFP in a downstream expression cassette. The clues in one case we studied were that the transgene ran too high on a protein gel and GFP became nuclear. Subsequent investigation revealed that the vector had found a splice donor site a few nucleotides before the 3′-end of the transgene, and spliced it to a cryptic acceptor site in the murine PGK promoter that was used to express GFP.

Acknowledgments

I thank my present and former colleagues in Switzerland, France, and Australia for help in developing these protocols. I thank the French Cancer League and the Fondation pour la lutte contre le cancer et pour des recherches médico-biologiques for funding work in my lab.

References[1]

1. Naldini L, Blomer U, Gallay P, Ory D, Mulligan R, Gage FH, Verma IM, Trono D (1996) In vivo gene delivery and stable transduction of nondividing cells by a lentiviral vector. Science 272(5259):263–267
2. Naldini L, Blomer U, Gage FH, Trono D, Verma IM (1996) Efficient transfer, integration, and sustained long-term expression of the transgene in adult rat brains injected with a lentiviral vector. Proc Natl Acad Sci U S A 93(21):11382–11388
3. Zufferey R, Dull T, Mandel RJ, Bukovsky A, Quiroz D, Naldini L, Trono D (1998) Self-inactivating lentivirus vector for safe and efficient in vivo gene delivery. J Virol 72(12): 9873–9880
4. Dull T, Zufferey R, Kelly M, Mandel RJ, Nguyen M, Trono D, Naldini L (1998) A third-generation lentivirus vector with a conditional packaging system. J Virol 72(11): 8463–8471
5. Duss S, Andre S, Nicoulaz AL, Fiche M, Bonnefoi H, Brisken C, Iggo RD (2007) An oestrogen-dependent model of breast cancer created by transformation of normal human mammary epithelial cells. Breast Cancer Res 9(3):R38
6. Hickey TE, Selth LA, Chia KM, Laven-Law G, Milioli HH, Roden D, Jindal S, Hui M, Finlay-Schultz J, Ebrahimie E, Birrell SN, Stelloo S, Iggo R, Alexandrou S, Caldon E, Zwart W, Palmieri C, Sartorius CA, Swarbrick A, Lim E, Carroll JS, Tilley WD (2021) The androgen receptor is a tumor suppressor in estrogen receptor positive breast cancer. Nature Med 27: 310–320. https://doi.org/10.1038/s41591-020-01168-7
7. de Felipe P, Luke GA, Hughes LE, Gani D, Halpin C, Ryan MD (2006) E unum pluribus: multiple proteins from a self-processing polyprotein. Trends Biotechnol 24(2):68–75
8. Ryan MD, Drew J (1994) Foot-and-mouth disease virus 2A oligopeptide mediated cleavage of an artificial polyprotein. EMBO J 13(4): 928–933
9. Jordan M, Schallhorn A, Wurm FM (1996) Transfecting mammalian cells: optimization of critical parameters affecting calcium-phosphate precipitate formation. Nucleic Acids Res 24(4): 596–601. https://doi.org/10.1093/nar/24.4.596
10. Escarpe P, Zayek N, Chin P, Borellini F, Zufferey R, Veres G, Kiermer V (2003) Development of a sensitive assay for detection of replication-competent recombinant lentivirus in large-scale HIV-based vector preparations. Mol Ther 8(2):332–341
11. Hacein-Bey-Abina S, Hauer J, Lim A, Picard C, Wang GP, Berry CC, Martinache C, Rieux-Laucat F, Latour S, Belohradsky BH, Leiva L, Sorensen R, Debre M, Casanova JL, Blanche S, Durandy A, Bushman FD, Fischer A, Cavazzana-Calvo M (2010) Efficacy of gene therapy for X-linked severe combined immunodeficiency. N Engl J Med 363(4):355–364. https://doi.org/10.1056/NEJMoa1000164
12. Bridge AJ, Pebernard S, Ducraux A, Nicoulaz AL, Iggo R (2003) Induction of an interferon response by RNAi vectors in mammalian cells. Nat Genet 34(3):263–264
13. Pertel T, Hausmann S, Morger D, Zuger S, Guerra J, Lascano J, Reinhard C, Santoni FA, Uchil PD, Chatel L, Bisiaux A, Albert ML, Strambio-De-Castillia C, Mothes W, Pizzato M, Grutter MG, Luban J (2011) TRIM5 is an innate immune sensor for the retrovirus capsid lattice. Nature 472(7343): 361–365. https://doi.org/10.1038/nature09976
14. Lahouassa H, Daddacha W, Hofmann H, Ayinde D, Logue EC, Dragin L, Bloch N, Maudet C, Bertrand M, Gramberg T, Pancino G, Priet S, Canard B, Laguette N, Benkirane M, Transy C, Landau NR, Kim B, Margottin-Goguet F (2012) SAMHD1 restricts the replication of human immunodeficiency virus type 1 by depleting the intracellular pool of deoxynucleoside triphosphates. Nat Immunol 13(3):223–228. https://doi.org/10.1038/ni.2236
15. Mangeat B, Turelli P, Caron G, Friedli M, Perrin L, Trono D (2003) Broad antiretroviral defence by human APOBEC3G through lethal editing of nascent reverse transcripts. Nature 424(6944):99–103. https://doi.org/10.1038/nature01709
16. Kane M, Yadav SS, Bitzegeio J, Kutluay SB, Zang T, Wilson SJ, Schoggins JW, Rice CM, Yamashita M, Hatziioannou T, Bieniasz PD (2013) MX2 is an interferon-induced inhibitor of HIV-1 infection. Nature 502(7472): 563–566. https://doi.org/10.1038/nature12653

[1] Additional useful information can be found on http://tronolab.epfl.ch/lentivectors and http://www.addgene.org.

17. Pebernard S, Iggo RD (2004) Determinants of interferon-stimulated gene induction by RNAi vectors. Differentiation 72(2–3):103–111
18. An J, Woodward JJ, Sasaki T, Minie M, Elkon KB (2015) Antimalarial drugs inhibit IFN-beta production through blockade of cyclic GMP-AMP synthase-DNA interaction. J Immunol 194(9):4089–4093. https://doi.org/10.4049/jimmunol.1402793
19. Balak JRA, de Graaf N, Zaldumbide A, Rabelink TJ, Hoeben RC, de Koning EJP, Carlotti F (2019) Highly efficient ex vivo lentiviral transduction of primary human pancreatic exocrine cells. Sci Rep 9(1):15870. https://doi.org/10.1038/s41598-019-51763-z
20. Hines WC, Yaswen P, Bissell MJ (2015) Modelling breast cancer requires identification and correction of a critical cell lineage-dependent transduction bias. Nat Commun 6:6927. https://doi.org/10.1038/ncomms7927
21. Moiani A, Paleari Y, Sartori D, Mezzadra R, Miccio A, Cattoglio C, Cocchiarella F, Lidonnici MR, Ferrari G, Mavilio F (2012) Lentiviral vector integration in the human genome induces alternative splicing and generates aberrant transcripts. J Clin Invest 122(5):1653–1666. https://doi.org/10.1172/JCI61852

Chapter 9

Modification of Single Cells Within Mouse Mammary Gland Derived Acini via Viral Transduction

Lucia Garcia del Valle, Marta G. Montero, and Martin Jechlinger

Abstract

The growth of organoid cultures from primary donor tissue is able to recapitulate the original tissue morphology, heterogeneity, and characteristics. Close study of these cultures grants a deeper understanding of the chain of events occurring during disease progression and healthy tissue development. While patient derived organoids are particularly suited to assay for novel treatment options, organoids obtained from model organisms are perfectly suited to establish in-depth analysis technology, including longitudinal imaging approaches, as well as proof of principle studies that rely on a steady source of primary tissue. All these approaches profit from advancements in technology to manipulate cells within an organoid.

Here we present an optimized protocol to generate, culture, and transduce 3D acini obtained from mouse primary mammary epithelial cells via viral vectors. Applying this method, a few cells within the preserved organoid can be marked, changed, and tracked within an unaltered neighboring environment of non-transduced cells to better understand processes like, for instance, tumor initiation.

Key words 3D cell culture, Mammary gland, Organoids, Primary cell growth, Viral transduction of single cells in a multicellular structure, Stochastic cell system

1 Introduction

Modern media and matrix technology permit the establishment of three-dimensional (3D) culture systems from primary cells to generate so-called organoids [1]. Organoids have gained importance for the biomedical field, since they can be assayed for responses to candidate drugs while recapitulating the cellular heterogeneity as well as growth and polarization characteristic of the original primary tissue [1, 2]. In general, multicellular structures grown in the appropriate 3D architectural context in vitro enable the study of important aspects of cell behavior, which are partially lost in 2D culturing conditions [3, 4], this includes features like correct

Lucia Garcia del Valle and Marta G. Montero contributed equally with all other contributors.

Maria dM. Vivanco (ed.), *Mammary Stem Cells: Methods and Protocols*,
Methods in Molecular Biology, vol. 2471, https://doi.org/10.1007/978-1-0716-2193-6_9,

control and timing of cell division [5, 6] as well as correct cell polarization and cell orientation [7, 8] which all impinge on tissue functionality [9, 10].

Mammary gland organoids derived from mouse models can be used to recapitulate the intricated, polarized network of the alveolar structures, including the study of the interactions of mammary cells with the cellular microenvironment and matrix components, all of which govern cell behavior in the breast tissue [11–14]. Hence, the development of in vitro-derived mammary gland acini-like structures has improved the study and understanding of the factors involved in the establishment and maintenance of both healthy and disease stages [15–17].

One limitation of organoids derived from genetically modified model organisms—like transgenic mice—is rooted in the fact that all cells within a genetically modified tissue will express a certain gene of interest in all cells. Therefore, important contextual information, for instance of the process of tumor initiation within a healthy epithelium, is lost. Here, we present a protocol that allows to modify a fraction of cells within an organoid via viral transduction to monitor the behavior and cell fate of transformed cells within a healthy, "normal" epithelial cell layer [18]. In addition, we envisage this protocol to be easily further adapted towards the study of other organoid model systems and patient derived organoids.

2 Materials

1. Culture medium: 500 mL Mammary Gland Basal Medium, Mammary Epithelial Growth Factor 5 mL mix, filter and store at 4 °C for a month max.
2. STOP media: DMEM F12 with 3.151 g/L glucose and L-glutamine, 10% Fetal Bovine Serum, 1% HEPES, and 1% Penicillin/streptomycin. Store at 4 °C (*see* **Note 1**).
3. Digestion media: DMEM F12 with 3.151 g/L glucose and L-glutamine, 1% HEPES, and 1% Penicillin/streptomycin. Store at 4 °C.
4. 3D Matrix Growth Factor Reduced Phenol-Free. Store at 4 °C (*see* **Note 2**).
5. 3D Collagen I Rat Tail. Store at 4 °C.
6. Phosphate-Buffered Saline (PBS).
7. Collagenase (stock): 75,000 units/mL in digestion media. Store at 4 °C.
8. Liberase TM (stock): 10 mg/mL in digestion media. Store at 4 °C.

9. DNase I from bovine pancreas (stock): 1 mg/mL in digestion media. Store at 4 °C.
10. Trypsin-EDTA 0.25%. Store at 4 °C.
11. Collagen I-coated 6-well plates.
12. 12-well plates.
13. Viral particles containing the genetic construction of interest. We use a lentivirus carrying the pLove-pGK-rtTA-p2A-H2b-GFP construct, manufactured by the Genetic & Viral Engineering Facility, EMBL Rome.

3 Methods

3.1 Organotypic Culture of Primary Mammary Epithelial Cells

All used components were pre-warmed at 37 °C, unless otherwise specified.

1. Sacrifice an 8 to 10 weeks old female virgin mouse and harvest all ten mammary glands. A detailed explanation describing mammary gland collection can be found in Nguyen-Ngoc et al. [14]. Distribute the glands equally in four loosely capped 50-mL Falcon tubes, each one filled with 5 mL of digestion media supplemented with 450 units of Collagenase Type 3 and 60 μg of Liberase. Let the tissue digest inside an incubator for 14–16 h at 37 °C, 5% CO_2 (*see* **Note 3**).
2. Mechanically disrupt the tissues by using a 5 mL pipet. Combine the content of two Falcon tubes into one, ending up with just two 50 mL Falcon tubes in total.
3. Add up to 40 mL of PBS to the suspension and centrifuge at 190 × *g* for 5 min at room temperature to wash the tissue fragments.
4. Remove the liquid interphase without disturbing the lipidic layer nor the cell pellet (*see* **Note 4**). Leave around 5 mL of solution and resuspend the pellet in that volume. Add 5 mL of 0.25% trypsin-EDTA to each loosely capped 50-mL tube, mix it, and incubate it for 20–30 min at 37 °C, 5% CO_2.
5. Add up to 40 mL of STOP media. Treat each suspension with 5–8 μg of DNase I for 5 min at room temperature to avoid cell aggregation due to sticky, free DNA from apoptotic cells that causes clumping. Centrifuge the tubes at 190 × *g* for 5 min at room temperature.
6. Discard all the supernatant, removing the fat layer as well and leaving just the pellet (*see* **Note 5**). Combine the two pellets using 12 mL of culturing media.
7. Seed the cell suspension into 4 wells of collagen I-coated 6-well. Distribute 3 mL of the cell suspension into each well.

Incubate them at 37 °C, 5% CO_2 for 24 h to allow brief propagation of adherent primary cells.

8. Wash the cells with 3 mL of PBS to remove cellular debris. Treat the attached cells with 0.5 mL of 0.25% trypsin-EDTA per well for 3–5 min. After cell detachment, inactivate the trypsin with STOP media.
9. Centrifuge the cells at 190 × *g* for 5 min at room temperature. Discard the supernatant and resuspend the pellet in 200–500 mL of culturing media. Mix 1:1 10 μL of cell suspension with 10 μL of Trypan Blue dye to color dead cells and exclude from the count. Take 10 μL from the mixture and count the number of single cells using a Neubauer chamber or a cell counter (*see* **Note 6**).
10. Working on ice, prepare a seeding master mix of 3D matrix composing of ice-cold PBS, Collagen I, and Matrigel in a 1:1:4 ratio, pipetting on that order (*see* **Note 7**). Calculate a total volume of 100 μL per seeded gel, always taking surplus volume into account to make up for pipetting losses due to the viscosity of the master mix (*see* **Note 8**).
11. Working quickly, mix 15,000 mammary epithelial cells with 100 μL of 3D matrix on ice. Dispense 95 μL into one well of a 12-well tissue-culture plate (*see* **Note 9**).
12. Let the gels solidify on a level surface in the incubator at 37 °C, 5% CO_2 for 30 min.
13. Add 1.5 mL of culturing media to each gel, and incubate the plates at 37 °C in a CO_2 incubator. Replace the medium every 2 days.
14. Let the organoids develop until they reach the desired size and age, changing culturing media every 48 h. For this protocol, we usually let the organoids grow for 4 days to establish structures with an already developed lumen, and then we perform the viral transduction (*see* Subheading 3.2). Growth period and resulting size of the organoids before starting the transduction protocol will depend on the experimental purpose, and therefore should be adjusted. Alternatively, organoids can be harvested and analyzed via immunostaining as described in [19].

3.2 Lentiviral Transduction

Carry out this procedure in a SP2 culture flow hood. All used components are pre-warmed at 37 °C unless otherwise specified. Starting from 6 initial 100 μL gels—with 15,000 cells seeded per gel—, 3 new 100 μL gels are seeded after recovery of structures and addition of lentivirus. Adjust the number of initial gels according to your planned follow-up experiments.

1. Digest the 3D matrix by adding 300 units of Collagenase Type 3 into each well (containing 1 mL of culturing media and a

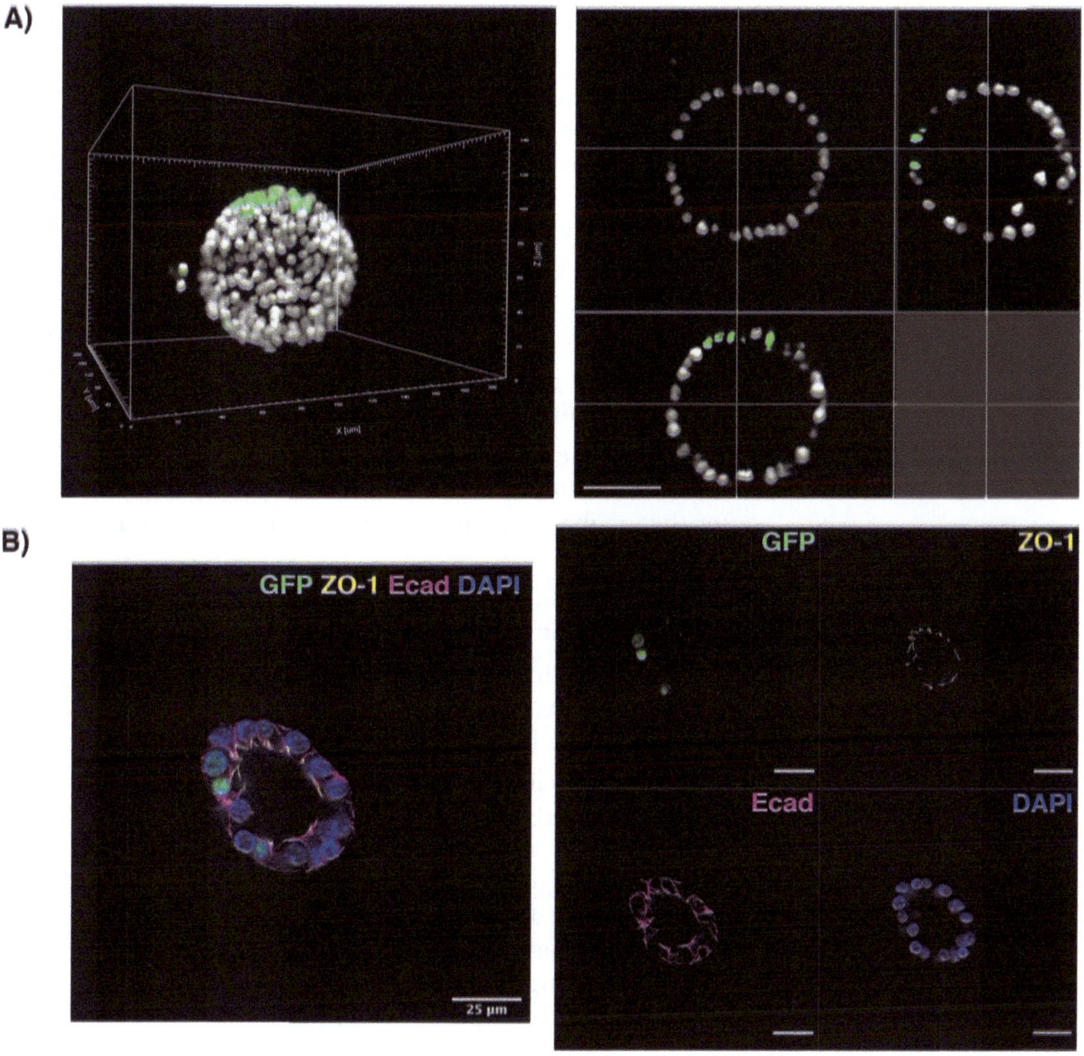

Fig. 1 Examples of potential experimental applications of the explained protocol on mammary gland organoids. (**a**) Organoids transduced with a lentivirus which induces the co-expression of GFP together with rtTA were imaged using an inverted Luxendo InVi light-sheet microscope. On the left, a 3D reconstruction of a transduced mammary gland organoid can be observed. On the right, 2D sections through the middle area of the same organoid. The main panel shows a Z section of the middle area of the organoid (*x*/*y*). This organoid is further sectioned in *x*/*z* and *y*/*z* (marked by the two axes shown in the main panel); the resulting sections can be observed on the right and lower sub-panels. In green, H2B-GFP, nuclei from the transduced cells can be observed; in gray, nuclei from all non-transduced cells that conform the organoid. Scale bar, 50 µm. (**b**) Example of an immunostaining assay performed on one transduced organoid with the same lentiviral construct. Few cells within the organoid are infected by the virus. H2B-GFP (green), introduced by the viral construct; ZO-1 (yellow), a tight junction protein expressed in the apical membranes. E-cad (magenta); DAPI nuclear stain (blue). Scale bar, 25 µm

100 µL gel with organoids). Resuspend the gels to ensure proper digestion by using mechanical force with a P1000 pipet. Vigorously pipet up and down 6–8 times each gel to

ensure proper gel disruption (*see* **Note 10**). Incubate the enzyme for 50 min at 37 °C, 5% CO_2 (*see* **Note 11**).

2. Using mechanical disruption again, collect and transfer the contents of the wells into a 15-mL Falcon tube. Typically, a maximum of 6 starting 3D 100 μL gels are transferred into a 15-mL Falcon tube (*see* **Note 12**). Fill up the tube(s) up to 14 mL with PBS.
3. Centrifuge the samples for 5 min at 190 × *g* (*see* **Note 13**).
4. Carefully aspirate the supernatant, leaving around 600–1000 μL of culturing media to not disturb the pellet. Resuspend the pellet vigorously in the remaining supernatant with a P1000 pipet, rinsing the walls of the tube (*see* **Note 14**).
5. Fill up the tube(s) with STOP media. Add 5–8 μg of DNase and let it work for 3–5 min at room temperature. Immediately after the incubation, centrifuge the samples as in **step 3** of this section.
6. Repeat **step 4** but using PBS instead of STOP media, and centrifuge the samples as in **step 3**.
7. Aspirate as much supernatant as possible this time, without disturbing the cell pellet (*see* **Note 15**). Resuspend the pellet vigorously in the remaining supernatant.
8. Add the lentiviral particles to the organoid suspension. We usually add 5 μL of lentiviral particles per 2 starting gels (which amounts to an MOI of ~17 TU/cell); hence, for a starting amount of 6 gels/15-mL Falcon tube, 15 μL of lentiviral particles are added (*see* **Note 16**). The suspension is left at room temperature while the master mix of 3D matrix is prepared, allowing the viral particles to interact with the resuspended organoids.
9. Working on ice, prepare a seeding master mix of 3D matrix composing of ice-cold PBS, Collagen I, and Matrigel in a 1:1:4 ratio, pipetting on that order. Calculate a total volume of 100 μL per seeded gel, always taking surplus volume into account (*see* **step 9** in Subheading 3.1).
10. Take 300 μL of the master mix and add it to the cell suspension, mixing carefully to avoid bubble formation. Working quickly, seed 100 μL of the suspension into the 12-well culture plate. Plate well size can be adapted based on the follow-up experiment.
11. Let the gels solidify on a level surface for 30 min at 37 °C, 5% CO_2.
12. Add 1.5 mL of culturing media to each gel, and let the gels rest inside an incubator at 37 °C, 5% CO_2.

13. The day after, change the media to remove any possible excess of lentiviral particles.
14. 48–72 h after—depending on the type of viral construction used—, the lentiviral construct should be fully integrated and expressed in the infected cells within an organoid. Therefore, the cultured organoids could potentially be ready to be used in the desired experiment. For instance, if the lentiviral particles include a reporter gene, organoids could be used in longitudinal live imaging (Fig. 1a) or fixed and used for immunofluorescence analysis of the desired targets (Fig. 1b).

4 Notes

1. In our specific case, we work with mammary glands from genetically modified mice containing doxycycline-dependent transgenes. Therefore, we use STOP media containing a Tet-approved FBS. Standard FBS contains enough traces of tetracycline and tetracycline analogs to interfere with a doxycycline-dependent regulation system.
2. Matrigel vials are typically stored at −80 °C. Thawing a new vial takes long time; we recommend letting the Matrigel thaw at 4 °C overnight prior usage.
3. It is important that cells are exposed to sufficient CO_2 during digestion for buffering and increasing cell viability. This is achieved in the 5% CO_2 incubator by loosely capping the digestion tubes.
4. After centrifugation, three different phases can be easily observed: a fat layer at the top, an intermediate liquid interphase where no cells can be found; and a clear cell pellet. Some undigested cells may still be found in the lipidic layer; that's why we keep it during the second digestion step.
5. Tip: start by aspirating the fat layer; if it enters in contact with the pellet, it will be harder to remove.
6. At this point, surplus cells can be cryopreserved for future experiments. When thawing a cryopreserved vial, we let the cells recover by doing a 2D culturing step in collagen I-coated plates prior to 3D seeding.
7. We work with a 3D matrix composed by Matrigel and Collagen I. Matrigel is routinely used to enable mammary gland organoid growth and development; the addition of Collagen I facilitates other relevant processes such as cell migration, tumorigenesis, or branch formation [20, 21].
8. When preparing the 3D matrix made of Matrigel, it is important to avoid solidification of the Matrigel and Collagen I by

working on ice. Always pre-hydrate the tips with ice-cold PBS before pipetting Collagen I or Matrigel.

9. The ratios of these compounds for the master mix vary with the original density of the Matrigel and number of cells seeded as well as cell type. We normally use Matrigel that has a protein concentration of around 10.5 mg/mL. For 100 μL gels, we use 12,000–15,000 cells; different volumes of gels can be used, adjusting the number of cells seeded on them.
10. Mechanical disruption of the 3D matrix before the incubation with the enzyme fastens and ensures a complete digestion of the 3D matrix. The gel should appear completely dissolved in the media before proceeding with the protocol.
11. Shorter incubation times or reduced enzymatic concentration results in an incomplete digestion of the 3D matrix. It is possible to estimate if the 3D matrix is being completely degraded by observing the obtained pellets in the following step of the protocol; a typical cell pellet should be observed when the 3D matrix is being properly digested, whereas a larger, less concentrated, whiter pellet is found when not. Incomplete removal of the 3D matrix drastically compromises the efficacy of the virus particles to infect the organoids, as they cannot move through it. The 3D matrix we are digesting is composed of PBS, Collagen Type I, and Matrigel, in a 1:1:4 ratio; other concentrations might need adjusted enzymatic concentration and incubation times. We have observed that using non-enzymatic approaches result in an incomplete 3D matrix disruption, leading to inefficient transductions with very low numbers of infected cells.
12. We find that starting with 6 initial 100 μL gels in a 15-mL Falcon tube is the most optimal way of performing this protocol. A larger number of initial gels will suppose more matrix and enzymatic content, which will not be wash out as nicely. This has the risk of leaving some enzymatic traces within the pellet, which can degrade the newly seeded 3D matrix, disrupting the future cultured gels. A smaller number of initial gels can suppose a more drastic loss of organoids along the process due to the instability of the pellet, yielding lower number of organoids of what it would be expected.
13. Alternatively, centrifuge for 4 min at 230 × *g*.
14. We have noticed that organoids tend to attach to the walls of the 15 mL tubes. By rinsing them carefully every time, we recover higher yields of organoids after the whole process.
15. To avoid disturbing the pellet, you can first aspirate as much supernatant as possible, leaving enough of it to not alter the cell pellet. Then, try to remove the excess with a P200. Depending on the degree of digestion of the 3D matrix, the

pellet will have a different volume (larger when incomplete digestion). In optimal conditions, a small cell pellet will be found, meaning that you can leave around 20–30 μL of supernatant. If larger pellets are obtained, you can leave around 50–80 μL of supernatant without affecting the outcome.

16. Depending on the type of viral particles used, its MOI and the experimental outcome, the amount of used viral particles can change and should be adjusted.

References

1. Clevers H (2016) Modeling development and disease with organoids. Cell 165(7): 1586–1597. https://doi.org/10.1016/j.cell.2016.05.082
2. Drost J, Clevers H (2018) Organoids in cancer research. Nat Rev Cancer 18(7):407–418. https://doi.org/10.1038/s41568-018-0007-6
3. Muthuswamy SK (2011) 3D culture reveals a signaling network. Breast Cancer Res 13(1): 103. https://doi.org/10.1186/bcr2800
4. Grunert S, Jechlinger M, Beug H (2003) Diverse cellular and molecular mechanisms contribute to epithelial plasticity and metastasis. Nat Rev Mol Cell Biol 4(8):657–665. https://doi.org/10.1038/nrm1175
5. Knouse KA, Lopez KE, Bachofner M, Amon A (2018) Chromosome segregation fidelity in epithelia requires tissue architecture. Cell 175(1):200–211.e213. https://doi.org/10.1016/j.cell.2018.07.042
6. Rowald K, Mantovan M, Passos J, Buccitelli C, Mardin BR, Korbel JO, Jechlinger M, Sotillo R (2016) Negative selection and chromosome instability induced by Mad2 overexpression delay breast cancer but facilitate oncogene-independent outgrowth. Cell Rep 15(12): 2679–2691. https://doi.org/10.1016/j.celrep.2016.05.048
7. Jechlinger M, Grünert S, Beug H (2002) Mechanisms in epithelial plasticity and metastasis: insights from 3D cultures and expression profiling. J Mammary Gland Biol Neoplasia 7(4):415–432
8. Tanner K, Mori H, Mroue R, Bruni-Cardoso A, Bissell MJ (2012) Coherent angular motion in the establishment of multicellular architecture of glandular tissues. Proc Natl Acad Sci U S A 109(6):1973–1978. https://doi.org/10.1073/pnas.1119578109
9. Hillreiner M, Muller NI, Koch HM, Schmautz C, Kuster B, Pfaffl MW, Kliem H (2017) Establishment of a 3D cell culture model of primary bovine mammary epithelial cells extracted from fresh milk. In Vitro Cell Dev Biol Anim 53(8):706–720. https://doi.org/10.1007/s11626-017-0169-7
10. Duval K, Grover H, Han LH, Mou Y, Pegoraro AF, Fredberg J, Chen Z (2017) Modeling physiological events in 2D vs. 3D cell culture. Physiology (Bethesda) 32(4):266–277. https://doi.org/10.1152/physiol.00036.2016
11. Shaw KRM, Wrobel CN, Brugge JS (2004) Use of three-dimensional basement membrane cultures to model oncogene-induced changes in mammary epithelial morphogenesis. J Mammary Gland Biol Neoplasia 9(4):297–310. https://doi.org/10.1007/s10911-004-1402-z
12. Sumbal J, Chiche A, Charifou E, Koledova Z, Li H (2020) Primary mammary organoid model of lactation and involution. Front Cell Dev Biol 8:68. https://doi.org/10.3389/fcell.2020.00068
13. Jarde T, Lloyd-Lewis B, Thomas M, Kendrick H, Melchor L, Bougaret L, Watson PD, Ewan K, Smalley MJ, Dale TC (2016) Wnt and Neuregulin1/ErbB signalling extends 3D culture of hormone responsive mammary organoids. Nat Commun 7:13207. https://doi.org/10.1038/ncomms13207
14. Nguyen-Ngoc KV, Shamir ER, Huebner RJ, Beck JN, Cheung KJ, Ewald AJ (2015) 3D culture assays of murine mammary branching morphogenesis and epithelial invasion. Methods Mol Biol 1189:135–162. https://doi.org/10.1007/978-1-4939-1164-6_10
15. Jechlinger M, Podsypanina K, Varmus H (2009) Regulation of transgenes in three-dimensional cultures of primary mouse mammary cells demonstrates oncogene dependence and identifies cells that survive deinduction. Genes Dev 23(14):1677–1688. https://doi.org/10.1101/gad.1801809
16. Havas KM, Milchevskaya V, Radic K, Alladin A, Kafkia E, Garcia M, Stolte J, Klaus B, Rotmensz N, Gibson TJ, Burwinkel B,

Schneeweiss A, Pruneri G, Patil KR, Sotillo R, Jechlinger M (2017) Metabolic shifts in residual breast cancer drive tumor recurrence. J Clin Invest 127(6):2091–2105. https://doi.org/10.1172/JCI89914

17. Mroue R, Bissell MJ (2013) Three-dimensional cultures of mouse mammary epithelial cells. Methods Mol Biol 945:221–250. https://doi.org/10.1007/978-1-62703-125-7_14

18. Alladin A, Chaible L, Garcia Del Valle L, Sabine R, Loeschinger M, Wachsmuth M, Heriche JK, Tischer C, Jechlinger M (2020) Tracking cells in epithelial acini by light sheet microscopy reveals proximity effects in breast cancer initiation. elife 9. https://doi.org/10.7554/eLife.54066

19. Jechlinger M (2015) Organotypic culture of untransformed and tumorigenic primary mammary epithelial cells. Cold Spring Harb Protoc 2015(5):457–461. https://doi.org/10.1101/pdb.prot078295

20. Kai F, Drain AP, Weaver VM (2019) The extracellular matrix modulates the metastatic journey. Dev Cell 49(3):332–346. https://doi.org/10.1016/j.devcel.2019.03.026

21. Spill F, Reynolds DS, Kamm RD, Zaman MH (2016) Impact of the physical microenvironment on tumor progression and metastasis. Curr Opin Biotechnol 40:41–48. https://doi.org/10.1016/j.copbio.2016.02.007

Chapter 10

Surgical Procedure for Implantation of Human Tumor Tissue into the Epithelium-Free Mammary Fat Pad of Immunocompromised Mice to Generate Patient-Derived Xenografts (PDX)

Lacey E. Dobrolecki and Michael T. Lewis

Abstract

Implantation of cancerous tissue obtained from patient biopsies or surgical procedures to create patient-derived xenograft models (PDX) has greatly enhanced the ability of scientists and physicians to perform translational research involving pre-clinical testing. PDX models are more representative of patient cancer tumors than traditional cell lines as they retain the three-dimensional architecture of the tumor, and are genomically, transcriptomically, and proteomically analogous. These models can be used for a variety of experiments to evaluate response of tumors to experimental therapeutic agents and to aid in the elucidation of response and resistance mechanisms of existing therapies. Orthotopic transplantation of breast cancer tissue into the mammary fat pad has been proven to be the most robust method of establishing breast cancer PDX models. Here we describe the procedure for implanting patient tumor tissue, or existing PDX tissue, into the mammary fat pad of immunocompromised mice, with notes, where necessary, for transplantation of normal or preneoplastic tissue.

Key words Cleared mammary fat pad, Epithelium-free mammary fat pad, Orthotopic transplantation, Patient-derived xenograft, PDX, Breast cancer

1 Introduction

The mammary fat pad transplantation technique is used for analyses of growth and behavior of both normal and neoplastic mammary tissue. The technique was developed in the late 1950s by Dr. Kenneth DeOme and colleagues [1], and fundamentally transformed the fields of normal mammary gland development and breast cancer. The original technique made use of the epithelium-free ("cleared") mammary fat pad by making an incision in the shape of an inverted "Y" on the ventral side of the mouse. Briefly, a first cut is made by lifting the skin mid-abdomen and snipping a small hole into which surgical scissors can be inserted. Thereafter,

Maria dM. Vivanco (ed.), *Mammary Stem Cells: Methods and Protocols*,
Methods in Molecular Biology, vol. 2471, https://doi.org/10.1007/978-1-0716-2193-6_10,

an approximately 1.5 cm incision is made from that site toward the sternum. Subsequently, two incisions are made from the initial point toward the hind limbs to complete the inverted "Y" form. The shape of the incision leaves two triangular skin flaps that can be separated from the peritoneum using mouse-tooth forceps and a cotton swab. Once completed, the skin flap can be affixed to the surgical board with a small pin. Major blood vessels are cauterized, as well as the nipple to prevent ingrowth of host epithelium into the portion of the fat pad that remains at the transplantation site. The fat pad is separated from the skin moving from the nipple toward the dorsal side of the animal, and the separated fad pad is cut on the dorsal side of the large lymph node near the center of the #4 (inguinal) fat pad. The remaining fat pad can be used immediately for transplantation as below, or the mouse can be closed for transplantation at a later date, depending on the experimental goals.

This technique remained unchanged for over five decades, when a "sparing" technique was developed for clearing the fat pad in 2008 by Brill et al. [2]. In this procedure, the #4 nipple is grasped, raised with forceps, and resected with a small cut. The fat pad is pulled out through the small, circular incision and the portion of the glad containing the lymph node is excised. Neoplastic or normal tissue can then be implanted into the remaining fat pad. The wound is closed with one wound clip at the incision site. As compared to the "Y" incision, the sparing technique is much less invasive and can be performed more rapidly. Hence, there are many benefits for the animal including decreased time under anesthesia, enhanced movement, and rapid recovery.

The technique described below, and in another recent publication [3], is modified from the sparing technique by making a straight line incision medial to the #4 nipple as opposed to resection of the nipple area. The incision is still minimal but allows the skin to be pinned back without stretching which increases visualization of the mammary fat pad to allow for precise placement of the transplanted tissue. Proper placement is essential for ease of caliper measurements for tumor models, to ensure proper movement of the animal, and to decrease tumor ulceration. This method retains the benefits of the sparing technique in that it can be performed rapidly, is minimally invasive, and promotes rapid healing of the mouse.

2 Materials

In this protocol, two autoclaved surgical packs are described for the sterile tools and supplies needed for transplantation. This may be modified to accommodate the use of instrument sterilization trays so long as all supplies are autoclaved and can be placed in the surgical space in a sterile manner.

2.1 Equipment

1. Anesthesia system with a dual port tubing system.
2. Animal cages for post-surgery housing.
3. Electric shaver.
4. Glass bead sterilizer.
5. Oxygen tank with wrench.
6. Scavenging pump system (*see* **Note 1**).
7. Warming devices (*see* **Note 2**).

2.2 Supplies

1. 50 mL conical tube.
2. Absorbent Underpads—polyethylene backing with spunbound top layer (*see* **Note 3**).
3. Alcohol Wipes (*see* **Note 4**).
4. Betadine Scrub.
5. Cauterizer.
6. Gloves.
7. Isoflurane.
8. Labeling Tape—white, 60 yd., ½″ thick.
9. Ophthalmic ointment.
10. Petri dishes—sterile.
11. Rescue™ solution or similar (Accelerated Hydrogen Peroxide).
12. Sterile surgical gloves.
13. Tissue fragments in media (DMEM or similar).
14. Vet bond.

2.3 Autoclaved Surgical Pack #1

1. Instant Sealing Sterilization Pouch—10 × 15″.
2. OK™ Steam Sterilization Monitor Strips, 4″.
3. Silicone surgical board—6 × 8″, ½″ thick, -60 - 500 °F, 35–45 Shore A, 650 psi tensile strength (*see* **Note 5**).
4. Sterile surgical drape, full or half size.
5. Underpad—white spunbound polypropylene facing sheet with blue polypropylene backing sheet (*see* **Note 6**).

2.4 Autoclaved Surgical Pack #2

1. 27G x ½" Needles (*see* **Note 7**).
2. Instant Sealing Sterilization Pouch—10 × 15″.
3. Cotton swabs—6″, wood handle, small tip cotton (*see* **Note 8**).
4. Knapp Scissors 4″ straight, sharp/blunt.
5. Micro Dissecting Forceps, 1 × 2 teeth, angular tip width 0.8 mm, length 4″ (claw).
6. Micro Dissecting Forceps, serrated, angular tip width 0.8 mm, length 4″ (regular).

7. Micro Dissecting Forceps, McPherson, angled, smooth typing platform, tip 0.4 × 5 mm, length 4″ (tissue placing forceps—*see* **Note 9**).
8. Micro Dissecting Spring Scissors (Castroviejo), angled, one sharp point and one 1.1 mm ball tip, 9 mm cutting edge, 0.15 mm tip width, length 3.5″.
9. OK™ Steam Sterilization Monitor Strips, 4″.
10. Sterile surgical drapes (*see* **Note 10**).
11. Tin foil—4 × 6″ and 2 × 2″, autoclaved (*see* **Note 11**).
12. Wound clip applicator.
13. Wound clip remover.
14. Wound clips, 9 mm (*see* **Note 12**).

3 Methods

Place all items from Subheading 2.3, Autoclaved Surgical Pack #1, into the Instant Sealing Sterilization Pouch. Seal the pouch and label with the date and lab identification. Autoclave on Gravity setting for 40 min at 121.0 °C, 15 psi.

From Subheading 2.4, Autoclaved Surgical Pack #2, take one of the two half-sheets of surgical drape and lay it out on the bench top. Center the surgical tools, cotton swabs, needles, wound clips, and OK™ strip on the drape and fold up into a pouch to contain the supplies. If there are too many cotton swabs, they may be autoclaved in their own pouch or in their original packaging if it has an autoclave indicator. Place pouch and all other items listed into the Instant Sealing Sterilization Pouch. Seal the pouch and label with the date and lab identification. Autoclave on Gravity setting for 40 min at 121.0 °C, 15 psi.

This method describes the transplantation procedure assuming there are two people working together as a team, one animal "prepper" and one surgeon. The procedure can be done with one individual, but it is recommended that the same setup be used so that sterility around the surgical board can be maintained. If an individual is working alone, new sterile surgical gloves will need to be donned prior to initiating the transplantation of tissue for each animal.

Follow institutional guidelines for handling immunocompromised animals. All supplies should be sprayed or wiped down with disinfectant (e.g., Rescue™) prior to placing them into the surgical space. All animals should be given pre-operative analgesic medication per institutional guidelines.

3.1 Preparation of the Non-sterile Surgical Area (Can Be Done Concurrently by the Prepper and Surgeon in Regular Gloves)

1. Upon entering the surgical space, spray all working surfaces with Rescue™ solution and allow the solution to sit for at least 1 min. This includes all bench tops, cage-warming devices, supplies, spaces around the surgical setup, etc. (*see* **Note 13**).
2. Bring the animals from their housing room into the surgical space.
3. Gather the correct number of clean, autoclaved cages needed for post-surgery housing.
4. Place the cages on the slide warmers located on the bench. The slide warmers should be set to 37–40 °C. The cage should be positioned transversely on the warmer (half on and half off—*see* **Note 14**).
5. Set up the prepping area on the left side of the surgical space by placing a heat source down and covering with an absorbent underpad (*see* **Note 15**).
6. Place the following supplies in the prepping area: alcohol wipes, sterile petri dishes (one-half for the ophthalmic ointment and one-half for each tissue being transplanted), a 50 mL conical tube and holder, betadine scrub, and ophthalmic ointment.
7. Fill the 50 mL tube with enough betadine scrub to cleanse each mouse 3 times (about 5–15 mL) and place it into the tube holder (*see* **Note 16**).
8. Squeeze out enough ophthalmic ointment into one-half of a petri dish to cover both eyes for each mouse. Use a cotton swab for application.
9. The electric shaver should be wiped down with Rescue™ and placed into the prepping area (*see* **Note 17**).
10. Spray the rubber tubing that is connected to the anesthesia machine with Rescue™ and arrange it in the surgical space so that the longer tube is on the prepping side and the smaller, flexible tubing with the nose cone is on the surgeon's side.
11. Place the glass bead sterilizer in the surgical space so that it can be easily reached by the surgeon and turn it on.
12. Put the applicator tube in the vet bond and place it next to the surgical area with a cauterizer.
13. Check the isoflurane and oxygen levels in the anesthesia machine:
    (a) When refilling isoflurane always be sure to use a scavenging pump system. Unscrew the metal lid and fill the reservoir to the top indicator line.
    (b) When replacing the oxygen tank, make sure the oxygen is off and the valve has been flushed. Remove the blue seal from the valve post and place the tank in the holder on the anesthesia machine. Next, unscrew the regulator from the old tank and place it onto the new one. Line up the pins correctly and tighten.

3.2 Preparation of the Sterile Surgical Area (Prepper and Surgeon Will Work Together While Maintaining Sterility)

1. The surgeon can now don surgical gloves but must keep them sterile. If sterility is broken, remove gloves and don a new pair.
2. The prepper will open Surgical Pack #1 and allow the surgeon to remove the items. The surgeon will check the OK™ strip for proper sterilization and then lay out the sterile drape on the left side of the surgical area.
3. The prepper will take the silicone board and blue-backed underpad and wrap the board with the underpad (blue plastic side out). Secure with tape on the backside.
4. Next, the prepper will tape down the nose cone in the center of the surgical board leaving enough space to tape down the mouse (*see* **Note 18**).
5. The prepper will open Surgical Pack #2 and the surgeon will remove the half-sheet surgical drape and lay it on the right side of the surgical board.
6. The surgeon should then remove the supply pouch, open it on top of the drape on the right side, and check the OK™ strip for proper sterilization. The foil and small square surgical drapes can then be removed and placed on a sterile surface.
7. The surgeon can place the appropriate number of cotton swabs in the prepping area (*see* **Note 8**).
8. The surgeon should take the sterile foil pieces and carefully wrap the vet bond and cauterizer. They can then be placed in the sterile surgical area for use.

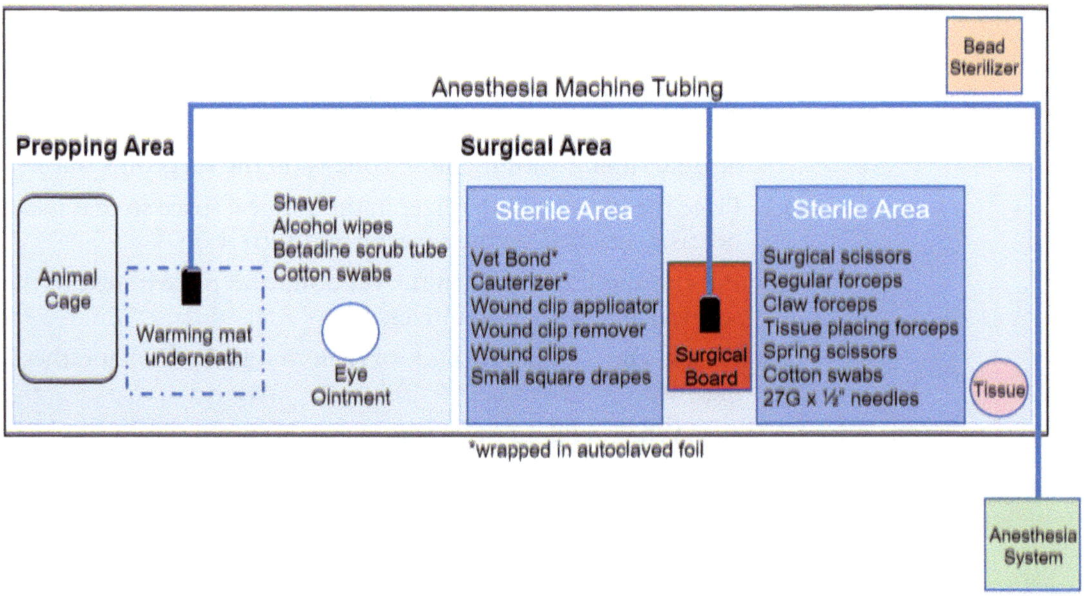

Fig. 1 Schematic layout of a surgical area for a mammary fat pad transplantation surgery

9. The prepper will spray the first tissue specimen tube into the surgical space, pour it into a petri dish, and place it on the surgeon's side making sure not to touch any sterile surfaces (*see* **Note 19**).
10. The prepper will turn on the anesthesia machine:
    (a) Open the oxygen flow using the oxygen tank wrench.
    (b) Ensure the oxygen gauge is at 2.5 LPM.
    (c) Turn the isoflurane gauge to 2.5 LPM.
    (d) Check both ports on the tubing to make sure they are in the "open" position.
11. The prepper will put the cage of mice into the hood on the far left side of the prepping area, spray with Rescue™, and remove the first mouse (Fig. 1).

3.3 Preparing the Mouse for Surgery (Prepper)

1. Shave the area over and around the #4 nipple (from below the #4 nipple to the rib cage), and place the mouse's nose into the nose cone.
2. Apply betadine scrub in a circular motion going outward from the center of the shaved area using a cotton swab, then remove using an alcohol wipe. Repeat two more times using a new cotton swab and new alcohol wipe each time.
3. Once the mouse is fully under anesthesia, gently apply ophthalmic ointment to completely cover each eye. Be careful not to scratch the eye of the mouse.
4. Transfer the mouse onto the surgery board, do a final toe pinch to ensure the mouse is in the appropriate plane of anesthesia, put the nose into the nose cone, and tape down the mouse's limbs.

3.4 Surgical Procedure to Implant the Tissue Fragment (Surgeon)

1. At this time the surgeon can begin the transplantation procedure.
2. Place a sterile drape with cutout over the mouse, only exposing the surgical area.
3. Using the regular forceps, pinch and lift up the skin at the #4 nipple.
4. With the blunted side down, use the Knapp scissors to make a short (about 1 cm), parasagittal incision, from the #4 nipple toward the head.
5. Hold the edge of skin up on the medial side of the incision with regular forceps and use a cotton swab to separate the skin from the peritoneum.
6. Pull the lateral side of the incision away from the body and use the same swab to gently peel the #4 mammary fat pad off the skin. It will take a couple of sweeps of the swab to completely separate the fat pad from the skin.

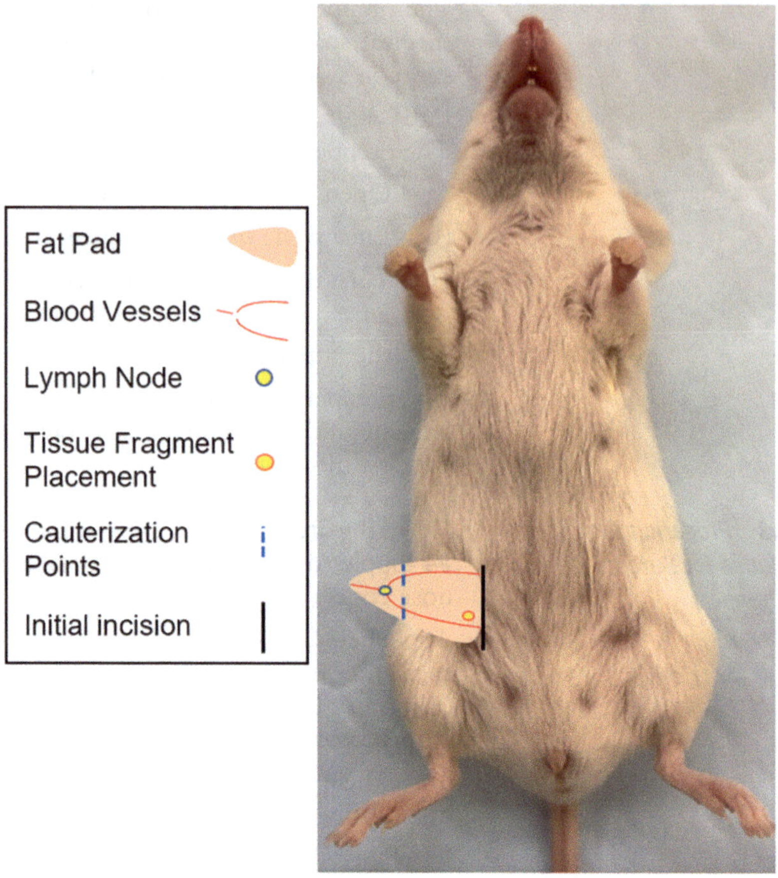

Fig. 2 Schematic image of the #4 mammary fat pad indicating the incision site, anatomical landmarks for clearing, and implantation site

7. Once the fat pad is separated and pushed toward the body, pin down the lateral skin flap using a 27G needle placed close to the animal's body (*see* **Note 20**).
8. Use the regular forceps to gently hold and pull out the fat pad and perform the clearing technique if indicated (Fig. 2) (if clearing is not indicated proceed to **step 9**):
   (a) The goal of this step is to remove the area of the fat pad containing the lymph node and all mammary epithelium.
   (b) Temporarily turn off the oxygen during this step (fire hazard). Cauterize medial to the lymph node, making sure to cauterize each vessel to prevent bleeding. If transplanting normal/non-neoplastic tissue (i.e., non-neoplastic tissue such as from a different wild-type or mutant mouse, or preneoplastic immortalized tissue), the nipple must also be cauterized to prevent ingrowth of nipple-associated epithelium into the remaining fat pad,

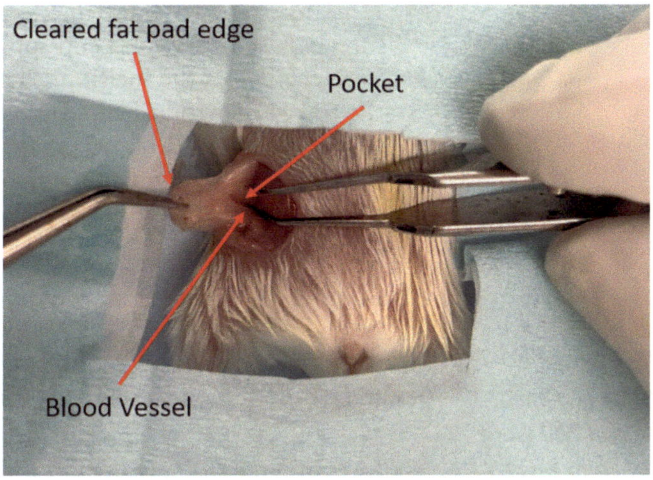

Fig. 3 Cleared #4 mammary fat pad. Surgeon demonstrating creation of the pocket near the body wall of the mouse and just above the main vessel in the fat pad

which can interfere with experimental conduct and interpretation.

(c) Using the Micro Dissecting Spring Scissors cut each cauterized vessel from the bottom to the top (to make sure that there is no bleeding) until the section of the fat pad that contains the lymph node and epithelium is removed (*see* **Note 21**).

(d) Place the cleared fat pad on the preppers side mat without touching the forceps to the mat [3].

9. Gently pull out the fat pad and push back the peritoneum with the jeweler's forceps or a cotton swab.

10. Locate the vessel in the #4 fat pad and use the tip of the jeweler's forceps to poke a hole just above the vessel and close to the body wall.

11. Take the jeweler's forceps and slightly open the hole into a small pocket (Fig. 3).

12. With the jeweler's forceps still in your hand, pick up a fragment of tissue from the petri dish and place it in the small pocket. Make sure that it goes completely into the pocket and does not fall out when the forceps are removed (*see* **Note 22**).

13. Release the fat pad from the forceps and unpin the skin.

14. Starting at the base of the incision collect the skin on each side using your regular forceps. While still holding the skin, use the clawed forceps to bring the two sides together and lift slightly to prepare the skin for wound clip application. Uncurl the edges of the incision and make sure they form a continuous surface at the cut edges.

15. Holding the two sides of the incision together with regular forceps, use the wound clip applicator to place one wound clip in the center of the incision. The edges of the skin should be at the top of the wound clip. If the wound clip is too low, is loose, or has gathered too much skin, use the wound clip remover to remove the clip and place another.
16. Apply vet bond above and below the wound clip to seal then ends of the incision (*see* **Note 23**).
17. Remove the drape from the mouse.

3.5 Post-Surgical Procedures (Prepper and Surgeon Will Work Together While Maintaining Sterility)

1. The prepper will now remove the tape on the legs of the mouse and transfer the mouse to a clean cage on the warming device.
   (a) Check the incision for proper wound clip placement and absence of bleeding.
   (b) Place the mouse in a prone position and be sure not to touch the incision site.
   (c) Ensure there is bedding under the mouse and that the mouse is in a section of the cage over the warmer.
2. The surgeon will clean the tools for 10 s in the glass beads sterilizer before continuing with the next transplant. Wait for the tools to cool before proceeding.
3. Repeat **step 1** of Subheading 3.3 to **step 2** of Subheading 3.5 until all mice are transplanted.
4. If more than one PDX model is transplanted in the same session, remember to do the following:
   (a) Remove and discard the petri dish with tissue from the previous transplant.
   (b) Discard the 27G needle used to pin back the skin and use a new one.
   (c) Clean tools thoroughly with alcohol wipes and bead sterilizer or have an additional sets of tools.
   (d) Place the petri dish with new tissue in the hood after validating and logging the tissue identification.
5. Check mice for bleeding and make sure they have resumed normal activities. Animals should be fully awake and alert within 5–10 min after the procedure.
6. Take cages back to the housing room and add fresh water and food (*see* **Note 24**).
7. Follow your protocol's post-surgical care plan.

4 Notes

1. Needed if transplantation is not performed in a ducted biological safety cabinet.
2. To place under the prepping mat in the surgical prepping area and to warm the cages—slide warmer, surgical board, warming disc, warming mat, etc.
3. This underpad is NOT autoclavable but it is ideal for the prepping area as it absorbs fluid well and the animal's nails do not get caught in the surface.
4. Need three per mouse to be transplanted plus extra for cleaning tools if desired.
5. Autoclavable, allows 27G needle to be placed securely and removed easily, tape does not stick to the surface, may come as a larger piece that needs to be cut to the appropriate size.
6. This underpad is autoclavable and is used to wrap the rubber surgery board. The blue backing will allow tape to be applied and peeled off if done carefully.
7. One per model to be transplanted; remove outer wrapping.
8. 4X the number of animals to be transplanted +5–10 extra; 3 per animal will be given to the prepper.
9. The small tip size is necessary for creating the pocket in the fat pad.
10. Two half-sheets plus smaller pieces to place over the mouse that have a square hole in the middle that is slightly larger than the surgical site.
11. To wrap around the vet bond tube and cauterizer to maintain sterility.
12. 2X the number of animals to be transplanted plus extra.
13. Surgical procedures may be done in a dedicated surgical space or inside a biological safety cabinet. If a shared surgical suite is used, care must be taken to keep the room free from contamination from immune-competent animals.
14. Alternate warming sources such as heating discs may be used in place of a slide warmer. Follow warming instructions for proper heating time so that the temperature is accurate. A portion of the cage should not be over the heat source so animals can retreat to that area if they become too hot.
15. A warming disc, heated surgical board, or warming mat may be used as a heat source.
16. Disposable betadine swab sticks may also be used.

17. Some researchers prefer to use Nair™ to remove the remaining hair after shaving. If that is desired, an additional petri dish, cotton swab, and alcohol wipe (one per mouse) will be needed to remove the Nair™.
18. Most tape will not stick directly to the rubber surgery board. The surgical board is not considered a sterile surface since the mouse will be taped to it.
19. Patient tissue is routinely placed in Matrigel™ prior to implantation [4, 5] and must be kept on ice. Spray the ice bucket and tissue tube with Rescue™ and place into a non-sterile surgical area close to the surgeon.
20. Keep the pinned skin flap close to the body to avoid stretching of the incision, which can cause bleeding at the edges of the skin. Turn off the oxygen and cauterize any areas showing signs of bleeding during the procedure. The needle may be bent to keep the hub out of the way during the procedure.
21. If the fat pad is cut off in one motion, it will retract back into the body and vessels that were not cauterized completely will bleed quickly obscuring where the bleeding is coming from and making it very difficult to find and cauterize the vessel. When the fat pad is cut slowly, vessel by vessel, if bleeding occurs, it is much easier to cauterize as the fat pad is still extended and being held in the forceps.
22. When implanting soft tissues, be careful not to squeeze the fragment into multiple pieces with the forceps. This could lead to small nodular tumors. Conversely, hard tumor fragments might have to be scooped up with the forceps, dropped on top of the pocket and pushed in with the forceps.
23. Two would clips may be used to secure the incision if desired.
24. Estrogen supplementation is required for some ER+ breast cancer PDX models. This may be accomplished by supplementing drinking water with 17-β-estradiol or estrogen pellet implants [4, 5].

Acknowledgments

This work is supported by the Patient-derived Xenograft and Advanced In Vivo Models (PDX-AIM) Core Facility at Baylor College of Medicine with funding from NIH grant P30CA125123 (to C. Kent Osborne), and a Core Facility Core Support Grant from the Cancer Prevention and Research Institute of Texas (CPRIT) grant 170691 (to MTL).

References

1. DeOme KB, Faulkin LJ Jr, Bern HA, Blair PB (1959) Development of mammary tumors from hyperplastic alveolar nodules transplanted into gland-free mammary fat pads of female C3H mice. Cancer Res 19(5):515–520
2. Brill B, Boecher N, Groner B, Shemanko CS (2008) A sparing procedure to clear the mouse mammary fat pad of epithelial components for transplantation analysis. Lab Anim 42(1): 104–110
3. Lv X, Dobrolecki LE, Ding Y, Rosen JM, Lewis MT, Chen X (2020) Orthotopic transplantation of breast tumors as preclinical models for breast cancer. J Vis Exp (159). https://doi.org/10.3791/61173
4. Dobrolecki LE, Airhart SD, Alferez DG, Aparicio S, Behbod F, Bentires-Alj M, Brisken C, Bult CJ, Cai S, Clarke RB, Dowst H, Ellis MJ, Gonzalez-Suarez E, Iggo RD, Kabos P, Li S, Lindeman GJ, Marangoni E, McCoy A, Meric-Bernstam F, Piwnica-Worms H, Poupon MF, Reis-Filho J, Sartorius CA, Scabia V, Sflomos G, Tu Y, Vaillant F, Visvader JE, Welm A, Wicha MS, Lewis MT (2016) Patient-derived xenograft (PDX) models in basic and translational breast cancer research. Cancer Metastasis Rev 35(4): 547–573
5. DeRose YS, Gligorich KM, Wang G, Georgelas A, Bowman P, Courdy SJ, Welm AL, Welm BE (2013) Patient-derived models of human breast cancer: protocols for in vitro and in vivo applications in tumor biology and translational medicine. Curr Protoc Pharmacol Chapter 14:Unit14.23

Chapter 11

The Chicken Embryo Chorioallantoic Membrane (CAM): A Versatile Tool for the Study of Patient-Derived Xenografts

Hugo Villanueva and Andrew G. Sikora

Abstract

Patient-derived xenografts represent the gold standard in pre-clinical research models. The chicken embryo chorioallantoic membrane (CAM) is used in functional studies for studying biological processes such as blood vessel development and embryogenesis, biocompatible material testing, and more recently three-dimensional patient-derived xenograft (PDX) tumor modeling. We describe here a detailed method used to readily engraft established mouse PDX and primary patient tumor specimens on the CAM with as little as 25 mg of tissue per embryonated egg.

Key words Chorioallantoic membrane, Patient-derived xenograft, PDX, Three-dimensional tumor model, Chicken egg model, CAM xenograft model

1 Introduction

The chicken embryo chorioallantoic membrane (CAM) represents a highly versatile, scalable, and economic xenograft growth platform. Since its utility as a self-contained in vivo model for cancer research was first realized [1], the CAM has been leveraged extensively to generate various tumor models. The rich vascular properties of the CAM have been extensively studied and provide the necessary support in addition to nourishment of the developing embryo and can likewise support patient-derived xenograft tumors and cancer cell lines [2, 3]. In addition to rich vasculature, the CAM is easily accessible and is naturally immune deficient during the majority of chick embryonic development. These characteristics make it an ideal platform to grow cell lines or primary human tumor tissue until the adaptive immune system matures at approximately embryonic day 18 (E18) and rejects the xenograft [4]. Cell and primary tumor xenografts on the CAM form three-dimensional, vascularized tumors which maintain properties of cancer cells growing in vivo that are often lost in two-dimensional or simple three-

Maria dM. Vivanco (ed.), *Mammary Stem Cells: Methods and Protocols*,
Methods in Molecular Biology, vol. 2471, https://doi.org/10.1007/978-1-0716-2193-6_11,

dimensional tissue culture models [4]. Such properties include epithelial-stromal interactions, tumor angiogenic properties, and complex three-dimensional cell to cell interactions. These properties make the CAM xenograft models ideal and well suited for studying cancer cell processes such as growth, invasion, angiogenesis, and metastasis of human tumor cells into the developing chick embryo visceral organs in a short period of time [4, 5].

Here we describe a detailed protocol for the engraftment of primary patient tumors or PDXs established in mouse models on the CAM of embryonated eggs. We have successfully used this method to engraft multiple cancer types including breast, head and neck, bladder, and prostate [6–10]. We outline a general method that can be applied for the engraftment of multiple primary tumors and PDXs on the CAM with defined procedures including CAM preparation (Figs. 1, 2, 3, 4, and 5), PDX engraftment, and tumor harvest.

2 Materials

2.1 Egg Housing and Preparation

1. Graefe angled serrated forceps.
2. Blunt-ended angled forceps.
3. 1.2 ml cryogenic vials (for 12 mm silicone rings).
4. Sterile rotary saw tool.
5. Egg candling lamp.
6. Sterile push pin.
7. Clear tape.
8. Cabinet egg incubator with rotating function.
9. Hova-bator tabletop egg incubator.
10. Egg crates.

2.2 PDX Engraftment

1. Sterile phosphate-buffered saline containing no calcium or magnesium ($PBS^{-/-}$).
2. Sterile phosphate-buffered saline containing calcium and magnesium ($PBS^{+/+}$).
3. Matrigel® Matrix.
4. P1000 and p200 pipettes and filtered tips.
5. 60 mm petri dishes.
6. Cold block rack.
7. Gel cold pack.
8. Sterile 1.5 ml Eppendorf tubes.
9. Sterile surgical scalpel.
10. Water bath set to 37 °C.

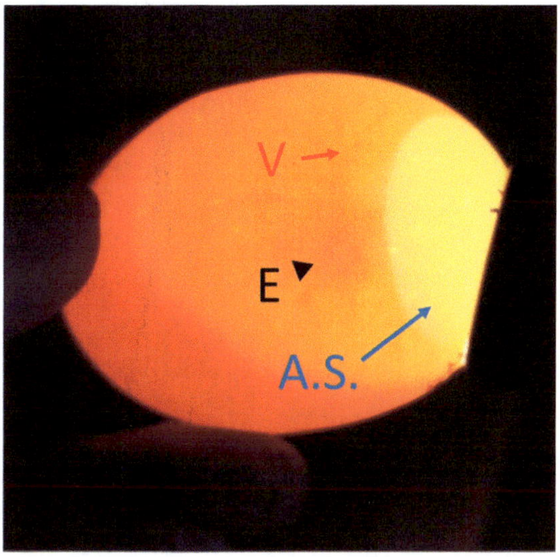

Fig. 1 A lamp is used to illuminate the interior of the fertilized egg for the identification of the embryo (E), vasculature (V), and the air sac (A.S.)

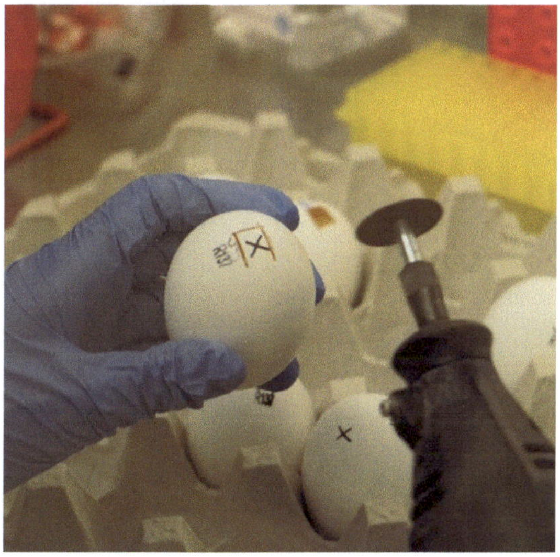

Fig. 2 Image of egg that has been drilled with a Dremmel saw tool. Parallel slits are shown, along with a third slit cut at the end and in between the other two slits

11. Spray bottle filled with 70% ethyl alcohol.
12. Analytical balance.

2.3 Tumor Harvest

1. Sterile phosphate-buffered saline containing no calcium or magnesium ($PBS^{-/-}$).
2. 10% Neutral Buffered Formalin.

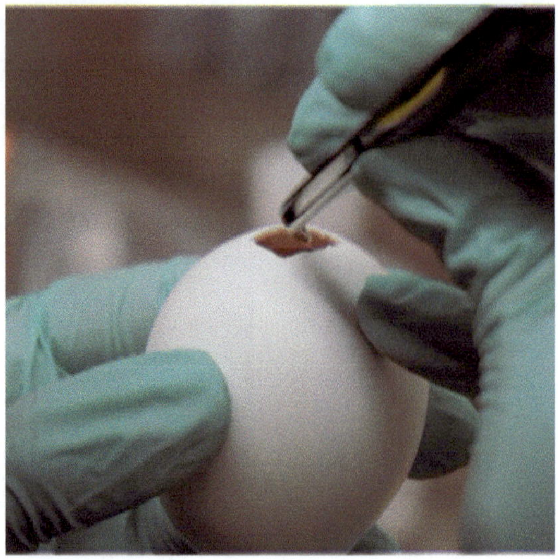

Fig. 3 Graefe angled forceps are used in an inverted orientation to pinch the inner shell membrane attached to the CAM

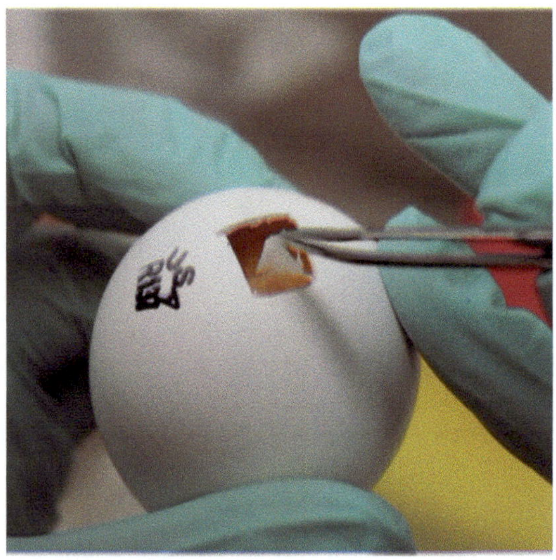

Fig. 4 Graefe angled forceps firmly grasping the inner shell membrane as it is gently detached from the CAM

3. Dissection scissors.
4. Graefe angled serrated forceps.
5. 60 mm petri dishes.

Fig. 5 CAM appearance immediately following inner shell membrane removal with vasculature exposed

3 Methods

3.1 Housing of Fertilized Eggs

1. Unpack, inspect, and maintain eggs at room temperature to allow acclimation for approximately 2 h (*see* **Note 1**).
2. Place fertilized eggs on crate with the narrow end pointing down in cabinet incubator at a constant temperature of 100° F and humidity levels of 50–60% for 7 days, ensuring the turning racks alternate positions once every hour.

3.2 Egg Preparation

1. Thaw Matrigel® (*see* **Note 2**) matrix on ice and pre-chill $PBS^{-/-}$ and $PBS^{+/+}$.
2. After 7 days of incubation in cabinet incubator, remove eggs from incubator and use a candling lamp to candle the eggs by placing them up against the light source (*see* **Notes 1** and **3**). Search for air sac, vascularization, and the embryo inside the egg on the wider, blunt end of the egg (Fig. 1, *see* **Note 4**). Mark the air sac location using a permanent marker for all eggs that will be prepared for engraftment.
3. Working inside of a biosafety cabinet, take the rotary saw tool and saw two parallel 2 cm slits that are 1.5 cm apart through the eggshell on the position previously marked as the air sac location. Ensure that the saw does not cut too deep into the egg in order to avoid damaging the CAM (*see* **Note 5**).
4. Use the saw to cut a third slit between and at the ends of the two parallel slits, using caution not to cut through the CAM

below the shell (Fig. 2). Repeat this process for all fertilized eggs.

5. Use a sterile pushpin to lift the shell off all drilled eggs.
6. Use sterile forceps to carefully peel away the inner shell membrane (Figs. 3, 4, and 5; *see* **Notes 6** and 7).
7. Use forceps to retrieve the sterile 12 mm silicone ring from a 1.2 ml cryogenic vial and gently drop it directly on top of the CAM (*see* **Note 8**).
8. Place a piece of clear tape over the newly cut shell opening and fold the tape at one end to facilitate removal later.
9. Repeat **steps 6** through **8** for all the eggs that will be engrafted.
10. Place eggs in the Hova-bator incubator on the flat egg tray for a minimum of 2 h before engraftment of cells (*see* **Note 9**).

3.3 PDX Engraftment

1. Prior to placing in biosafety cabinet, spray 70% ethyl alcohol on frozen gel cold packs, cold tube rack, sterile scalpels, sterile 60 mm petri dish sleeves, pre-chilled $PBS^{-/-}$ and $PBS^{+/+}$ bottles, bags containing sterile 1.5 ml Eppendorf tubes, sterile p1000 and p200 filtered pipette tip boxes, p1000 and p200 pipettes.
2. Spray a paper towel with 70% ethyl alcohol and wipe down the tube containing cold Matrigel® and place it on the cold block rack inside the biosafety cabinet, maintaining Matrigel® cold at all times before use in order to prevent its polymerization.
3. Retrieve viably cryopreserved patient tumor sample or PDX from deep freezer and quick thaw in water bath set to 37 °C (*see* **Note 10**).
4. While specimen is thawing, place a 60 mm petri dish directly on top of a gel cold pack and pour pre-chilled $PBS^{-/-}$ on the petri dish, place the petri dish lid on the gel cold pack, and place a sterile 1.5 ml Eppendorf tube on the cold block rack.
5. Weigh a separate, empty 60 mm sterile petri dish with lid on using an analytical balance and record the weight. Transfer the petri dish back into the biosafety cabinet.
6. Immediately after specimen is thawed, take the tube containing the specimen and spray it thoroughly with 70% ethyl alcohol and wipe it down with a clean paper towel before placing it in the biosafety cabinet.
7. Open the specimen tube and transfer the specimen using sterile forceps to the 60 mm petri dish containing cold $PBS^{-/-}$ to rinse the specimen.
8. Transfer the specimen to the pre-weighed 60 mm petri dish and weigh the specimen on the analytical balance, ensuring that the petri dish lid remains on during this process.

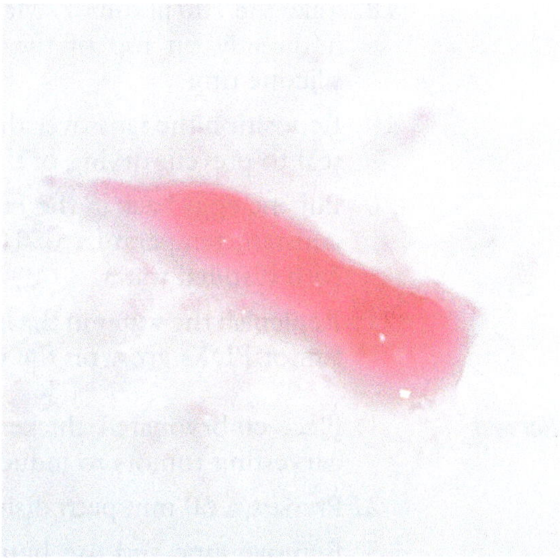

Fig. 6 Minced tumor prior to mixure with $PBS^{-/-}$ and Matrigel®

9. Record the specimen weight and immediately transfer the specimen to the petri dish lid that was placed on top of gel cold pack in **step 4**.
10. Take scalpel and begin mincing tumor specimen until a consistent homogenate is observed (Fig. 6, *see* **Note 11**).
11. Take tumor homogenate and transfer it to the pre-cooled 1.5 ml Eppendorf tube (*see* **Note 12**).
12. Calculate the total tumor weight and determine the number of prepared eggs that can be engrafted at 25–30 mg of tumor per egg.
13. Add adequate amount of $PBS^{+/+}$ and Matrigel® (1:1 ratio) to the tumor homogenate.

    For every chicken egg CAM to be engrafted, calculate a total of 100 μl volume mixture of tumor homogenate, $PBS^{+/+}$, and Matrigel® (*see* **Note 13**).
14. Cut off approximately 5 mm from a p200 tip and use shortened tip to mix the tumor homogenate, $PBS^{+/+}$, and Matrigel® by pipetting up and down several times while maintaining the tube on the cold block rack. Set pipette down while keeping the shortened tip on the pipette (*see* **Note 14**).
15. Retrieve eggs that are in the incubator and place them in the biosafety cabinet.
16. Peel the tape back enough to expose the opening that was created previously.
17. Repeat the previous step for all the eggs that will be engrafted.

18. Take the 100 μl tumor/Matrigel®/PBS$^{+/+}$ mixture and pipette it directly on top of the CAM and into the center of the silicone ring.
19. Reposition the tape over the opening and ensure it forms a full seal to prevent drying of the CAM.
20. Put the eggs back in the Hova-bator incubator and maintain a constant temperature of 100° F and fill the bottom reservoirs with distilled water.
21. Replenish the water in the incubators every 2–3 days and let the tumor PDXs grow on the CAM for 5–7 days.

3.4 Tumor Harvest

1. Place embryonated chicken eggs at 4 °C for 30 min prior to harvesting tumors to induce hypothermia sedation.
2. Prepare a 60 mm petri dish with 3–4 ml of PBS.
3. Remove tape and use blunt-ended, curved forceps to gently break off small fragments of the shell starting from the opening and working towards the outer edge, ensuring care is taken not to pull or breach the CAM that is adhered to the shell. Break off enough shell to give ample room to cut out the CAM, tumor, and silicone ring (Fig. 7).
4. Use light photography or a stereoscope equipped with a camera to image tumors on the CAM.
5. Using dissection scissors, cut around the silicone ring and include tumor, CAM, and silicone ring while firmly holding them with forceps (Fig. 8a–d, *see* **Note 15**).
6. Place the tumor and CAM on a 60 mm dish with PBS$^{-/-}$ to wash excess blood off the CAM.
7. While in the PBS$^{-/-}$ solution, use forceps to separate the silicone ring from the CAM and tumor (Fig. 9).
8. Transfer the CAM with tumor to a 10% Neutral Buffered Formalin solution for tissue fixation and subsequent histology. Alternatively, tumor can be flash frozen after trimming the CAM for downstream molecular biology assays.
9. After tumor removal, place embryos in a high CO_2 chamber as a primary method of euthanasia and follow this with a secondary method such as decapitation to ensure euthanasia [11]. Place euthanized embryos in freezer for 24 h and discard frozen embryos as biohazardous waste.

4 Notes

1. Always wear laboratory gloves prior to handling eggs to prevent hand oils coming in contact with eggs, which can clog

Fig. 7 Image of egg shell removed prior to dissecting out the tumor on the CAM

porous shell and promote suffocation, or introduce contamination.

2. It is recommended to aliquot Matrigel® matrix stock vial into 1.5 ml tubes for subsequent use prior to beginning with the engraftment procedure.
3. Ensure room lights are turned off when candling eggs to make identification of vascularization, embryo, and air sac easier.
4. Egg candling may not always result in visualization of embryo; however, the air sac and vascularization will always be easily identified.
5. Do not allow the saw blade to go deeper than 3 mm below the eggshell in order to avoid breaching the CAM.
6. As the membrane begins to detach from the CAM, it helps to gently rotate the egg with your non-dominant hand while firmly holding the forceps that are pinching the inner shell membrane. This allows the rotation motion to promote the removal of the inner shell membrane.
7. Remove enough inner shell membrane from the CAM to expose as much vasculature as possible (i.e., a minimum of 24 mm diameter). Some bleeding may be observed after this process, this is completely normal (Fig. 5).
8. Place the ring directly on top of the CAM that contains a well-vascularized area.
9. The 2 h of incubation following the CAM preparation procedure ensure that any embryos that will not survive post-preparation procedure are identified and not used.

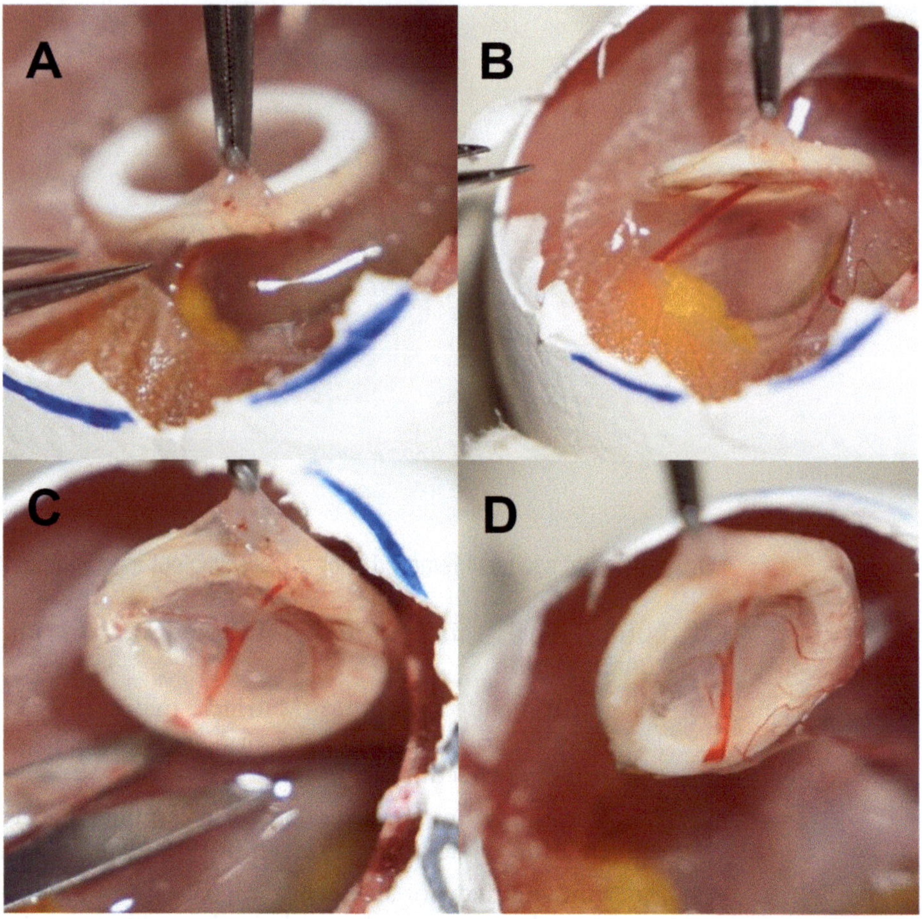

Fig. 8 (**a–d**) CAM and tumor removal from the embryonated egg

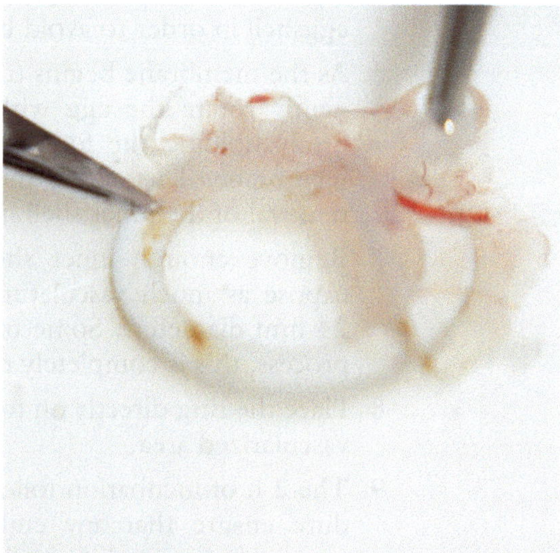

Fig. 9 Separation of CAM and tumor from silicone ring

10. If engrafting freshly obtained specimens, take specimen from storage medium and directly rinse in cold $PBS^{-/-}$.
11. Tumor mincing procedure is considered complete when tumor homogenate lacks larger fragments (i.e., no larger than 1 mm). Additionally, tumor homogenate completion can be checked with fine-tip angle forceps by pinching homogenate and lifting. If homogenate appears as a mucous-like consistency, then the mincing is complete.
12. Transfer tumor homogenate to a cold 1.5 ml Eppendorf tube by using the scalpel blade.
13. When calculating how much $PBS^{+/+}$ and Matrigel® to add to the minced tumor homogenate, make sure you calculate a volume of 100 μl of tumor homogenate mixture per egg (i.e., if a tumor weighs 120 mg, then you would add 200 μl of PBS +/+ and 200 μl of Matrigel® to the tumor homogenate and mix well).
14. Ensure that the cut tip does not touch any surface in the biosafety cabinet while you set it aside in order to keep it sterile.
15. Take care not to depress the forceps while excising the CAM to prevent loss of tumor into the albumin.

Acknowledgments

This work is supported by the Patient-Derived Xenograft and Advanced in vivo Models Core Facility at Baylor College of Medicine with funding from NIH grant 2P30CA125123-14 and Cancer Prevention and Research Institute of Texas (CPRIT) grant #170691.

References

1. Murphy JB (1913) Transplantability of tissues to the embryo of foreign species : its bearing on questions of tissue specificity and tumor immunity. J Exp Med 17:482–493. https://doi.org/10.1084/jem.17.4.482
2. Wilting J, Weich HA, Christ B (1993) Effects of vascular endothelial growth factor and basic fibroblast growth factor: application with corneal grafts on the chorioallantoic membrane. Acta Anat (Basel) 147:207–215. https://doi.org/10.1159/000147506
3. Lucarelli E, Sangiorgi L, Benassi S, Donati D, Gobbi GA, Picci P, Vacca A, Ribatti D (1999) Angiogenesis in lipoma: an experimental study in the chick embryo chorioallantoic membrane. Int J Mol Med 4:593–596. https://doi.org/10.3892/ijmm.4.6.593
4. Ribatti D (2014) The chick embryo chorioallantoic membrane as a model for tumor biology. Exp Cell Res 328:314–324. https://doi.org/10.1016/j.yexcr.2014.06.010
5. Deryugina EI, Quigley JP (2008) Chick embryo chorioallantoic membrane model systems to study and visualize human tumor cell metastasis. Histochem Cell Biol 130:1119–1130. https://doi.org/10.1007/s00418-008-0536-2
6. Vantaku V, Dong J, Ambati CR, Perera D, Donepudi SR, Amara CS, Putluri V, Ravi SS, Robertson MJ, Piyarathna DWB, Villanueva M, von Rundstedt FC, Karanam B, Ballester LY, Terris MK, Bollag RJ, Lerner SP, Apolo AB, Villanueva H, Lee M, Sikora AG, Lotan Y, Sreekumar A, Coarfa C, Putluri N

(2019) Multi-omics integration analysis robustly predicts high-grade patient survival and identifies CPT1B effect on fatty acid metabolism in bladder cancer. Clin Cancer Res 25(12):3689–3701. https://doi.org/10.1158/1078-0432.CCR-18-1515

7. Vantaku V, Putluri V, Bader DA, Maity S, Ma J, Arnold JM, Rajapakshe K, Donepudi SR, von Rundstedt FC, Devarakonda V, Dubrulle J, Karanam B, McGuire SE, Stossi F, Jain AK, Coarfa C, Cao Q, Sikora AG, Villanueva H, Kavuri SM, Lotan Y, Sreekumar A, Putluri N (2020) Epigenetic loss of AOX1 expression via EZH2 leads to metabolic deregulations and promotes bladder cancer progression. Oncogene 39(40):6265–6285. https://doi.org/10.1038/s41388-019-0902-7

8. Arnold JM, Gu F, Ambati CR, Rasaily U, Ramirez-Pena E, Joseph R, Manikkam M, San Martin R, Charles C, Pan Y, Chatterjee SS, Den Hollander P, Zhang W, Nagi C, Sikora AG, Rowley D, Putluri N, Zhang XH, Karanam B, Mani SA, Sreekumar A (2020) UDP-glucose 6-dehydrogenase regulates hyaluronic acid production and promotes breast cancer progression. Oncogene 39(15):3089–3101. https://doi.org/10.1038/s41388-019-0885-4

9. San Martin R, Pathak R, Jain A, Jung SY, Hilsenbeck SG, Piña-Barba MC, Sikora AG, Pienta KJ, Rowley DR (2017) Tenascin-C and integrin α9 mediate interactions of prostate cancer with the bone microenvironment. Cancer Res 77(21):5977–5988. https://doi.org/10.1158/0008-5472.CAN-17-0064

10. DeBord LC, Pathak RR, Villaneuva M, Liu HC, Harrington DA, Yu W, Lewis MT, Sikora AG (2018) The chick chorioallantoic membrane (CAM) as a versatile patient-derived xenograft (PDX) platform for precision medicine and preclinical research. Am J Cancer Res 8(8):1642–1660. eCollection 2018

11. Leary SU, Anthony W, Cartner R, Grandin S, Greenacre T, Gwaltney-Brant TC, McCrackin S, Meyer MA, Miller R, Shearer D, Turner J, Yanong R (2020) AVMA guidelines for the euthanasia of animals: 2020 edition. American Veterinary Medical Association, Schaumburg

Chapter 12

Intraductal Injections into the Mouse Mammary Gland

Erik Oliemuller, Richard Newman, and Beatrice A. Howard

Abstract

The mammary intraductal xenografting technique has been established to inject cells or other substances directly into the mammary ducts of female mice. Using this refined xenografting method provides the possibility of mimicking the normal microenvironment of preinvasive breast lesions including, ductal carcinoma in situ (DCIS), to study of the progression of DCIS to invasive breast cancer in a more relevant manner than with other mammary xenografting methods. Xenografting into the mammary fat pad delivers cells directly into the stroma and bypasses the occurrence of invasive transition, during which cells invade through the basement membrane. Either breast cancer cell lines or patient-derived breast cancer cells can be injected into the mammary duct using this protocol to model breast cancer progression. This protocol will cover the procedures required to perform this technique.

Key words Intraductal mammary injection, Xenograft, Preclinical model

1 Introduction

Despite significant progress in breast cancer research and clinical advances, breast cancer remains the most commonly diagnosed cancer (one in eight women will develop this disease during their lifetime and men can also develop breast cancer), which highlights the unmet clinical need for improved prevention strategies, early detection, and more efficient and specific treatments that are needed to accelerate progress and see more people surviving the disease. Mouse models of breast cancer are invaluable tools for use in preclinical studies. Xenograft models are widely used for preclinical studies of breast cancer but have limitations since it has been well documented that the mammary fat pad provides a microenvironment that will strongly influence the phenotypes of both normal mammary epithelial cells and breast cancer cells [1–4]. It is now evident that intraductal injections of mammary epithelial cells directly into the mouse mammary duct is superior to conventional subcutaneous and mammary fat pad xenografting when modeling breast cancer progression, since this technique grafts cells into a

Maria dM. Vivanco (ed.), *Mammary Stem Cells: Methods and Protocols*,
Methods in Molecular Biology, vol. 2471, https://doi.org/10.1007/978-1-0716-2193-6_12,

location with a microenvironment that is more similar to the sites that breast cancers arise in the human breast. Intraductal grafting of cells permits a progression of cell growth within the duct, followed by microinvasion, invasion, and systemic dissemination which is what occurs during breast cancer progression in patients [5].

Ductal carcinoma in situ (DCIS) are "pre-cancerous lesions" that form within the ductal component of the breast lobule. Some DCIS will progress to invasive breast cancer and this issue has considerable clinical consequences as diagnoses of DCIS have increased as a consequence of population-based mammographic screening. DCIS is usually treated similar to invasive breast cancer with surgery, radiation, and hormone-blocking medicines. While early cancer detection may offer a survival benefit, increasing numbers of lesions are also being identified that will never develop into lethal disease. To be able to accurately distinguish indolent from aggressive DCIS (determine when DCIS will not become a cancer) in order to reduce overtreatment and life-long side effects remains a major scientific challenge [6]. A number of researchers now routinely use intraductal xenografting to model both DCIS and breast cancer progression using cell lines and PDXs [7–11]. Here, we present a protocol to inject DCIS.com cells into the mouse mammary gland. Other cells including those derived from primary patient samples can be engrafted using the same protocol.

2 Materials

1. ISO Class 3 airflow workstation, laminar air flow cabinet, or other suitable location to perform surgery.
2. Stereoscope (Zeiss Stemi 2000 CS and LED dual lights, GT vision).
3. Light source.
4. Hamilton syringe with 30- or 33-gauge fixed needle (Hamilton Cat. No. 80508, 705 SN SYR).
5. Dumont fine forceps (Fine Science Tools Cat. No. 12200-14).
6. Vannas spring scissors straight (Fine Science Tools Cat. No. 15000-00).
7. 0.2% Evan's blue dye in phosphate buffered saline (PBS).
8. Microcapillary pipette (Sigma Lot 3110) fitted with pre-pulled glass microcapillary (70 μM) (FIVEphoton Biochemicals, Lot: MF-10505003).
9. Graefe angled serrated forceps (Fine Science Tools Cat. No. 11049-10).
10. Non-slip rubber mat (Vet Tech Cat. No. FM009L).
11. Surgical tape.

12. Cotton buds.
13. Veet Hair Removal Cream.
14. 250 mL beaker to autoclave cotton buds.
15. Medi-wipe tissue.
16. Warm ultra-filtrated water, sterile (warm up water in kettle or microwave).
17. 0.3 mL Insulin syringes for analgesics (BD Micro Fine+0.3 mL 30 gauge, Medisave Cat. No. ND890).
18. Cells for inoculation.
19. Metacam at 0.5 mg/mL (or suitable alternative analgesic).
20. Vetergesic at 0.3 mg/mL (buprenorphine hydrochloride or suitable alternative analgesic).
21. PBS.
22. 70% Ethanol.
23. Ultra-filtrated water, sterile to dilute Metacam.
24. Eye lubricant (Xailin night).
25. Isoflurane for anesthesia (*see* **Note 1**).
26. Anesthesia rig.
27. Luciferin dissolved in sterile water (15 mg/mL).
28. IVIS system (or other suitable in vivo imaging system).

3 Methods

3.1 Preparation of Mice/Removal of Fur Surrounding Nipples (See Note 2)

One or 2 days prior to the procedure apply an over-the-counter hair removal cream to the nipple area of the female mouse (Figs. 1 and 2). Wait 2–3 min and gently remove loose hair with a cotton-tip applicator using a circular motion. Remove all the cream (as failure to do so will cause skin irritation) using damp paper towels wetted with warm ultra-purified autoclaved water. Shaving is not recommended due to the risk of damaging the nipples but is also possible, if preferred. E45 dermatological cream may be used to aid recovery from skin irritation and redness that may occur after depilation.

3.2 Preparation of Materials and Equipment

1. Setup equipment: Prepare stereoscope and set appropriate focus (recommended 0.65–1.00 viewing) and lighting (Fig. 3).
2. Fill microcapillary pipette with Evan's blue dye.

3.3 Preparation of Mice (Pre-surgical)

On the day of the study:

1. Record the body weight of each mouse. As with all preclinical studies, animal weights should be monitored regularly (twice a

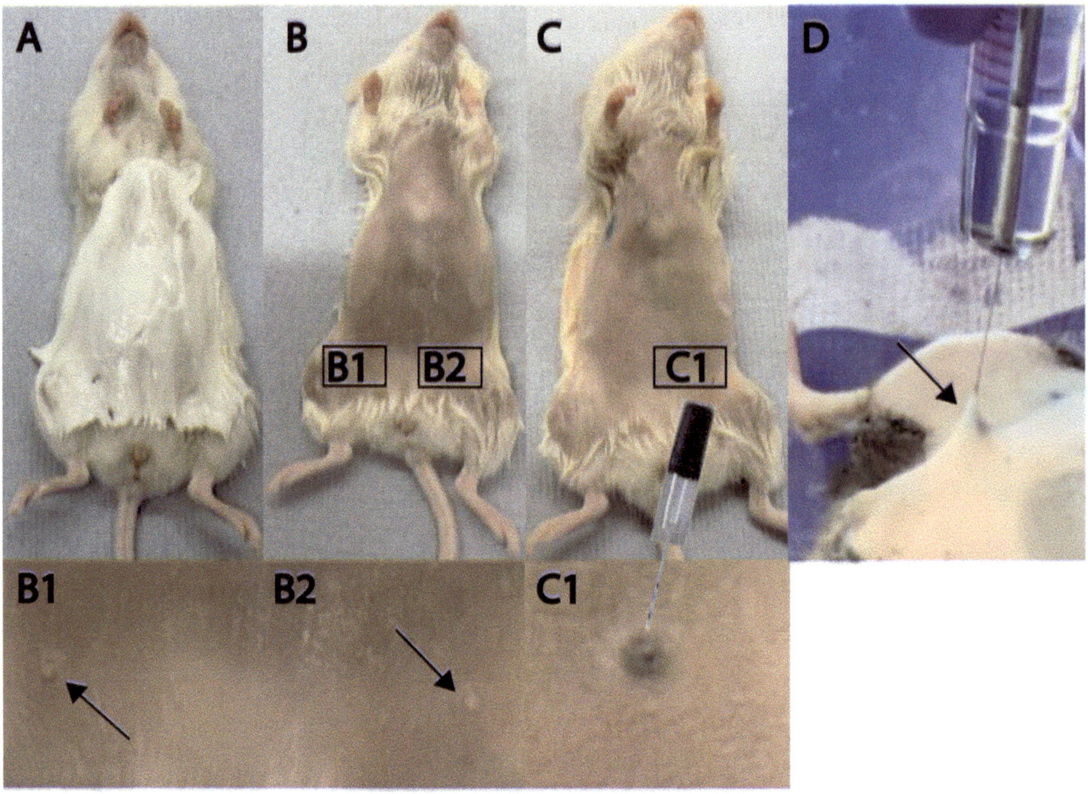

Fig. 1 Intraductal mammary injection. (**a**) Mouse with hair removal cream over ventral surface over nipples. (**b**) Mouse after hair removal exposing the nipples. Region with nipples of fourth mammary glands are marked with squares. (b1, b2) Close-up magnification of left and right nipples of fourth mammary glands shown in (b). (**c**) Left fourth mammary nipple is marked with square (c1) after being cut under the microscope and stained with Evan's Blue using a pre-pulled glass microcapillary. (c1) Close-up magnification of area marked with square in (c) displaying the ductal opening of nipple indicated by the darker blue area that is surrounded by lighter blue area. (**d**) Hamilton syringe needle in the ductal opening of nipple during intraductal injection

week or greater) as part of protocol to assess animal health and welfare.

2. Inject Metacam (200 μL for 20 g mouse) subcutaneously prior to the procedure as analgesia.
3. Anesthetize the mouse using an isoflurane chamber and apply eye lubricant. During the procedure, the mouse will continue to be anesthetized using inhaled 2–4% concentration of isoflurane in oxygen via a nose cone. Carefully monitor the mouse for changes in respiratory rate, adjusting the level of isoflurane accordingly.
4. Secure the mouse under the stereoscope by gently taping down the extremities.
5. Clean injection sites with alcohol swabs.

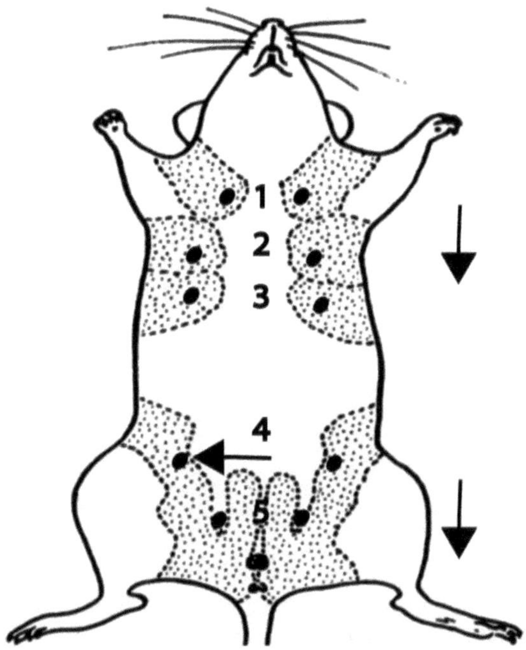

Fig. 2 Diagram of ventral surface of female mouse. Nipples are shown as filled black circles on left and right ventral surface of mouse labeled from 1 to 5 with the mammary fat pad location marked by dotted lines filled with dots. Arrows indicate needle orientation and angle for each nipple. The direction of needle entry for nipples 1, 2, 3, and 5 are directly down, while for nipple 4 the needle is oriented to enter away from the mouse midline at a 45° angle

3.4 Intraductal Injection

1. Load 5–10 μL of cells for injection into a 50 μL syringe with a 30*–33* gauge metal hub needle affixed (*see* **Note 3**).
2. Locate appropriate nipple to be injected under the stereoscope (Figs. 1 and 2).
3. Use fine forceps to remove any dead skin that covers the nipple opening.
4. Using Vannas spring scissors gently snip the top off the nipple taking care not to remove too much or the nipple will slide under the skin making the procedure more challenging. This will expose the mammary duct (*see* **Note 4**).
5. Using Graefe forceps, gently pull the skin back and up to make the skin taut around the nipple area and then, gently insert the tip of a pre-pulled glass microcapillary filled with blue dye into the mammary duct (Fig. 3).
6. Once tip with dye is in the duct, dispense a little of the blue dye. This will act as a guide for the Hamilton syringe as it should clearly outline the mammary duct.

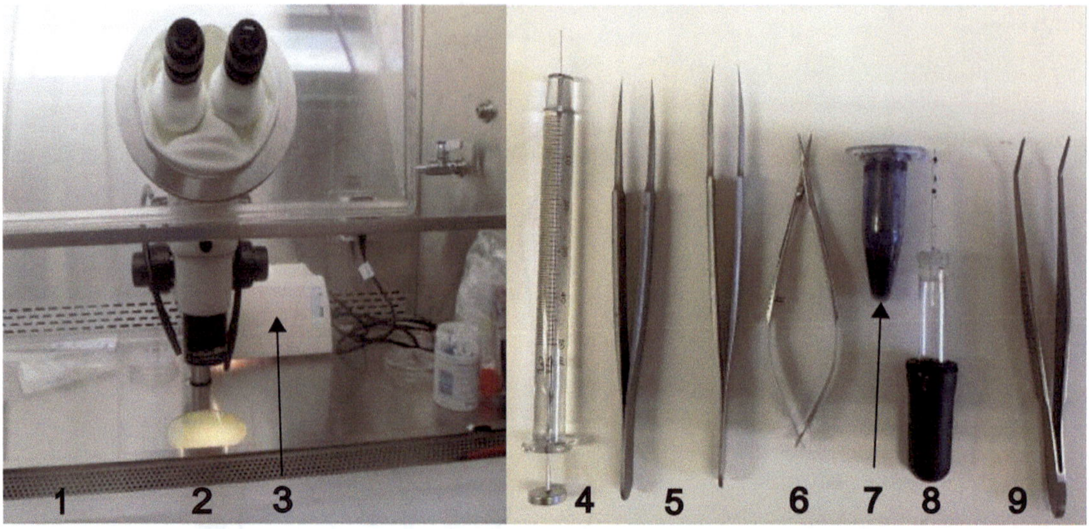

Fig. 3 Equipment used for MIND procedure. (1) Phantom cabinet. (2) Stereomicroscope (e.g., Zeiss Stemi 2000 CS). (3) Light source. (4) Hamilton syringe with 32-gauge fixed needle. (5) Dumont fine forceps. (6) Vannas spring scissors, straight. (7) Eppendorf tube containing 0.2% Methylene blue dye in PBS. (8) Microcapillary pipette with pre-pulled micro glass capillary inserted. (9) Graefe angled serrated forceps

7. Once the duct is defined with blue dye, it is now time to inject the cells into the duct using a 30- to 33-gauge Hamilton syringe (*see* **Note 5**).
8. Using Graefe forceps, gently pull the skin back and up to make the skin taut around the nipple area and then, using the fixed needle Hamilton syringe containing the cells, carefully insert this into the mammary duct. The angle and direction the needle enters the mammary duct is dependent on which mammary duct is being injected (Fig. 1). Once in the duct, release the skin and very gently pull back the needle from the nipple. Suction from the nipple should keep the needle within the duct. Inject the solution slowly to minimize potential damage caused by rapidly moving fluid within the ductal lumens. The injection rate should be maintained at approximately 40 μL per minute. Once completed keep the needle in the duct for a few seconds before withdrawing the needle to prevent leakage.
9. Observe the injection site. There should be no signs of trauma to the nipple region or surrounding tissue. A successful intraductal injection is shown in Fig. 4a–d. Swelling in the area surrounding the nipple is likely to indicate a mammary fat pad injection (Fig. 4e, f). Make an experimental record (*see* **Note 6**).

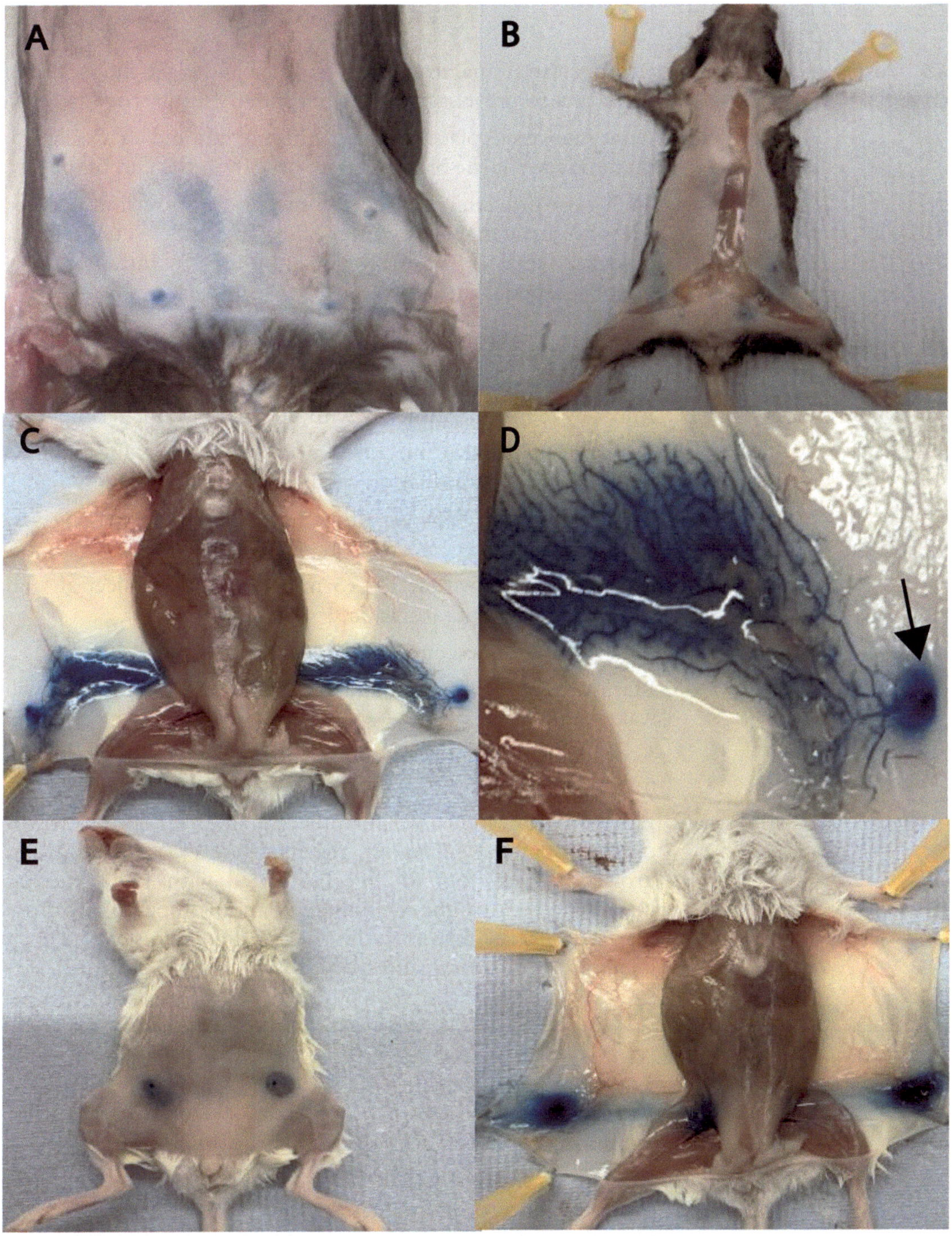

Fig. 4 Intraductal mammary injection dissection. (**a**) Ventral surface of mouse after intraductal mammary injection with 0.2% Evan's blue dye into left and right fourth and fifth mammary glands. (**b**) An inverted Y section is performed through the skin. Needles are used to pin out the mouse skin to expose the fourth and fifth mammary glands. (**c**) Skin is deflected from the abdominal wall to expose the left and right fourth mammary gland which is filled with Evan's Blue dye after a successful intraductal injection. (**d**) Close-up of fourth mammary gland injected intraductally with Evan's Blue dye highlighting the ductal tree. Arrow indicates nipple. (**e**) Mammary fat pad injection after failed left and right fourth mammary MIND injections. Note the dark blue dye concentrated around the nipple region. Swelling will also be evident around the nipple area. Note the difference between (**a**) and (**e**). (**f**) Left and right fourth mammary gland after unsuccessful intraductal injections of Evan's Blue dye

3.5 Postoperative Care

1. Remove the animal from the nose cone and place the mouse in a hot box to prevent hypothermia and assist in recovery.
2. Once recovered, place back in the home cage and check all mice that underwent the procedure are healthy at the end of the day.
3. 24 h post-procedure, check the mouse for any clinical signs, and as analgesia, administer 0.3 mg/mL Vetergesic at 0.05 mg/kg dose (65 μL for 20 g mouse) by intraperitoneal or subcutaneous injection.

3.6 Monitoring Cell or Tumor Growth

1. In vivo imaging should be performed to monitor cell or tumor growth throughout duration of experiment (*see* **Note** 7).
2. Inject mice intraperitoneally with 100–200 μL luciferin.
3. 7 min after luciferin injection, mouse should be anesthetized with isoflurane in induction chamber for 5 min.
4. Transfer mice to IVIS (or other suitable in vivo imaging system) chamber to receive isoflurane via nose cone.
5. Start measuring within 14–15 min of luciferin injection and avoid measuring at saturation.
6. Record bioluminescence in mammary glands.
7. Remove mice from IVIS and return to cage where they will wake and recover quickly.
8. Repeat at regular intervals and quantify region of interest in the mammary glands.
9. Plot bioluminescence per mammary gland to monitor growth over time. Figure 5 shows results from IVIS imaging and growth curves from MCF cells tagged with Luc2 injected intraductally into the mammary gland compared to those injected into the mammary fat pad. IVIS imaging results from luciferase-positive explanted organs are shown.

4 Notes

1. Use of injectable anesthesia (such as ketamine/xylazine or hypnorm/hypnovel) is also possible.
2. Practice on cadavers should be undertaken prior to doing as an experimental procedure. When training, to confirm a MIND injection, cut and deflect the skin from around the nipples to expose the mammary fat pads. For observing mammary fat pads 4 and 5, an inverted Y dissection would be performed (Fig. 4b). The ductal tree should be blue (Fig. 4c, d). We assess technical performance and make sure that >90% of injections are successful on cadavers prior to permitting a researcher to perform as a surgical procedure.

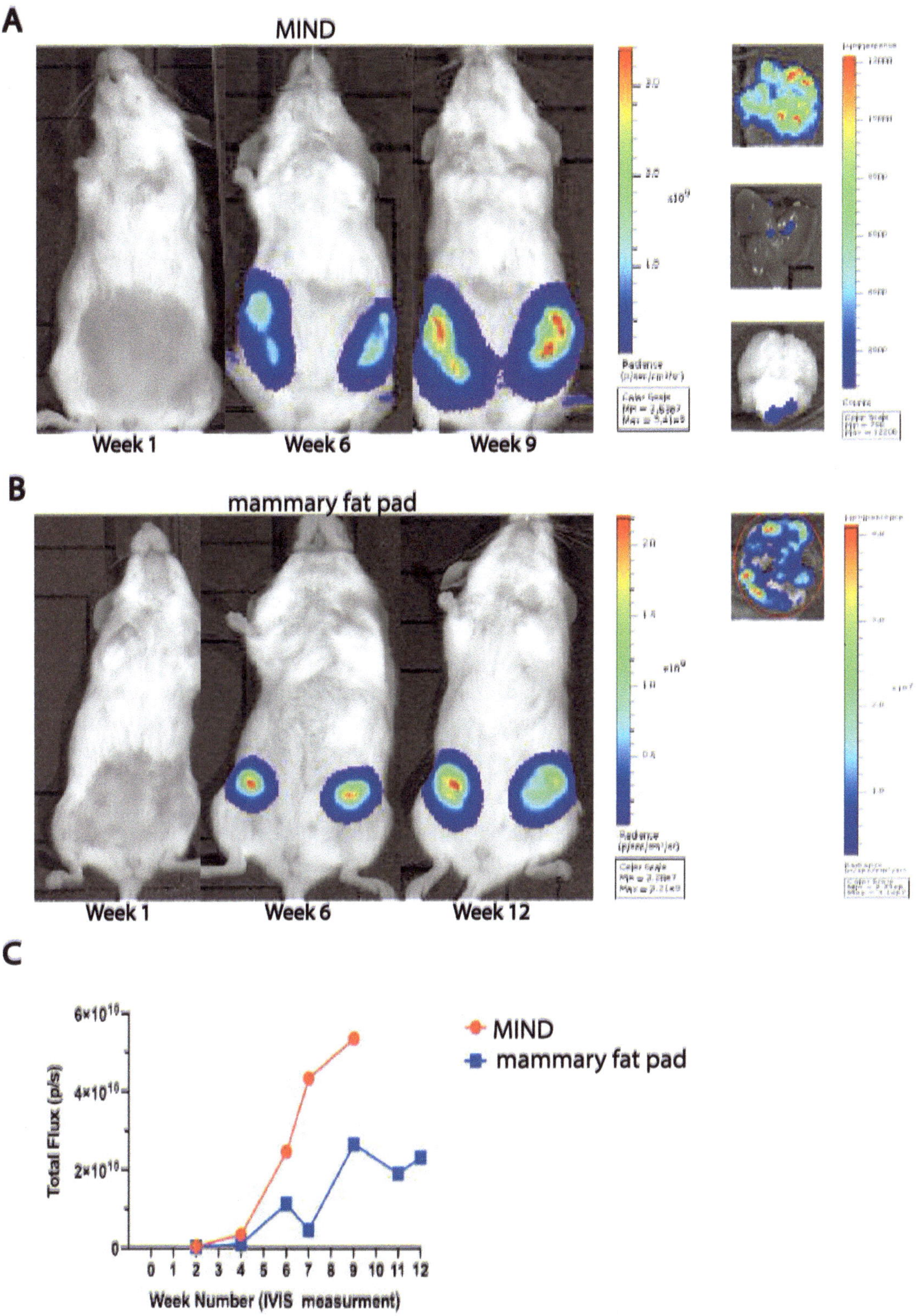

Fig. 5 IVIS imaging. (**a**) IVIS images after MIND injection of MCF7-luc2 cell line. IVIS images represent mouse at start, mid-way, and termination of experiment with organ IVIS imaging, removed directly after whole body

3. Mammary ducts 1–5 can be injected with a maximum volume of 10 μL. After injection, small amounts can leak out of the nipple upon pulling out the needle. Therefore, plan to inject 1 μL extra to account for potential loss.
4. Cutting the nipple is not strictly required and is dependent on the mouse strain. Removing the keratin plug may also increase the nipple size, sometimes substantially. If you cut the nipple, execute this procedure with care and cut straight across, not at an angle, as if you cut too much you risk losing non-surgical access to the duct and decrease the likelihood of performing a successful intraductal injection without surgically cutting a flap.
5. Accessing the mammary duct and delivery of substance such as cells, virus, or dye into mammary duct. This can be challenging, and each mouse/strain may present a different challenge. The needle gauge used will depend on the size of the nipples; for example, in our experience, C57/BL6 mice have small nipples and a 33-gauge needle would be used while the NSG mice generally have larger nipples and therefore a 30-gauge needle would usually suffice. Make sure the mammary duct opening is marked with the blue dye prior to needle insertion, if necessary, reapply the blue dye as this will significantly increase the chances for a successful injection.
6. Usually the experimental plan will dictate which nipples to inject. However, if the mammary duct injection was unsuccessful document this and, if possible, move on to another "replacement" nipple and again document this so a protocol amendment/deviation can be written and added to the implemented protocol after the surgery. If the substance is not entering the mammary duct, check the direction of injection needle (Fig. 1). Even if this is correct, it may be beneficial to slightly change the direction the needle is entering the duct. Another cause of a failed intraductal injection is being too heavy-handed when dispensing the plunger. The recommended flow rate is approximately 40 μL per minute. A good indicator you have a successful ductal entry in live mice is to lightly pull back the syringe and the nipple should pull back with the needle. Small or inverted nipples can also be a problem; ways around this issue is to order more mice than you need (but you must have a valid use for the spare mice prior to ordering) and removing the fur on all mice and pre-select mice based on nipple size prior to surgery. Another option is

Fig. 5 (continued) imaging shown at right. (**b**) IVIS images after mammary fat pad injection of MCF7-luc2 cell line. IVIS images represent mouse at start, mid-way through and termination of experiment with organ IVIS imaging, removed directly after whole body imaging shown at right. (**c**) Graph showing bioluminescence as total flux from MIND and mammary fat pad experiments. Week represents time post cell implantation

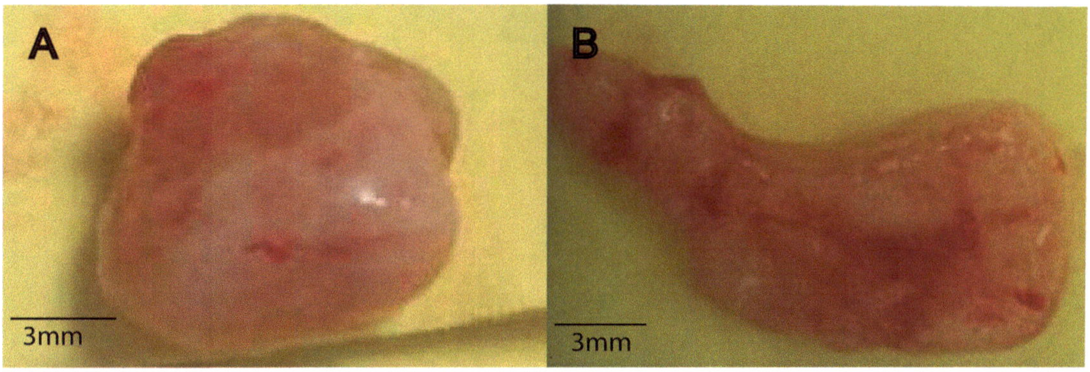

Fig. 6 Mammary tumor after MIND and mammary fat pad injections. (**a**) Tumor 3 months after mammary fat pad injection into the fourth mammary gland. Note the round shape compared with the tumor formed after MIND injection. (**b**) Tumor 3 months after MIND injection. Note the shape remains similar to the mammary fat pad with tumor growth evident throughout ductal system

to start with mice with bigger nipples and end with more challenging ones when you are steadier. It is helpful to be adaptable with a variety of different techniques to hand, which can be applied as and when necessary to gain a successful injection.

7. With the MIND method, accurate tumor measurements using the traditional caliper are difficult due to the dispersed character of the MIND tumor growth throughout the entire mammary ductal tree (Figs. 6 and 7). Unlike mammary fat pad tumors which are superficial and grow outwards and can be felt through gentle palpation of the tumor and easily measured, in contrast, the MIND tumor grows and fills the entire ductal tree, making it difficult to accurately measure with calipers. This can make it difficult to allocate animals to experimental groups based on tumor size. To overcome this issue, tumor cells should be luciferin and/or mCherry tagged, so tumor size can be accurately recorded throughout the experiment duration using IVIS imaging. IVIS imaging is performed at regular intervals to monitor cell growth and frequency will depend on the growth kinetics of each cell line but weekly or twice-monthly imaging is typically used in our own experiments.

Acknowledgments

We thank Breast Cancer Now for funding this work as part of Program Funding to the Breast Cancer Now Toby Robins Research Centre.

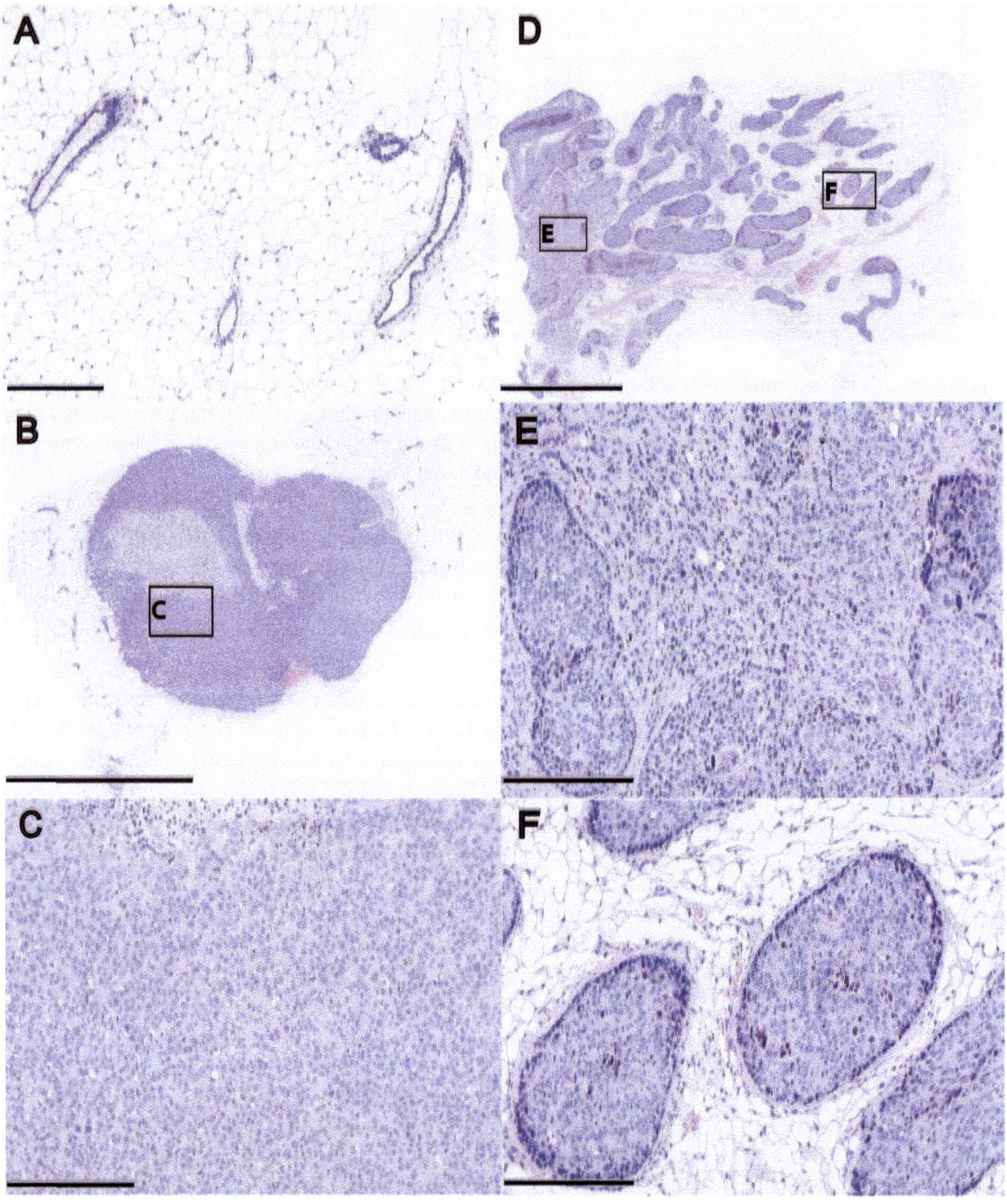

Fig. 7 Mammary tumor histology. (**a–c**) H&E staining of normal mammary ducts (**a**) and tumor (**b**, **c**) after mammary fat pad injection into fourth mammary gland. (**a**, **c**) magnification, ×10, scale bar 250 μm; (**b**) magnification, ×0.75, scale bar, 2.5 mm. (**d–f**) H&E staining after MIND injection into mammary gland 4 (**d**). Fully invasive tumor cells outside ducts (**e**) and tumor cells within ducts (**f**) are observed. (**d**) magnification 0.97, scale bar 2.5 mm; (**e**, **f**) magnification ×10, scale bar 250 μm

References

1. Sflomos G, Dormoy V, Metsalu T, Jeitziner R, Battista L, Scabia V, Raffoul W, Delaloye JF, Treboux A, Fiche M, Vilo J, Ayyanan A, Brisken C (2016) A preclinical model for ERalpha-positive breast cancer points to the epithelial microenvironment as determinant of luminal phenotype and hormone response. Cancer Cell 29(3):407–422. https://doi.org/10.1016/j.ccell.2016.02.002
2. Bussard KM, Boulanger CA, Booth BW, Bruno RD, Smith GH (2010) Reprogramming human cancer cells in the mouse mammary gland. Cancer Res 70(15):6336–6343. https://doi.org/10.1158/0008-5472.CAN-10-0591
3. Boulanger CA, Smith GH (2009) Reprogramming cell fates in the mammary microenvironment. Cell Cycle 8(8):1127–1132. https://doi.org/10.4161/cc.8.8.8189
4. Visvader JE, Stingl J (2014) Mammary stem cells and the differentiation hierarchy: current status and perspectives. Genes Dev 28(11):1143–1158. https://doi.org/10.1101/gad.242511.114
5. Simpson PT, Reis-Filho JS, Gale T, Lakhani SR (2005) Molecular evolution of breast cancer. J Pathol 205(2):248–254. https://doi.org/10.1002/path.1691
6. Groen EJ, Elshof LE, Visser LL, Rutgers EJT, Winter-Warnars HAO, Lips EH, Wesseling J (2017) Finding the balance between over- and under-treatment of ductal carcinoma in situ (DCIS). Breast 31:274–283. https://doi.org/10.1016/j.breast.2016.09.001
7. Oliemuller E, Kogata N, Bland P, Kriplani D, Daley F, Haider S, Shah V, Sawyer EJ, Howard BA (2017) SOX11 promotes invasive growth and ductal carcinoma in situ progression. J Pathol 243(2):193–207. https://doi.org/10.1002/path.4939
8. Fiche M, Scabia V, Aouad P, Battista L, Treboux A, Stravodimou A, Zaman K, Rls DV, Ayyanan A, Sflomos G, Brisken C (2019) Intraductal patient-derived xenografts of estrogen receptor alpha-positive breast cancer recapitulate the histopathological spectrum and metastatic potential of human lesions. J Pathol 247(3):287–292. https://doi.org/10.1002/path.5200
9. Behbod F, Kittrell FS, LaMarca H, Edwards D, Kerbawy S, Heestand JC, Young E, Mukhopadhyay P, Yeh HW, Allred DC, Hu M, Polyak K, Rosen JM, Medina D (2009) An intraductal human-in-mouse transplantation model mimics the subtypes of ductal carcinoma in situ. Breast Cancer Res 11(5):R66. https://doi.org/10.1186/bcr2358
10. Valdez KE, Fan F, Smith W, Allred DC, Medina D, Behbod F (2011) Human primary ductal carcinoma in situ (DCIS) subtype-specific pathology is preserved in a mouse intraductal (MIND) xenograft model. J Pathol 225(4):565–573. https://doi.org/10.1002/path.2969
11. Elsarraj HS, Hong Y, Limback D, Zhao R, Berger J, Bishop SC, Sabbagh A, Oppenheimer L, Harper HE, Tsimelzon A, Huang S, Hilsenbeck SG, Edwards DP, Fontes J, Fan F, Madan R, Fangman B, Ellis A, Tawfik O, Persons DL, Fields T, Godwin AK, Hagan CR, Swenson-Fields K, Coarfa C, Thompson J, Behbod F (2020) BCL9/STAT3 regulation of transcriptional enhancer networks promote DCIS progression. NPJ Breast Cancer 6:12. https://doi.org/10.1038/s41523-020-0157-z

Chapter 13

Modeling Breast Cancer in Organoid and Intraductal Models

Richard Iggo

Abstract

We present protocols to create estrogen receptor positive (ER+) and androgen receptor positive (AR+) breast cancer models by combining organoid culture with mammary intraductal injection.

Key words Lentivirus, Transfection, Infection, Intraductal injection

1 Introduction

1.1 Hormone Receptor Positive Breast Cancer Models

We describe the creation of breast cancer models by transducing tumor cells and normal mammary epithelial cells with lentiviral vectors in organoid culture and injecting them into the mammary ducts of mice. Previous approaches failed to generate ER or AR positive tumors because the cells did not receive the correct micro-environmental signals. The approach presented here generates models that express hormone receptors and FOXA1, the hallmark of mammary hormone sensing cells. In vitro culture in Clevers 3D culture system [1] preserves hormone sensing cell identity, as shown by FOXA1 expression, but injection of the cells into the mammary ducts ("MIND" injection [2]) is required to see robust ER expression. Both Clevers culture and MIND injection are necessary to preserve hormone sensing cell identity: a single round of in vitro culture on plastic can abolish FOXA1 expression, and injection into the mammary stroma promotes squamous metaplasia or epithelial-to-mesenchymal transition (EMT, [3]).

1.2 Culture Systems for Breast Tumors and Normal Mammary Epithelial Cells

The idea behind Clevers culture is to mimic the natural microenvironment by activating the signal transduction pathways that should be active and blocking the rest. This means it contains a daunting mixture of growth factors and inhibitors [4, 5]. It is a step up from fetal bovine serum (FBS) that contains a random assortment of growth factors released by platelets, whose real function is to

Maria dM. Vivanco (ed.), *Mammary Stem Cells: Methods and Protocols*,
Methods in Molecular Biology, vol. 2471, https://doi.org/10.1007/978-1-0716-2193-6_13,

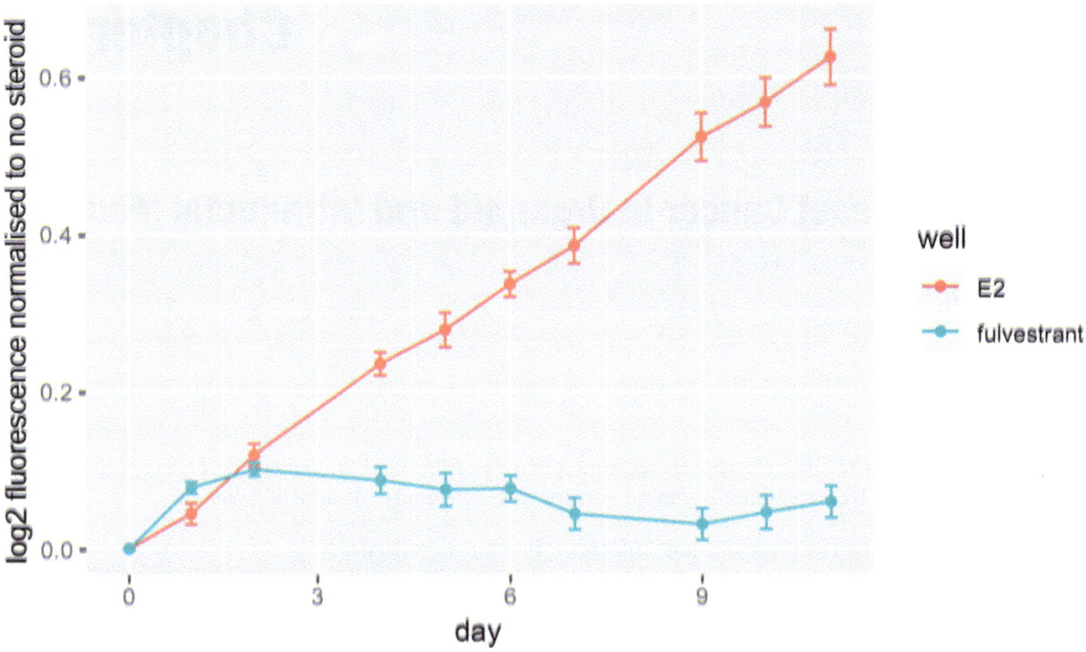

Fig. 1 LA-PDX1 growth curve in organoid culture. A lobular ER+ PDX [6] was put into organoid culture and treated with 1 nM E2 or 10 nM fulvestrant. The cells were infected with a lentiviral vector expressing tdTomato before the first graft, passaged five times in vivo by intraductal injection and four times as organoids before plating for the growth curve. Fluorescence of the entire Matrigel dot was captured with a 2× objective on a standard tissue culture microscope (IX71) (*See* **Note 1**)

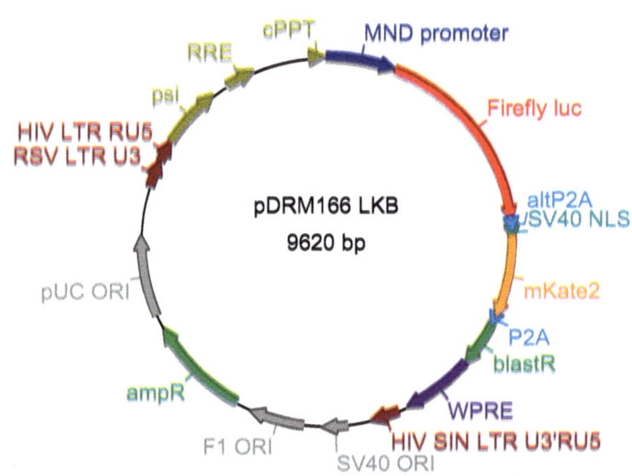

Fig. 2 LKB (pDRM166) is a lentiviral vector expressing firefly luciferase, mKate2 and the blasticidin resistance gene separated by 2A sites. Luciferase is used for bioluminescence imaging. mKate2 is useful when harvesting tissue from mice and makes it easy to quantify growth of organoid cultures. The blasticidin resistance gene (*bsd*) makes it easy to eliminate murine cells from organoid cultures by adding blasticidin

stimulate wound healing by fibroblasts. Cells in normal mammary ducts exchange paracrine factors in a cascade triggered by stimulation of hormone sensing cells with estrogen. The growth factors added to Clevers medium should render most of this regulation superfluous, potentially masking the effect of exogenous steroids. That said, some models do show strong dependence on estradiol, for example a luminal A PDX we established from a lobular tumor (LA-PDX1, [6]). This can be tested by simply adding or removing estrogen from the medium (Fig. 1)—there is no need to use stripped serum because there is no serum. The cells are grown in dots (or domes) of Matrigel, a floppy matrix that elicits more physiological behavior than plastic, but falls short of reproducing the microenvironment in human tissue. It is possible to generate bilayer ducts from a single stem cell grown in very simple medium provided it is embedded in a floating collagen gel, presumably because the cells synthesize and exchange the growth factors they want to see. For this to work the myoepithelial cells must pull on the collagen and feel just the right resistance [7]. This points to Matrigel being a suboptimal matrix, but incorporating patient-derived tumor cells into ducts formed in floating collagen gels would be a tall order for even the most gifted student.

The urgency scientists feel is not necessarily shared by tumor cells. A lobular tumor that doubles once every 100 days in vivo is probably growing at its normal rate, no matter how inconvenient that may be for scientists. When it doubles every 10 days in organoid culture it is making a major concession. Forcing it to grow even faster is probably not wise. The risk is that the cells will revert to their ancestral fate, which is to be skin. Pathologists call this squamous metaplasia, and they lose all interest when they see it in breast cancer models. It was the fate of almost all traditional human mammary epithelial cell (HMEC) cultures [8]. It is a common artifact when too much growth factor is added (for example, too much EGF or Wnt [9, 10]). Some of the supplements in Clevers medium are specifically included to prevent it (in particular, SB202190 to inhibit p38 kinase). Anything that promotes stem cell identity or epithelial-to-mesenchymal transition will probably increase the risk of squamous metaplasia and decrease the chance of luminal cells maintaining their sensor cell identity. The Ronnov-Jessen group showed that TGFBR inhibitors preserve ER expression in their culture system [11], and the Brisken lab showed that TGFβ in the mammary stroma forces grafted cells to undergo EMT and lose ER expression [3]. Clevers medium contains A83-01 to inhibit TGFBR but ER expression is still lower than in the original tumors or after intraductal injection, so this problem is not completely solved. Rho kinase (ROCK) inhibitor (Y-27632) is used to promote survival when cells are passaged but it is also commonly included in stem cell media. Together with irradiated feeders, it forms the basis of a culture system in which the cells are reprogrammed to a more proliferative state [12]. An attractive feature of Clevers culture is that the Matrigel dots are plated in 24 well plates, usually one dot per well. Since you can replace the medium without disturbing the cells, it is easy to test the requirement for different growth factors. The critical growth factor that maintains luminal identity is probably neuregulin [10, 13]. Many groups are experimenting with different medium formulations, for example, to adapt the signals to the expectations of individual tumors [14] and to reduce the cost (*see* **Note 2**). It is well worth reading some recent papers from these groups before embarking on your own experiments (for example, [5, 10–12, 14]).

Lentiviral infection of mammary epithelial cells in Clevers culture is straightforward. Cells are more infectable immediately after digestion of human tissue than after in vitro culture. Hence, we recommend inserting an imaging vector like LKB (Fig. 2) as soon as you receive the tissue, and definitely before injecting the cells into the mammary ducts. Making genetically defined models from normal cells is more complicated because single oncogenes often trigger oncogene-induced apoptosis or senescence, so it makes sense to do simultaneous infections with complementing oncogenes in the first round of infection. We commonly insert four vectors, starting with a CCND1-MYC vector and a PI3K vector because the former given alone kills the cells. The remaining oncogene we would like to express is ERBB2 so we add this in a second round of infection that includes an imaging vector like LKB (Fig. 2).

1.3 Mammary Intraductal Injection (MIND)

The only graft site that maintains the hormone sensing cell identity of human mammary epithelial cells is the mammary duct. This makes sense because the cells receive the correct microenvironmental signals at this site. Dan Medina's group originally showed that intraductal injection maintains ER expression by DCIS [2]. Our group then showed that it maintains ER expression in genetically defined xenografts from mammoplasty tissue (GDX) [13]. Finally, Cathrin Brisken's group and our group showed that it maintains ER expression in PDX [3, 6], including AR expression in molecular apocrine/HER2-E PDX [6].

Mice have five mammary glands on each side. The cervical glands are not normally injected, although it is certainly possible. The gland most commonly used is the fourth (inguinal gland). It is the easiest to inject and harvest. From an ethical perspective it is considered a good thing to inject multiple glands because it reduces the number of mice required, and the wounds are so tiny that giving multiple injections is not perceived to cause undue suffering. This is most relevant when you want to graft a PDX with a low take rate, in which case the mouse may only develop one or two tumors despite receiving injections in eight glands; it should not be done just to generate more material by giving a mouse eight tumors.

Bioluminescence scanning is highly recommended [3]. It allows you to check after a few days whether your grafts have taken. This allows you to terminate experiments early if tumors fail to engraft. That said, it can be quite deceptive because it only takes a handful of cells to generate a strong signal. Hence you should not rush to sacrifice mice just because you have a strong signal. Signal strength depends on the multiplicity of infection and the activity of the promoter driving luciferase in your particular tumor cell. This means you should base welfare decisions on normal criteria (mouse weight, appearance, and behavior) not bioluminescent signal strength. We use a French IVIS scanner (Biospace PhotonImager) that performs kinetic scans. The advantage is that you can see when the signal reaches a plateau, so you do not count the photons too early. Intraductal injections often give a bimodal peak, probably generated by slow diffusion of luciferin into necrotic tissue (intraductal injections commonly give comedo necrosis). If the imaging vector also expresses a fluorescent protein (as in the LKB vector in Fig. 2) you can identify tumor foci at tissue harvest with a fluorescence stereomicroscope. This allows you to enrich for tumor cells by manually dissecting out the fluorescent foci while still in the animal house. By including an antibiotic resistance gene in the vector (as in LKB), you can also easily eliminate murine cells when you put them back into Clevers culture. Note that NSG mice contain the neoR gene so you can not use G418 to eliminate mouse cells from organoid cultures.

2 Materials

2.1 Components for Isolation of Mammary Epithelial Cells

Advanced DMEM/F12.

Primocin 50 mg/ml or penicillin/streptomycin.

HEPES 1 M.

Glutamax 100×.

Phosphate-buffered saline.

Collagenase type IV: make 30,000 U/ml (~100 mg/ml) in PBS, spin to avoid clogging filter then filter sterilize (0.22 μm) and store aliquots at −20 °C (100× stock).

Hyaluronidase: make 10,000 U/ml (~10 mg/ml) in PBS, filter sterilize (0.22 μm) and store aliquots at −20 °C (100× stock).

Deoxyribonuclease 1 (DNase): make 10 mg/ml in PBS, filter sterilize (0.22 μm) and store aliquots at −20 °C.

Trypsin Solution 0.25%.

Dispase: make 5 U/ml in Advanced DMEM/F12, filter sterilize (0.22 μm) and store aliquots at −20 °C.

2 ml plastic pipettes.

EasySep Human EpCAM Positive Selection Kit II.

EasySep Magnet (STEMCELL Technologies 18000).

EPCAM (CD326) antibody 5E11.3.1-FITC.

Hyflex gloves.

Sterile forceps.

Sterile scalpels.

Sterile chopping board.

AdDF+: Advanced DMEM/F12, 1× Glutamax, 20 mM HEPES, 2× pen/strep, 10 μM Y-27632. Digestion medium: AdDF+, 300 U/ml collagenase, 100 U/ml hyaluronidase.

Complete digestion medium: CBOM (*see* Subheading 2.3), 10 μM Y-27632, 300 U/ml collagenase, 100 U/ml hyaluronidase.

AdDF+/DNase: AdDF+, 0.1 mg/ml DNaseI.

15 cm cell culture dishes.

15 and 50 ml conical base tubes.

FACS tubes (polystyrene).

2.2 Components for Lentiviral Transduction of Mammary Epithelial Cells

Cells from digestion of reduction mammoplasty or tumor.

Viral stock concentrated to at least 1×10^7 infectious units/ml (preferably $>5 \times 10^7$/ml). (See Chapter 3 for instructions on how to produce lentiviral vectors).

Phosphate-buffered saline.

0.05% Trypsin/EDTA

6-well plate.

96-well plate U-shaped well.

Y-27632 2HCl.

2.3 Components for Clevers Breast Organoid Medium (CBOM)

Advanced DMEM/F12.

Primocin 50 mg/ml.

HEPES 1 M.

Glutamax.

B27 Supplement 50×.

293T cells expressing Rspo1 and 293 cells expressing Noggin.

Rspo conditioned medium[c].

Noggin conditioned medium[c].

Nicotinamide.

N-Acetyl-L-cysteine (NAC).

Heregulinβ-1 (NRG1).

Amphiregulin (AREG).

FGF-7.

FGF-10.

PGE2.

A83-01.

SB202190.

E2 stock 25 μM in ethanol.

Y-27632 2HCl.

Matrigel: Test a specific batch then order enough of that batch to complete the experiment.

Freezing medium: 10% DMSO, 90% Cryo3 (www.stemalpha.com), 10 μM Y-27632.

Recipe for 50 ml CBOM

Product	Initial	Diluent to make stock	Conc stock	Conc final	Vol for 50 ml
Ad DMEM/F12[a]	1×	Na	1×	~1×	38 ml
Primocin[b]	50 mg/ml	Na	50 mg/ml	100 μg/ml	100 μl
HEPES	1 M	Na	1 M	10 mM	500 μl
Glutamax	100×	Na	100×	1×	500 μl
B27	50×	Na	50×	1×	1 ml
Rspo1[c]	10×	Na	10×	1×	5 ml
Noggin[c]	10×	Na	10×	1×	5 ml
FGF7	10 μg	100 μl PBS 0.1% BSA	100 μg/ml	5 ng/ml	2.5 μl
FGF10	25 μg	250 μl PBS 0.1% BSA	100 μg/ml	20 ng/ml	10 μl
Heregulinβ-1	50 μg	500 ul PBS 0.1% BSA	100 μg/ml	37.5 ng/ml	18.75 μl
AREG	50 μg	500 μl PBS 0.1% BSA	100 μg/ml	10 ng/ml	5 μl
A83-01	10 mg	2.4 ml DMSO	10 mM	0.5 μM	2.5 μl
SB202190	5 mg	3.0 ml DMS0	5 mM	0.5 μM	5 μl
PGE2[d]	10 mg	2.8 ml DMSO	10 mM	1 μM	5 μl
Nicotinamide[e]	1.2 g	10 ml PBS	1 M	10 mM	500 μl
N-acetyl-L-cysteine[e]	0.8 g	10 ml PBS	0.5 M	1.25 mM	125 μl

(continued)

Product	Initial	Diluent to make stock	Conc stock	Conc final	Vol for 50 ml
E2	136 mg	20 ml ethanol	25 mM	–	–
E2	25 mM	5 μl in 5 ml ethanol	25 μM	1 nM	2 μl
Y-27632[e]	10 mg	3.1 ml TC water	10 mM	10 μM	50 μl

[a] To make phenol red-free CBOM, order SILAC Advanced DMEM/F-12 Flex Media no glucose no phenol red (Thermo A2494301) and add back arginine, lysine, and glucose (this is much cheaper than ordering custom media)

[b] Broad-spectrum antibiotic. Not needed unless you have just made the organoids from human or mouse tissue. Add pen/strep if you always use it in tissue culture

[c] R-spondin1 final 500 ng/ml, Noggin final 100 ng/ml. Rather than buy pure factors we make conditioned medium. The Rspo1 cells are available from Trevigen. To request the Noggin cells, contact Hans Clevers (h.clevers@hubrecht.eu). To skip over MTAs, you can make your own cells by cloning gBlocks into your favorite lentiviral vectors. Checking the amount of Rspo1 and Noggin in conditioned medium is nontrivial (we use a Wnt luciferase reporter cell line, TOP293, for Rspo and a BRE-luc reporter plasmid for Noggin, both from Susie Woods, SAHMRI, Adelaide)

[d] COX2 mimetic, optional for mammary cells

[e] Filter sterilize

2.4 Components for Injection of Mammary Ducts (MIND Protocol)

NSG mice (Jackson strain 005557).

Vannas Spring Scissors (FST 15000-00).

Angled tip fine forceps (FST 11063-07).

Hamilton syringe 705 SN Volume 50 μL Needle point style 3, gauge 30, length 13 mm.

Hair removal cream (e.g., Veet).

Trypan blue 0.4% (filter sterilize).

Culture medium with 2× pen/strep to rinse syringe (e.g., Advanced DMEM/F12).

Ethanol 70% to rinse syringe.

Emla cream (lidocaine 2.5% and prilocaine 2.5%).

Luciferin sodium salt 1 g: dissolve in 33 ml PBS, filter sterilize and freeze in 500 μl aliquots.

Solutions for tissue harvest (e.g., neutral buffered formalin, organoid culture medium).

2.4.1 General Animal House Equipment

Isoflurane anesthetic machine.

Basic equipment for mouse surgery (anesthetic nose piece, electric shaver, heat pad).

Bioluminescence scanner (Perkin-Elmer IVIS or Biospace PhotonIMAGER).

Basic equipment for tissue harvest (scissors, forceps, tubes, histology cassettes).

Simple stereomicroscope for injections.

Stereomicroscope with fluorescence to visualize tumor at harvest (optional).

3 Methods

3.1 Prepare Mammary Epithelial Cells from Tumors and Mammoplasties

We commonly split the protocol into two parts separated by the freezing of the tissue after it has been digested with collagenase. Note that the term organoid now has two meanings: tissue fragments released by collagenase digestion of reduction mammoplasties or human tumors, and clumps of cells growing in Clevers medium. To avoid confusion, we sometimes use the term "tissue organoids" for the former. We use the terms PDO/PDX (patient-derived organoids/xenografts) for organoids/xenografts grown from tumors, and GDO/GDX (genetically defined organoids/xenografts) for organoids/xenografts grown from lentivirally transduced reduction mammoplasty cells.

1. Place reduction mammoplasty tissue or tumor fragments on a sterile chopping board.
2. Remove yellow fatty tissue from mammoplasties by scraping with a scalpel.
3. Shear and chop the remaining tissue using two scalpels until it reaches a mushy consistency. Wear Hyflex gloves to avoid cutting yourself while chopping.
4. Add 1 g tissue to 10 ml digestion media.
5. Incubate at 37 °C on a roller for 3–16 h.

   Mammoplasties are commonly digested overnight in collagenase. Tumor samples require closer attention. Some will be completely digested after 3 h. If they require longer, it is easier to monitor progress if they are spun down and the digestion is completed in a 6-well plate in complete digestion medium. Put the plate at 37 °C, checking them every few hours with a tissue culture microscope and pipetting up and down with a wide bore pipette to mix. It is time to stop when you can see mainly little clumps containing 5–20 cells (*see* **Note 3**).
6. Pellet the cells at 1200 rpm/270 × *g* for 4 min.
7. Wash the pellet twice by resuspending in 5 ml warm AdDF+ and spinning down at 1200 rpm. Add 0.1 mg/ml DNase I to the first wash buffer to ensure you do not throw away any cells with the supernatant. Remove the supernatant with a pipette rather than a suction line to reduce the risk of losing everything (*see* **Notes 4** and **5**).

   You can pause at this point and freeze the tissue organoids. Given the large amount of material commonly obtained from reduction mammoplasties, you will normally freeze down most of the material at this stage. You can freeze tumor biopsies at this stage, but it is better to continue without freezing.

8. **To freeze**: Resuspend the pellet in freezing medium. Aliquot into cryotubes and freeze in an isopropanol box (Mr Frosty) at −80 °C overnight then transfer to liquid nitrogen. Adjust the volumes and number of aliquots according to the amount of material (commonly 10–20 tubes for a mammoplasty, 3 tubes for tumours).

    To continue without freezing: Take up mammoplasty organoids in 5 ml AdDF+, and tumor organoids in 160 μl CBOM, 10 μM Y-27632.

To work with small volumes, put the cells in a 1.5 ml Eppendorf tube placed in a 50 ml Falcon tube in a normal tissue culture swing-out rotor and spin at 1200 rpm/270 × *g* for 4 min at 20 °C. If you use a fixed angle microfuge the cells may stick to the sides of the tube and be lost.

3.2 Purify EPCAM+ Mammoplasty Cells on Magnetic Beads

To make genetically defined models we normally purify the cells from mammoplasties on EPCAM beads at this stage. Skip this step for patient-derived tumor samples.

1. Thaw a tube of tissue organoids and transfer them to a 15 ml tube containing 5 ml prewarmed AdDF+ medium (*see* **Note 3**).
2. Spin at 1200 rpm (~270 × *g*) for 4 min.
3. Remove the supernatant with a pipette (the pellet may be sucked up by free DNA if you use an aspiration line).
4. Add 500 μl prewarmed Trypsin 0.25% and resuspend the organoids with a 2 ml pipette.
5. Incubate for 2 min at room temperature.
6. Add 5 μl DNase and pipette the organoid suspension repeatedly against the tube wall for 1 min with a 2 ml pipette.
7. Add 1 ml prewarmed Dispase and pipette repeatedly against the tube wall for 2 min (the suspension should become clear; add more DNase if still stringy).
8. Spin at 1200 rpm (~270 × *g*) for 4 min and remove the supernatant.
9. Resuspend cells in 5 ml AdDF+/DNase.
10. Mix 10 μl cell suspension with 10 μl trypan blue and examine on a Kova slide to count the cells, to make sure there is a mixture of single cells and clumps of <10 cells, and to confirm that most of the cells are alive (*see* **Note 5**).
11. Spin at 1200 rpm (~270 × *g*) for 4 min and remove the supernatant.

12. Resuspend cells in 150 μl AdDF+/DNase and transfer to a FACS tube on ice. Use a larger volume, as per the STEMCELL protocol, if you have more than 1×10^7 cells (since you don't need more for infection, start with fewer organoids next time if you end up with a ridiculous amount).
13. Add 15 μl STEMCELL EPCAM antibody (more if $>10^7$ cells), mix and incubate for 20 min at 2–8 °C.
14. Vortex to resuspend the STEMCELL magnetic beads, add 15 μl to the cell suspension, pipette to mix, and incubate for 15 min at 2–8 °C.
15. Add 2.5 ml AdDF+/DNase, pipette to mix, and place in the magnet (cool the magnet in the fridge before starting but do the incubations in the hood at room temperature).
16. Incubate 5 min.
17. Pour off the supernatant.
18. Resuspend in 2.5 ml AdDF+/DNase, pipette to mix and place back in the magnet for 5 min.
19. Repeat the washes in the magnet a total of four times.
20. Resuspend in 210 μl CBOM, 10 μM Y-27632.

    Optional: Test an aliquot by flow cytometry if you want to confirm that the beads have enriched for EPCAM+ cells (*see* **Note 6**).

3.3 Infect Mammary Epithelial Cells

If you froze the collagenase-digested tumor tissue while there were still lumps of collagen present, you should digest again with collagenase (as above) until you have eliminated the remaining collagen.

1. Put boxes of P20 and P200 tips and a tube of Matrigel in the fridge before you start so they will be cold and ready to use when you want to plate the cells.
2. You should now have 210 μl of EPCAM+ mammoplasty cells or 160 μl of tumor cells. Mix 10 μl of tumor cells or EPCAM+ mammoplasty cells with 10 μl trypan blue and count the cells on a Kova slide. Most of the cells should be alive. The cell count will be rather approximate because of the clumps. Do not use more than 10^6 cells per infection.
3. Perform each infection in a single well of a round bottom 96-well nonstick plate. For the patient-derived tumor cells put 150 μl of cells in the well and add 50 μl of virus. For the mammoplasty cells put 100 μl of cells in two wells and 50 μl of each virus to one well (the other well is an uninfected control). The CBOM contains 10 μM Y-27632 to discourage anoikis but this does not interfere with infection. If infecting multiple samples at the same time leave plenty of empty wells between the samples so you don't mix anything up (*see* **Note 7**).

4. Incubate overnight at 37 °C, 5% CO_2.
5. In the morning, pipette the cells up and down to break up the bigger clumps, then transfer them to a 1.5 ml Eppendorf containing 1 ml CBOM 10 μM Y-27632 and spin it in a swing-out rotor at 1200 rpm/270 × *g* for 4 min at room temperature (put the Eppendorf inside a 50 ml Falcon and use a normal TC centrifuge—don't use a fixed angle rotor as the cells may stick to the walls of the tube).
6. Remove the virus-containing medium, leaving behind about 100 μl, add 1 ml CBOM 10 μM Y-27632 and spin again. Remember that you are dealing with virus and inactivate the supernatant with disinfectant under the hood before disposing of it.

   Optional: If you want to skip organoid culture you can perform some additional washes and proceed directly to intraductal injection at this point.
7. Very carefully remove the supernatant, leaving only 10–20 μl over the cells. To avoid disaster when dealing with tiny pellets, it helps to put the supernatant in a spare 24-well plate in case you remove the cells by mistake (in which case you can just spin them down again). Pipette the cells up and down in the residual 10–20 μl to resuspend them, taking care not to create any bubbles.
8. Put the Eppendorf with the cells on ice, together with a tube of Matrigel that has been thawed overnight in the fridge. Take up 120 μl of Matrigel with a cold tip and resuspend the cells without creating bubbles.
9. Put 20 μl dots of cells in Matrigel in a 24-well plate, one dot per well. If in doubt, plate a single dot and check it under the microscope—if it looks very dense add more Matrigel to reduce the cell number in the remaining dots. Invert the plate and put it in the incubator for 15 min to let the Matrigel solidify.
10. Add 1 ml CBOM 5 μM Y-27632 per well (*see* **Note 2** if you want to adapt the medium to the cell type). There should be lots of individual cells and little clumps in each dot. If you started with 200 μl of bead prep you should now have two rows of six dots: infected cells and mock-infected control. If dealing with patient-derived tumor cells you will only have one row of six dots.
11. After 48 h, replace the medium with CBOM plus antibiotic. For the mammoplasty infections, if the cells received two viruses, one blast, the other hygro, add blasticidin 5 μg/ml to three wells, and hygromycin 200 μg/ml to three wells. Add antibiotics to two wells each in the mock-infected row (i.e., keep two wells without antibiotic). For patient-derived tumor cell infections with marker viruses, you can afford to wait

longer before adding antibiotic—if you used the LKB virus in Fig. 2, wait until you can see red fluorescence starting to appear then add blasticidin 5 μg/ml to 3 wells (i.e., keep 3 wells without antibiotic, just in case the infection efficiency was very low—you don't want to kill your precious cells for nothing).

12. Change the medium every 2 days, hoping to see the uninfected cells die and small clumps of transduced cells (Clevers organoids) appear in the infected wells. There will be a lot of dead cells in the mammoplasty wells, but this is expected because the EPCAM beads purify some mature luminal cells that are likely to die before the first passage.

    Optional: Given the difficulty of growing ER+ organoids it is prudent to inject 2 mice as soon as it is clear that tumor cells are starting to form organoids.

13. As soon as it is obvious that there are viable clumps of cells with single antibiotics in the mammoplasty infections, give both antibiotics to four wells (i.e. keep two infected wells in single antibiotics.) If one infection worked better than the other, let the less successful infection catch up for a few days in its own antibiotic before hitting it with the second antibiotic.

    Passage the mammoplasty wells that received both antibiotics after 1 week. Passage the patient-derived tumor infections when they have formed organoids (this may take more than 1 week) (*see* **Note 8**).

14. To passage organoids, remove the medium and pool identical conditions in a single well containing either 1 ml CBOM + 10 μM Y-27632 + 0.5 U Dispase or 1 ml Tryple + 10 μM Y-27632. If using dispase, put the cells back in the incubator for 15–30 min. If using tryple put the cells at room temperature for 2 min. Then look at the cells under the microscope. Tryple will reduce the organoids to single cells—this is fine for mammoplasty cells (GDOs) but may kill patient-derived organoids (PDOs). It is better to use dispase for the latter (the cells should remain in clumps). To avoid losing viable cells in the washes, add 0.1 mg/ml DNaseI if there is a lot of debris. Spin in a 1.5 ml Eppendorf at 1200 rpm/270 × *g* for 4 min at room temperature after digestion with tryple. Spin at 1500 rpm/420 × *g* for 6 seconds after digestion with dispase. Take up the cell pellet in 1 ml CBOM 10 μM Y-27632 and spin again, this time leaving only 10–20 μl supernatant over the cells.

15. Take up the cells in Matrigel and make dots as above. Depending on the amount of viable cells, split 1:2 or 1:3 (so for each single dot you split you should allow 40–60 μl Matrigel). Let the Matrigel set for 15 min at 37 °C then add CBOM 5 μM

Y-27632 with antibiotics. The cells should look bright and happy at plating, preferably in clumps of 5–10 cells.

16. Change the medium every 2 days.
17. You should now have multiple wells growing strongly in the presence of antibiotics. The transformed mammoplasty cells can tolerate splitting to single cells, making it possible to expand them rapidly. Passage them once to make frozen stocks then proceed to mammary intraductal injection. The patient-derived tumor cells will show much greater variation in growth rate.

    Optional: Infect the mammoplasty cells again after 1 week. Harvest them as for passaging but when you resuspend the cells in 1 ml CBOM, split it into 200 μl in one tube and 800 μl in another. Spin down the 200 μl tube and plate the cells in 4–6 dots of Matrigel, as for routine passage (you can use these singly infected cells to make stocks). Spin down the 800 μl tube and resuspend the cells in 300 μl CBOM 10 μM Y-27632 (assuming you want to do 2 infections: 100 μl for each infection, plus a control). Put 100 μl in round bottom wells in a 96-well nonstick plate and add 100 μl of concentrated viruses to the infection wells or 100 μl medium to the mock-infected control. Incubate overnight and process as for the first infection, plating each infection in 6 dots. Let the cells recover for 2 days then add the new single antibiotics to two wells each (for example, puromycin 2 μg/ml to first two wells, and G418 750 μg/ml to the next two wells), and the old double antibiotics (e.g., blasticiding+hygromycin) to the remaining two wells. Give the mock-infected control wells one antibiotic per well, making sure there is at least one well without antibiotics, so you can see when killing is complete in the new infections, and verify that the old infections are still resistant to the old antibiotics. As soon as you see that that they are resistant to the new single antibiotics, give the newly infected wells both new antibiotics (in this example, puromycin + G418). Continue changing the medium every 2 days until the controls are dead. Hopefully, everything except the controls will grow. At this point you can be pretty sure that both the first and second infections have worked, and the cells are resistant to all four antibiotics. To confirm this, passage the cells from the new infections to six wells, adding just the new antibiotics to two wells and all four antibiotics to the remaining four wells.

3.4 Inject the Mammary Ducts

The protocol below suggests removing the fur over the teats 1 or 2 days before injection. Experienced scientists can do everything on the day of injection, but it will greatly reduce the stress for beginners to prepare the mice in advance. Hair removal cream makes the teats very easy to see, so it is worth using cream initially, particularly if you want to inject 8 glands per mouse.

1. One or 2 days before injection, anesthetize the mice with isoflurane, remove the fur over the planned injection sites by shaving and/or hair removal cream, mark the mice, and insert a 0.5 mg estradiol pellet if required (*see* **Note 9**).
2. Allow at least one Matrigel dot per gland to inject. For GDOs that grow well this corresponds to about 10^5 cells per injection. For PDOs the amount of material may be limiting, so you may be forced to harvest more dots and inject fewer cells. Harvest the organoids in dispase and resuspend them in CBOM, 10 μM Y-27632, allowing 20 μl per injection to allow for losses. To avoid wasting material, do not filter the cells through cell strainers. This assumes the organoids are less than 159 μm in diameter, the internal diameter of the 30 gauge needle. If they are bigger than that, digest with tryple instead of dispase.
3. Rinse the Hamilton syringe with 70% ethanol before starting and between samples, to prevent cell carry-over, then rinse with tissue culture medium to remove the ethanol before aspirating the cell suspension.
4. Anesthetize the mice with isoflurane, gently squeeze the skin around the teat with the bent tip forceps and use the spring scissors to cut off the upper half of the teat (use a stereomicroscope so you can see exactly what you are doing). Dab the site with trypan blue to stain the exposed cells at the cut: you should see a perfect blue circle into which to insert the needle (*see* **Note 10**).
5. Inject 10–15 μl of cell suspension into the duct (*see* **Note 11**).
6. Put a dab of local anesthetic cream on the skin around the teat and put the mice in a recovery cage on a heated pad.
7. Scan the mice by IVIS/photonImager 1 week after injection to see if the glands were successfully injected and the cells have survived. To do this, place the mice in an isoflurane chamber, shave the abdominal skin, inject 100 μl of luciferin 30 mg/ml ip, and transfer them to the scanner when they are sleeping comfortably. Collect data when the signal has stabilized (usually after about 15 min). Repeat the scans as required by your experimental protocol.
8. Harvest glands containing tumors by making a midline incision, reflecting the abdominal skin, and dissecting in the tissue plane between the skin and the subcutaneous fat. If harvesting the fifth gland, make a single midline ventral incision from top to bottom—do not make the Y incision commonly used for harvesting only the fourth gland or you may cut through the fifth gland, reducing it to a pile of mush. Put tissue for culture

or reinjection in culture medium with pen/strep for transfer to the tissue culture lab. Depending on your experimental goals, put the remaining tissue in neutral buffered formalin for histology, or freeze for protein or RNA extraction. To find metastases, soak individual organs in luciferin and scan bioluminescence separately from the primary tumor.

9. The protocol for digestion of mouse glands is the same as for human tissue except that mouse glands are easier to digest. Chop the tissue with scalpels then incubate it in CBOM 10 μM Y-27632 plus collagenase/hyaluronidase for 1 h. Add 0.1 mg/ml DNaseI and spin at 1200 rpm for 4 min. The adipocytes will float to the top, leaving a cell pellet that is, hopefully, brightly fluorescent. If there are still large amounts of stringy fluffy stuff, continue the digestion with fresh CBOM 10 μM Y-27632 plus collagenase/hyaluronidase in a 6-well plate so you can see what is happening. When satisfied, wash with CBOM 10 μM Y-27632, resuspend in freezing medium and put the vials in a Mr. Frosty at −80 °C overnight then transfer to liquid nitrogen. Since there is no bead step you may need to use a cell strainer to remove stringy bits before injecting mice: they are not harmful in organoid culture and will disappear at the first passage, but you must get rid of anything that would block the needle if you are going to reinject mice immediately. Plate organoids or reinject ducts as above. If you start in the morning it should be possible to reinject on the same day (but it will be a long day). If making PDO from PDX containing the LKB vector shown in Fig. 2, add blasticidin 5 μg/ml to the CBOM and you will prevent overgrowth by mouse cells. If making GDO from GDX as described above (i.e., cells containing puro, hygro, blast and neo vectors), start by adding puromycin to get rid of the mouse cells quickly. Rather than adding all four antibiotics at once after the first passage, make sure you have at least one well with each antibiotic alone—that way you should be able to work out quickly what happened if the cells have lost one of the vectors.

4 Notes

Two issues we have not touched on are LDEV and cross-contamination. LDEV is a virus that contaminated Matrigel and led to widespread contamination of PDX. The problem has now been solved by the manufacturers but there is a legacy of contaminated PDX. You will probably be asked to screen for it before introducing new PDX into the animal house. Cross-contamination is as much a problem for PDX as for any cell lines and you should take the normal steps to detect and prevent it.

1. To make any headway with organoids, you need a simple way to measure growth. This is commonly done by Cell Titer Glo assay (Promega G9241) but this is a destructive, endpoint assay. If you have transduced the cells with a virus expressing a fluorescent marker (for example, mKate2 in LKB, Fig. 2), an easy way to quantify growth of live cells is to take daily pictures with a 2× objective. This lets you take a photo of the entire Matrigel dot in a single image. You can then count the total signal with an ImageJ macro and generate growth curves with very tight error bars (Fig. 1). The critical factor is the depth of field—most of the organoids will be out of focus but this hardly matters with very low power objectives. To make growth curves on the Incucyte, you can plate 2 μl Matrigel dots in 96-well plates and capture whole wells (the object of the exercise being to have such a thin dot that almost all of the organoids are in focus). You will also need a way to check that the histology is correct. This can be quite challenging if you have small amounts material because it is difficult to find the organoids when sectioning the paraffin blocks. One solution is to embed the organoids in low melting point agarose and embed that in paraffin. A more sophisticated solution is to embed Matrigel blocks in histogel [15].
2. An interesting variant of the Clevers protocol is the addition of Wnt3a to the medium for culture of normal cells [5]. Activating the Wnt pathway with Rspondin was a cornerstone of the original Clevers organoid approach [16]. Adding both Wnt3a and Rspondin leads to much stronger activation but carries the risk of inducing squamous metaplasia [10]. For normal cells, Clevers adds forskolin to elevate cAMP [16] (cholera toxin was commonly used for this in the past). Many organoid media contain B27, a medium supplement of known ingredients at unknown concentrations [17]. It improves organoid growth generally [18]. Unfortunately for mammary gland biologists, B27 contains progesterone, probably at about 1 μM in the 50× stock. Many of the other ingredients in B27 are now included in the base medium anyway (we are trying to define a cheaper substitute leaving out the unnecessary components). The Welm media have specific adaptations for particular cell types: 100 ng/ml FGF2 and 1 mM N-acetyl cysteine for ER+ PDOs; 10 nM neuregulin for HER2+ PDOs; all versions contain 10 μM ROCK inhibitor (Y-27632) and some of the more classic ingredients in mammary media (5% FBS, 10 ng/ml EGF, and 1 μg/ml hydrocortisone) [14]. It is worth testing different media to the find the one that best suits your cells; for example, the LA-PDX1 line in Fig. 1 can be passaged multiple times in CBOM but is killed quickly by the Welm luminal medium. The most controversial ingredient in Welm media is

FBS. From a philosophical standpoint it is a shame to add it to an otherwise fully defined medium. In our hands it promotes the growth of fibroblasts. More generally, we strongly recommend inserting a virus that can be selected with antibiotic immediately after receiving human tissue, and passaging the cells in mice before performing any complicated experiments. That way you can get rid of normal cells of both human and mouse origin simply by adding antibiotic. Without these steps there is no guarantee that your organoids will contain human tumor cells, as many people have found to their dismay.

3. Trypsin and dispase are no match for collagen I. If you skimped on collagenase, you cannot make up for it by digesting in trypsin and dispase. You will simply kill the cells and produce a horrible gloopy mess. It is much better to digest again in collagenase. Collagenase and dispase need calcium, DNase needs magnesium; both are present in tissue culture medium. You should not try adding standard tissue culture trypsin/EDTA to hasten the digestion—the EDTA will inhibit the other enzymes, and prolonged exposure to trypsin will kill the cells.
4. Many protocols call for fire-polished glass Pasteur pipettes at this stage. Surrounding yourself with broken glass is a bad idea in a virus lab. Two ml plastic pipettes (same as aspiration pipettes but with a cotton wool plug) have a larger opening than Gilson tips and work well for gently breaking up organoids. If you use normal Gilson tips at this stage, the partially digested organoids tend to form a gooey mass that sticks to the tips. Wetting tips with medium containing BSA or FCS may reduce the tendency to stick but using wide bore pipettes is safer.
5. Most protocols ask for the cell suspension to be passed repeatedly through a 40 or 70 μm cell strainer before purification on beads. This is pointless when purifying cells on EPCAM beads because stringy mush does not stick to the beads, so it gets poured away harmlessly during the washes. It is not essential, and probably harmful, to force the digestion to completion. The beads are strong enough to retain small clumps. Any cells within the clumps that are not infected will be eliminated as soon as antibiotics are added, and up to that point they will provide moral support to the other cells, so there is really nothing to be gained from digestion to completion unless you want to sort the cells.
6. To check the success of your EPCAM bead purification, label an aliquot with FITC-labeled anti-EPCAM antibody (StemCELL #5E11.3.1-FITC) according to your standard flow cytometry protocol. Before purification there should be a bimodal

peak; after purification you should only see the upper peak. If you want to infect specific subsets you can use Miltenyi enzymes (Miltenyi Biotec 130-095-929), digest to completion in a gentleMACS tissue dissociator, and use multicolor FACS to purify the cells you are interested in. Miltenyi enzymes do not cleave off extracellular epitopes (ectodomains) so FACS epitopes are better preserved, but FACS is much harsher for the cells. The main objection, however, is that some tumors never recover from being reduced to single cells. The EPCAM bead protocol is a good compromise: it cleans up the cells for very little effort and retains little clumps that survive and infect well.

7. The chance of getting ≥1 virus per cell at an moi of 1 is 63%, but the chance of getting at least one copy of two different viruses is only 40%. Infecting at anything less than an moi of 1 will give very low numbers of doubly infected cells. Trying to use more than two vectors at a time is unattractive because higher multiplicities are required leading to multiple integrations in some of the cells. If your viruses are at 10^7 infectious units/ml, 50 μl of virus will contain 5×10^5 infectious units. If you have 10^5 cells you will infect at a multiplicity of 5. If you use two viruses, 50 μl of each, you should be able to infect most of the cells with both viruses. Where this falls down is that some cells can be much more difficult to infect. This is particularly true of patient-derived tumor cells. If you need to use a larger volume of virus, you can use nonstick 24-well tissue culture plates, with the cells in 250 μl CBOM, and add 250 μl of virus. The cells will clump together during the night, but this is not a problem. Another way to increase the infection efficiency and save time is to perform spinfection: rather than putting the cells in an incubator overnight, put them in a closed tube and spin them at 2500 rpm in a swing-out rotor for 2 h at room temperature. If you enjoy making virus you can even add some virus during collagenase digestion of patient-derived tumors, but this should only be done if you have eliminated the bulk of the collagen and have transferred the prep to a 6 well plate for a second round of digestion (you will probably still need to do the formal infection step later on in the protocol).

8. The timing of the first passage of PDOs depends heavily on the state of the culture. There may be a large amount of debris; normal mammary epithelial cells may make extravagant attempts to form ducts; fibroblasts may try to contract the dots; and the tumor cells may break through the Matrigel and start adhering to the plastic. You can passage the cells 1:1 to get rid of most of these problems. To reduce adhesion, we use plastic that has not been tissue culture treated. Some groups

find that normal human cells can colonize the mouse ducts (Cathrin Brisken, pers. com.) but in our experience passage in mice is a good way to eliminate normal cells from PDOs.

9. Despite everyone's best efforts there are still conspicuously few ER+ models from primary breast tumors. Most are from metastases, particularly pleural effusions. Intraductal injection is a major advance in this respect. It has been shown that ER+ tumors do not require estrogen supplementation after intraductal injection [3], but that does not mean they will not grow better if you do provide additional estrogen. Given the generally poor outcome of ER+ grafts it is reasonable to test whether the cells will grow better with estrogen, for example by giving estrogen supplements to half the mice in the primary grafts. The two options are to give estradiol pellets, or to add estradiol to the drinking water [19]. The latter is much gentler for the mice. The classic side effect of estrogen pellets is urinary retention leading to renal failure. The move away from estrogen supplementation is driven primarily by a desire to avoid this. It is never seen when estradiol is added to the drinking water, perhaps because of a first pass effect. While this is laudable, it is probably also less effective at stimulating tumor growth, so you can't win.

   To add estradiol to the drinking water: make 2.7 mg/ml estradiol in ethanol and add this to water bottles at a final concentration of 8 μg/ml; change the bottles weekly.

   To make estradiol pellets: put 4.3 g Nusil part A and 0.45 g Nusil part B in a 50 ml Falcon tube, add 250 mg estradiol and mix with a sterile spatula (work in a tissue culture hood). When it is homogeneously mixed, spin 1 min at 4000 rpm to remove bubbles, aspirate the slurry into 1 ml syringes, put them at 37 °C overnight to harden, then store them at 4 °C. You should be able to fill 3 syringes. It is easier to measure out and maintain sterility if you use Nusil in syringe format (Nusil MED-4011, 37 ml SxS, 2 Part Kit 175068-57). To make pellets, wear Hyflex gloves and use a scalpel to cut half-way through the barrel of the syringe near the tip, then snap off the tip. You should then be able to pull out the entire sausage and cut it into 1 mm slices without drama. To get 0.5 mg pellets, cut the slices in half (if you want to be more precise you can weigh the pellets—a 10 mg pellet will contain 0.5 mg estradiol). To insert the pellet under isoflurane anesthesia, make a small incision at the back of the neck, create a tunnel with long thin forceps, insert the pellet with the forceps to a position level with the hips, and close the wound with a clip.

10. The level of analgesia required depends on the philosophy of your regulator. The size of the wound is comparable to that from a simple subcutaneous injection. For a maximalist

approach, you can give paracetamol or meloxicam in the drinking water starting the day before injection, buprenorphine at the time of injection, and apply local anesthetic cream to the cut site. The wound is barely visible to the naked eye, so it is hard to justify a maximalist approach. My personal view is that local anesthetic cream provides ample cover for the injection itself. By preparing the mice in advance it is possible to make the MIND procedure itself quick and minimally invasive, reducing to almost zero the risk that anything untoward can happen. If the procedure takes more than 5 min, for example when you are injecting multiple glands, use Viscotears (Novartis) to prevent the eyes drying out while the mice are anesthetized.

11. Cathrin Brisken runs a MIND injection course at the EPFL that is highly recommended if you want to learn the technique (https://www.epfl.ch/labs/brisken-lab/preclinicalmodelcourse/). There are two schools of thought when it comes to intraductal injection. In both cases you should lift the skin with forceps 5 mm from the gland on the side of the teat opposite the needle. Older scientists like me point the needle in the direction of the duct (for example, North-West if injecting the right fourth gland) and cannulate the duct (i.e., insert the needle ~1 mm into the duct). Once the needle is in the duct, let go with the forceps and use the hand that was holding the forceps to press on the syringe plunger. When the needle is removed it is usually obvious that it was in the duct because the duct is partially everted as the needle is withdrawn. Younger scientists with better motor skills should use the Audrey Hepburn technique: hold the syringe between the first and second fingers like a cigarette and put the thumb on the plunger. Rather than pointing the needle toward the gland, point it away from the gland. Slide it across the skin toward the opening until it drops into the orifice. The goal is to press the needle against the duct orifice to make a seal, and to inject without letting go with the forceps. The Audrey Hepburn technique is more elegant if you can master it. In particular, you need Hamilton syringes with plungers that still provide a little resistance—they are no good if the plunger falls under gravity because the cells will spill out uncontrollably at the slightest touch of your thumb.

Acknowledgments

I thank my present and former colleagues in Switzerland, France, and Australia for help in developing these protocols, and Valentina Scabia in the Brisken lab for showing me the Audrey Hepburn injection technique. I thank the French Cancer League and the Fondation pour la lutte contre le cancer et pour des recherches médico-biologiques for funding work in my lab.

References

1. Sachs N, de Ligt J, Kopper O, Gogola E, Bounova G, Weeber F, Balgobind AV, Wind K, Gracanin A, Begthel H, Korving J, van Boxtel R, Duarte AA, Lelieveld D, van Hoeck A, Ernst RF, Blokzijl F, Nijman IJ, Hoogstraat M, van de Ven M, Egan DA, Zinzalla V, Moll J, Boj SF, Voest EE, Wessels L, van Diest PJ, Rottenberg S, Vries RGJ, Cuppen E, Clevers H (2018) A living biobank of breast cancer organoids captures disease heterogeneity. Cell 172:373–386. https://doi.org/10.1016/j.cell.2017.11.010
2. Behbod F, Kittrell FS, LaMarca H, Edwards D, Kerbawy S, Heestand JC, Young E, Mukhopadhyay P, Yeh HW, Allred DC, Hu M, Polyak K, Rosen JM, Medina D (2009) An intraductal human-in-mouse transplantation model mimics the subtypes of ductal carcinoma in situ. Breast Cancer Res 11(5): R66
3. Sflomos G, Dormoy V, Metsalu T, Jeitziner R, Battista L, Scabia V, Raffoul W, Delaloye JF, Treboux A, Fiche M, Vilo J, Ayyanan A, Brisken C (2016) A preclinical model for ERalpha-positive breast cancer points to the epithelial microenvironment as determinant of luminal phenotype and hormone response. Cancer Cell 29(3):407–422. https://doi.org/10.1016/j.ccell.2016.02.002
4. Sachs N, Clevers H (2014) Organoid cultures for the analysis of cancer phenotypes. Curr Opin Genet Dev 24:68–73. https://doi.org/10.1016/j.gde.2013.11.012
5. Dekkers JF, van Vliet EJ, Sachs N, Rosenbluth JM, Kopper O, Rebel HG, Wehrens EJ, Piani C, Visvader JE, Verissimo CS, Boj SF, Brugge JS, Clevers H, Rios AC (2021) Long-term culture, genetic manipulation and xenotransplantation of human normal and breast cancer organoids. Nat Protoc. https://doi.org/10.1038/s41596-020-00474-1
6. Richard E, Grellety T, Velasco V, MacGrogan G, Bonnefoi H, Iggo R (2016) The mammary ducts create a favourable microenvironment for xenografting of luminal and molecular apocrine breast tumours. J Pathol 240(3):256–261. https://doi.org/10.1002/path.4772
7. Linnemann JR, Miura H, Meixner LK, Irmler M, Kloos UJ, Hirschi B, Bartsch HS, Sass S, Beckers J, Theis FJ, Gabka C, Sotlar K, Scheel CH (2015) Quantification of regenerative potential in primary human mammary epithelial cells. Development 142(18): 3239–3251. https://doi.org/10.1242/dev.123554
8. Ince TA, Richardson AL, Bell GW, Saitoh M, Godar S, Karnoub AE, Iglehart JD, Weinberg RA (2007) Transformation of different human breast epithelial cell types leads to distinct tumor phenotypes. Cancer Cell 12(2): 160–170
9. Pasic L, Eisinger-Mathason TS, Velayudhan BT, Moskaluk CA, Brenin DR, Macara IG, Lannigan DA (2011) Sustained activation of the HER1-ERK1/2-RSK signaling pathway controls myoepithelial cell fate in human mammary tissue. Genes Dev 25(15):1641–1653. https://doi.org/10.1101/gad.2025611
10. Jarde T, Lloyd-Lewis B, Thomas M, Kendrick H, Melchor L, Bougaret L, Watson PD, Ewan K, Smalley MJ, Dale TC (2016) Wnt and Neuregulin1/ErbB signalling extends 3D culture of hormone responsive mammary organoids. Nat Commun 7:13207. https://doi.org/10.1038/ncomms13207
11. Fridriksdottir AJ, Kim J, Villadsen R, Klitgaard MC, Hopkinson BM, Petersen OW, Ronnov-Jessen L (2015) Propagation of oestrogen receptor-positive and oestrogen-responsive normal human breast cells in culture. Nat Commun 6:8786. https://doi.org/10.1038/ncomms9786
12. Liu X, Krawczyk E, Suprynowicz FA, Palechor-Ceron N, Yuan H, Dakic A, Simic V, Zheng YL, Sripadhan P, Chen C, Lu J, Hou TW, Choudhury S, Kallakury B, Tang DG, Darling T, Thangapazham R, Timofeeva O, Dritschilo A, Randell SH, Albanese C, Agarwal S, Schlegel R (2017) Conditional reprogramming and long-term expansion of normal and tumor cells from human biospecimens. Nat Protoc 12(2):439–451. https://doi.org/10.1038/nprot.2016.174
13. Verbeke S, Richard E, Monceau E, Schmidt X, Rousseau B, Velasco V, Bernard D, Bonnefoi H, MacGrogan G, Iggo RD (2014) Humanization of the mouse mammary gland by replacement of the luminal layer with genetically-engineered preneoplastic human cells. Breast Cancer Res 16(6):504. https://doi.org/10.1186/s13058-014-0504-9

14. Guillen KP, Fujita M, Butterfield AJ, Scherer SD, Bailey MH, Chu Z, DeRose YS, Zhao L, Cortes-Sanchez E, Yang C-H, Toner J, Wang G, Qiao Y, Huang X, Greenland JA, Vahrenkamp JM, Lum DH, Factor RE, Nelson EW, Matsen CB, Poretta JM, Rosenthal R, Beck AC, Buys SS, Vaklavas C, Ward JH, Jensen RL, Jones KB, Li Z, Oesterreich S, Dobrolecki LE, Pathi SS, Woo XY, Berrett KC, Wadsworth ME, Chuang JH, Lewis MT, Marth GT, Gertz J, Varley KE, Welm BE, Welm AL (2021) A breast cancer patient-derived xenograft and organoid platform for drug discovery and precision oncology. bioRxiv
15. Pinto MP, Jacobsen BM, Horwitz KB (2011) An immunohistochemical method to study breast cancer cell subpopulations and their growth regulation by hormones in three-dimensional cultures. Front Endocrinol (Lausanne) 2:15. https://doi.org/10.3389/fendo.2011.00015
16. Sato T, Vries RG, Snippert HJ, van de Wetering M, Barker N, Stange DE, van Es JH, Abo A, Kujala P, Peters PJ, Clevers H (2009) Single Lgr5 stem cells build crypt-villus structures in vitro without a mesenchymal niche. Nature 459(7244):262–265. https://doi.org/10.1038/nature07935
17. Brewer GJ, Torricelli JR, Evege EK, Price PJ (1993) Optimized survival of hippocampal neurons in B27-supplemented neurobasal, a new serum-free medium combination. J Neurosci Res 35(5):567–576
18. Rosenbluth JM, Schackmann RCJ, Gray GK, Selfors LM, Li CM, Boedicker M, Kuiken HJ, Richardson A, Brock J, Garber J, Dillon D, Sachs N, Clevers H, Brugge JS (2020) Organoid cultures from normal and cancer-prone human breast tissues preserve complex epithelial lineages. Nat Commun 11(1):1711. https://doi.org/10.1038/s41467-020-15548-7
19. Levin-Allerhand JA, Sokol K, Smith JD (2003) Safe and effective method for chronic 17beta-estradiol administration to mice. Contemp Top Lab Anim Sci 42(6):33–35

Chapter 14

Single Organoids Droplet-Based Staining Method for High-End 3D Imaging of Mammary Organoids

Jakub Sumbal and Zuzana Koledova

Abstract

In the last decade, organoids became a tremendously popular technique in developmental and cancer biology for their high pathophysiological relevance to in vivo models with the advantage of easier manipulation, real-time observation, potential for high-throughput studies, and reduced ethical issues. Among other fundamental biological questions, mammary organoids have helped to reveal mechanisms of mammary epithelial morphogenesis, mammary stem cell potential, regulation of lineage specification, mechanisms of breast cancer invasion or resistance to therapy, and their regulation by stromal microenvironment. To exploit the potential of organoid technology to the fullest, together with optimal organoid culture protocols, visualization of organoid architecture and composition in high resolution in three dimensions (3D) is required. Whole-mount imaging of immunolabeled organoids enables preservation of the 3D cellular context, but conventional confocal microscopy of organoid cultures struggles with the large organoid sample size and relatively long distance from the objective to the organoid due to the 3D extracellular matrix (ECM) that surrounds the organoid. We have overcome these issues by physical separation of single organoids with their immediate stroma from the bulk ECM. Here we provide a detail protocol for the procedure, which entails single organoid collection and droplet-based staining and clearing to allow visualization of organoids in the greatest detail.

Key words 3D culture, Clearing, Confocal imaging, Microenvironment, Organoid, Staining

1 Introduction

Organoid cultures represent an important tool in mammary gland biology and breast cancer research [1]. For over two decades, they have helped to faithfully model complex processes of mammary epithelial tissue development and tumorigenesis in vitro. Among other studies, they have helped to dissect the signaling and cellular mechanisms of mammary epithelial branching morphogenesis [2–5], lineage specification and mammary stem cell potential [6], lactation and involution [7], mechanisms of breast cancer initiation [8], invasion [9–11], recurrence [12] or resistance to therapy [13]. Cocultures of mammary epithelial organoids with stromal

Maria dM. Vivanco (ed.), *Mammary Stem Cells: Methods and Protocols*,
Methods in Molecular Biology, vol. 2471, https://doi.org/10.1007/978-1-0716-2193-6_14,

cells or in different ECM have helped to untangle regulation of epithelial morphogenesis by stromal microenvironment [5, 14, 15]. Patient-derived breast cancer organoids and patient-derived xenograft organoids have been used to capture breast tumor heterogeneity, for biobanking and for drug screening [16].

Several approaches for mammary organoid culture have been developed, including derivation of organoids of primary origin (mammary epithelial or breast tumor explants) [7, 17–19], organoid culture from mammary stem cells/breast cancer cells in 3D ECM [16, 20, 21], and organoid coculture with mammary stromal cells [19, 22]. Regardless of the specific organoid culture techniques used, morphological and immunohistological analysis of organoids is an integral part of all studies performed on organoids and a source of invaluable information. Classic histology techniques employing sample sectioning are great approaches to visualize cellular and subcellular details of organoids (using light and electron microscopy, respectively), but these methods lack the valuable 3D information. Therefore, techniques for 3D imaging of whole mammary organoids have been developed, which often employ optical clearing to increase penetrance of light through the thick organoid structures [23]. To overcome other common difficulties of whole organoid staining and imaging, which are decreased antibody penetrance through bulk ECM and long distance of organoids from the objective imposed by organoid culture in 3D ECM, current protocols often extract the organoid from the ECM [23]. However, for studies focused on organoid interaction with its stromal microenvironment, including ECM and/or stromal cells, preservation of the 3D ECM (eventually containing the stromal cells) in the vicinity of the organoid is required.

Therefore, we developed a novel protocol for organoid staining for high-resolution 3D imaging, which preserves the proximate 3D stromal microenvironment of the organoid. In our method the bulk ECM distant from organoid of interest is mechanically removed by dissection. Using gentle manipulation under a stereoscope/dissection microscope, the organoid of interest and its proximate stroma are together removed from the 3D organoid culture and stained in small (microliter) volumes of antibody staining solutions. Thus, our method not only overcomes both problems of antibody penetration and long distance to organoid during imaging, but it is also cost-effective. With traditional whole 3D culture staining methods, which required overlaying whole 3D cultures with antibody staining solutions, the cost of antibodies has been a considerable issue.

Here, we provide a detail protocol for mammary organoid dissection and collection from bulk 3D cultures, and for their further processing for successful 3D imaging, including organoid staining, clearing and mounting (Fig. 1). The procedure can be completed in 3 days. This method is of elemental importance and

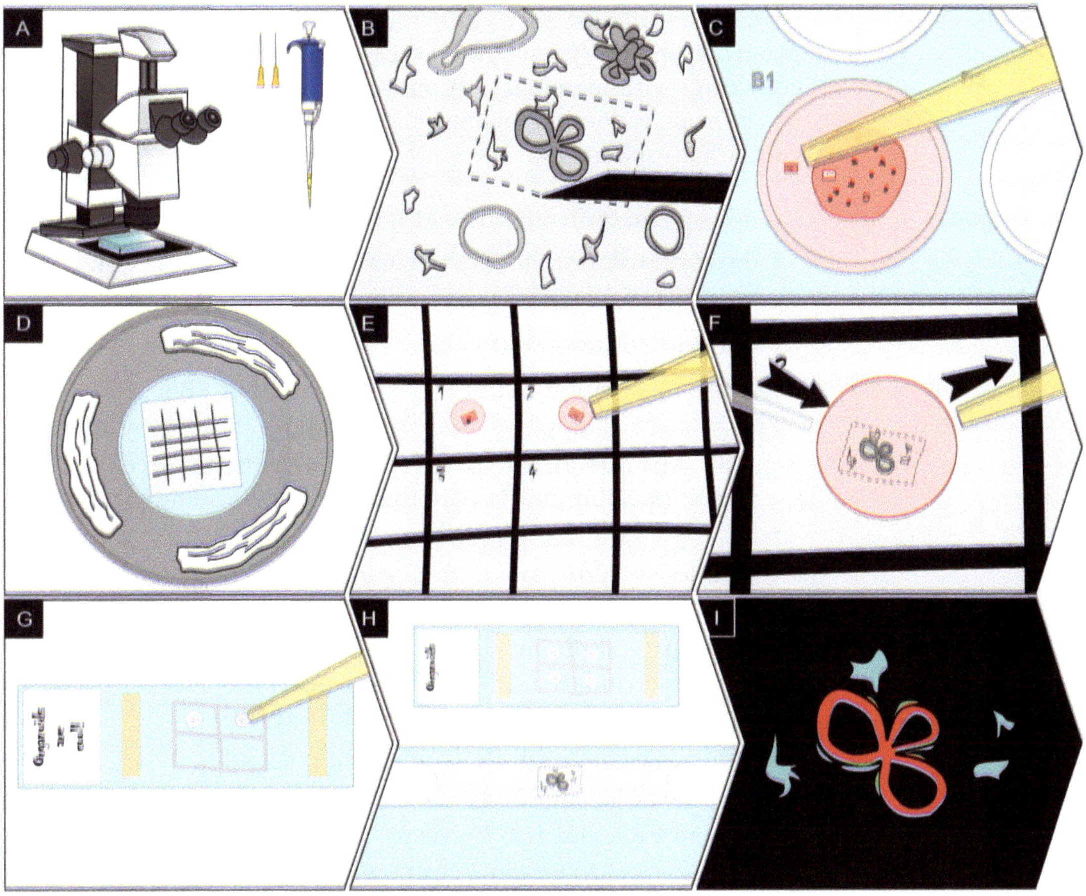

Fig. 1 Overview of the procedure. (**a**) A stereoscope, needles, and a micropipette are required for collection of organoids. (**b**) Using needles, an organoid is cut out from the 3D culture. (**c**) Organoid is transferred from a cell culture dish to a staining chamber using a micropipette. (**d**) Schematic representation of the staining chamber placed within a wet chamber. (**e**) Collected organoids are placed on the parafilm-coated surface of the staining dish at marked positions. (**f**) Organoid washing procedure. (**g**) After immunostaining, the organoids are transferred onto a coverslip in positions demarcated with pap pen. (**h**) Schematics of a top and side view of the organoid mounted on a coverslip. (**i**) Schematics of expected results after imaging of organoids processed by this method

usefulness for studies focusing on the role of stromal cells (such as fibroblasts, immune cells, endothelial cells) in mammary epithelial development or tumor growth, as well as for studies on ECM remodeling, either by stromal or epithelial cells, but can be very well applied also to epithelium-centric studies.

2 Materials

2.1 Biological Materials

Mammary and breast cancer organoids or spheroids and their cocultures with stromal cells (*see* **Note 1**). Using this protocol, we have successfully stained and imaged mouse primary mammary gland organoids (from 7–10 weeks old ICR females) and their cocultures with mammary fibroblasts, human primary breast organoids from reduction mammoplasty and MCF7 (human breast cancer line) spheroids.

2.2 Reagents

1. Phosphate buffered saline (PBS).
2. Fixation solution: 8% (v/v) paraformaldehyde (PFA) in PBS.
3. Permeabilization solution: 0.5% (v/v) Triton X-100 in PBS.
4. Blocking solution: 10% (v/v) FBS in PBS (*see* **Note 2**).
5. Washing solution: 0.05% (v/v) Tween 20 in PBS.
6. Staining solution: 8% (v/v) FBS, 0.1% (v/v) Triton X-100 in PBS. Mix 800 μL of blocking solution with 200 μL of permeabilization solution. Prepare fresh for each staining procedure.
7. Primary antibody staining solution: Usually a 1:100–1:1000 dilution of primary antibody/antibodies in staining solution (*see* **Note 3**).
8. Secondary antibody staining solution: Usually a 1:1000 dilution of secondary antibody/antibodies conjugated with a fluorophore (*see* **Note 4**) and 2 ng/mL DAPI.
9. (Optional) fluorophore-conjugated phalloidin.
10. Glycerol solutions: 50% (v/v) glycerol in PBS, 75% (v/v) glycerol in PBS, and 100% glycerol (*see* **Note 5**).
11. Immersion liquid, water-, oil- or glycerol-based, depending on the microscope objectives.

2.3 Equipment

1. Syringe needles, 28 gauge.
2. Micropipettes, 10 μL, 20 μL, 100 μL, 1000 μL, and sterile pipette tips.
3. Stereoscope (e.g., Leica M165 C; Leica Microsystems).
4. Plastic Petri dishes, 10 cm diameter (*see* **Note 6**).
5. Parafilm, 10 cm width.
6. Safe-lock tubes, 1.5 mL and 0.5 mL.
7. Coverslip glasses, 55 × 24 mm.
8. Glass microscope slides.
9. Double-sided tape.
10. Scalpel.
11. Confocal microscope (laser scanning or spinning-disc).
12. Glass Petri dish,15 cm diameter.
13. Hydrophobic barrier pap pen.
14. Wet chamber (*see* **Note 7**).

3 Methods

3.1 Organoid Culture Remarks

1. Culture organoids according to your standard culture protocol and experimental requirements (*see* **Note 8**).
2. Check organoid growth regularly using a bright-field microscope.
3. When the organoids achieve the desired morphology and/or at the desired timepoints as per experiment, the organoids are ready for collection, fixation, and subsequent staining and imaging.
4. Organoids can be collected from ongoing living cultures either at the endpoint of the culture, or at various timepoints as required by the experiment (*see* Subheading 3.2.2).
5. Alternatively, whole organoid cultures can be fixed with 4% PFA in PBS, washed twice with PBS and stored in PBS at 4 °C until organoid collection (*see* Subheading 3.2.3).

3.2 Organoid Collection

3.2.1 Preparation of a Staining Chamber Dish

1. Take a plastic 10 cm culture dish and lay it with parafilm.
2. Using a pen, draw a grid of lines approximately 1 cm apart on the parafilm.
3. Mark each square of the grid with a number to indicate positions.

3.2.2 Organoid Collection from Living Cultures, Followed by Fixation

1. Under a stereoscope, select an organoid for collection and using 28 G needles, cut the organoid out of the Matrigel (Fig. 1a, b) (*see* **Note 9**).
2. Using a micropipette with 10 μL tip, transfer the organoid (by aspirating it into the tip and then pushing it out of the tip) onto the parafilm in the staining chamber dish (Fig. 1c, d) (*see* **Note 10**). Add some extra culture medium to the organoid to the total volume of 20 μL.
3. Continue to collect as many organoids as needed (Fig. 1e) (*see* **Note 11**).
4. When organoid collection is complete, move the culture vessel back to the incubator (if later timepoints of the organoid culture are needed for collection or other analysis) or dispose of it according to appropriate procedures.
5. Add 20 μL of fixation solution (to the final volume of 40 μL and the final PFA concentration of 4%) to the drops with organoids on the parafilm and incubate for 30 min at room temperature (RT). Continue with steps described in Subheading 3.3.

3.2.3 Organoid Collection from Fixed 3D Cultures

1. Under a stereoscope, select an organoid for collection and using 28 G needles, cut the organoid out of the Matrigel.
2. Use the two needles as tweezers to carefully transfer the organoid onto the parafilm in the staining chamber dish (*see* **Note 12**).
3. Continue to collect as many organoids as needed.
4. When organoid collection is finished, place the fixed culture back to the fridge for further storage (if needed for further analysis) or dispose of it following the appropriate procedures.
5. To further process the collected organoids, proceed to Subheading 3.3.

3.3 Blocking and Permeabilization

1. From now on, work under the stereoscope. Prepare two pipettes, a 10 μL pipette to aspirate and dispose of solutions, and a 20 μL pipette to add new solutions, and a 1.5 mL tube for liquid waste collection.
2. Under the visual control of the stereoscope (usually 2× to 3× magnification is optimal) aspirate the fixation solution and dispose of it in the liquid waste tube.
3. Add 20 μL of washing buffer. To avoid drying of the sample, add washing buffer immediately after aspirating the previous solution, and process the drops one by one (Fig. 1f).
4. Incubate organoids in washing buffer for 30 min at RT.
5. Repeat two more times to wash the organoids three times in total after the fixation (*see* **Note 13**).
6. Aspirate washing buffer and add 20 μL of permeabilization solution. Incubate for 1 h at RT in a wet chamber to permeabilize the organoids.
7. Rinse once with 20 μL of washing solution.
8. Add 20 μL of blocking solution and incubate for 1 h at RT (or at 4 °C overnight) in a wet chamber.

3.4 Staining

1. Prepare 20 μL of primary antibody staining solution per each drop (*see* **Note 14**).
2. Aspirate blocking solution and add primary antibody staining solution to the organoids, 20 μL per each drop. Make a note of the antibodies used for each position. Incubate at 4 °C overnight in a wet chamber.
3. Wash organoids three times with washing solution, 30 min each wash.
4. Prepare secondary antibody staining solution, 20 μL per each drop (*see* **Note 15**). If desired, add fluorophore-conjugated phalloidin (at a 1:100 dilution) and/or nuclear dye to the secondary antibody staining solution as well.

5. Aspirate washing solution and add secondary antibody staining solution to the organoids, 20 μL per each drop. Incubate at 4 °C overnight in a wet chamber in dark (*see* **Note 16**).
6. Wash the organoids three times with washing solution, 20 μL and 30 min incubation per each wash.

3.5 Clearing

1. Aspirate the washing solution and add 20 μL of 50% glycerol to the organoids. Incubate for 30 min at RT in dark.
2. Aspirate the 50% glycerol solution and add 20 μL of 75% glycerol to the organoids (*see* **Note 17**). Incubate for 30 min at RT in dark.
3. Aspirate the 75% glycerol and add 20 μL of 100% glycerol to the organoids. Incubate for 30 min at RT in dark (*see* **Note 18**).

3.6 Mounting and Microscopy

1. Prepare a microscopy slide for mounting:
   (a) Use two layers of double-sided tape as spacers to create space between the slide and a coverslip (*see* **Note 19**).
   (b) Using a hydrophobic barrier pap pen, draw three horizontal lines and three vertical lines to create four squares of approximately 1 × 1 cm for individual samples.
2. Using a micropipette with cut tip, transfer the organoid with approximately 10 μL of glycerol on the microscopy slide in the middle of each square (Fig. 1g). Make a note of sample position on the slide. Continue transferring other organoids until all positions on the pap pen-drawn grid are occupied.
3. Carefully cover the samples with a coverslip. Do not apply direct pressure on the organoids, only gently attach the coverslip to the tape (Fig. 1h).
4. Proceed immediately to imaging or store the slides at 4 °C in darkness for a few days (*see* **Note 20**).
5. Use a confocal microscope with appropriate objectives and immersion liquid to image the organoids (Figs. 1i and 2). To achieve the best results, match the refractive index of glycerol with glycerol immersion.

4 Notes

1. Other types of organoids or spheroids can be processed using this method as well.
2. Due to the serum content, the blocking solution is prone to contamination. To avoid contamination of the solution, prepare it using aseptic techniques in a cell culture hood. Then, aliquot the solution per 1 mL into safe-lock 1.5 mL tubes and use a new aliquot for each staining. Store the unused aliquots at

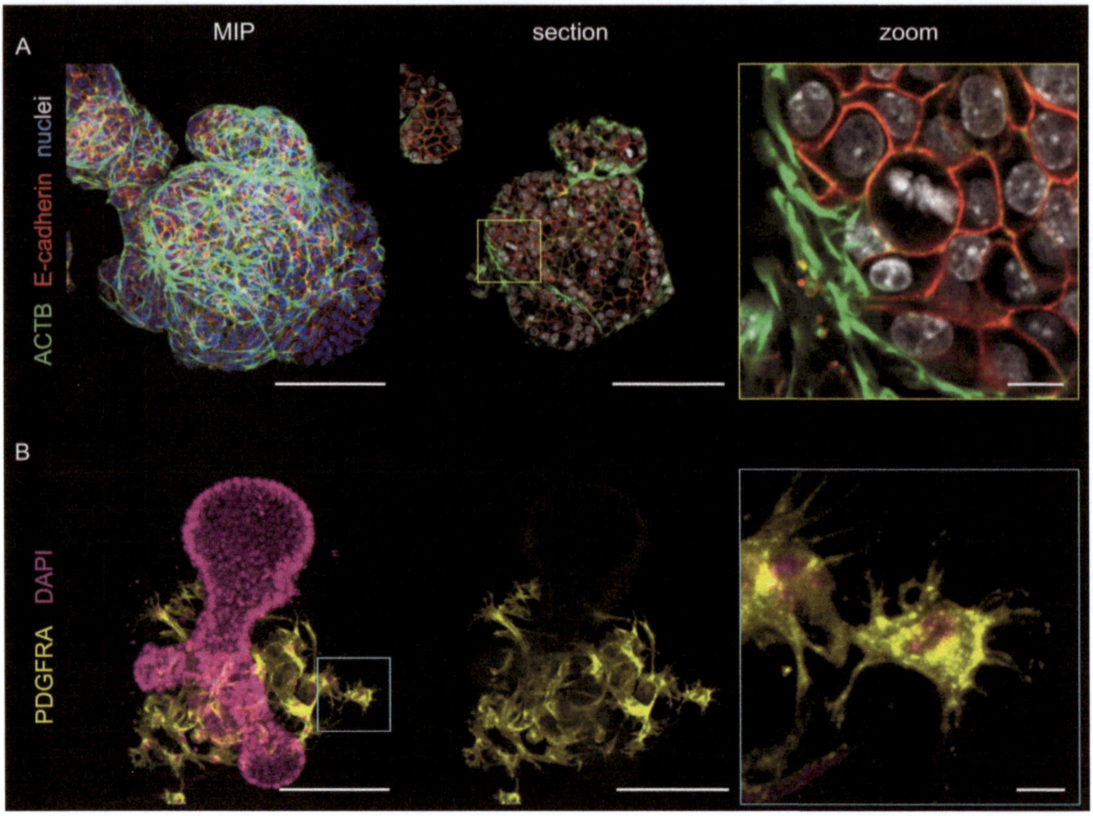

Fig. 2 Representative results of the staining procedure. The images show maximum intensity projection (MIP), optical section, and zoomed-in images from confocal imaging of organoids. (**a**) Lactating mammary organoid stained with a rat anti-E-cadherin antibody, an anti-rat secondary antibody Alexa Fluor 555, phalloidin-Alexa Fluor 488, and DAPI. (**b**) Mammary organoid co-cultured with mammary fibroblasts, stained with a rabbit anti-PDFRA antibody, anti-rabbit secondary antibody Alexa Fluor 568, and DAPI. Scale bars: 100 μm in MIP and section, 10 μm in zoom

4 °C for short-term storage or at −20 °C for long-term storage.

3. When performing double-staining, use a mixture of primary antibodies from different hosts, e.g., antibody against protein 1 raised in mouse, antibody against protein 2 raised in rabbit.
4. Make sure that the secondary antibodies are produced in the same species.
5. Glycerol is a viscous liquid, pipette it carefully to achieve exact volumes. To prepare a 50% glycerol solution, mix 25 mL of glycerol with 25 mL of PBS. To prepare a 75% glycerol solution, mix 32.5 mL of glycerol and 12.5 mL of PBS.
6. The Petri dishes will be used to build a staining chamber. Therefore, they do not need to be sterile or new/unused. Rather, it is convenient and eco-friendly to reuse Petri dishes

previously used for staining or for cell culture. Simply wash the used Petri dishes with water, clean them with 70% ethanol and let dry before using them to build the staining chamber.

7. To build a wet dark chamber, first wrap a 15 cm glass Petri dish (separately the bottom and the lid) in a tin foil. Lay the sides of the bottom dish with wet tissue paper, wetted with distilled water. To use the wet chamber, place the staining chamber dish in the bottom dish of the wet chamber, further cover the staining dish with a wet tissue paper, and enclose it by the cover part of the wet chamber.
8. To increase manual access to the organoids for collection during the following steps, it is proficient to culture the organoids in wells or dishes of the size no smaller than the well of a 12-well plate.
9. Work as clean as possible, wash instruments and gloves with 70% ethanol and wear a facemask, while collecting the organoids. Ensure that your organoid culture medium contains antibiotics.
10. When the organoid is large and would not fit through the opening of a 10 μL tip, use a cut 10 μL tip or a cut 20 μL tip with the appropriate micropipette.
11. When collecting organoids from different culture conditions, timepoints etc., make a list of positions with indication of organoid origin.
12. After fixation, Matrigel becomes sticky. Therefore, do not use a micropipette to transfer the organoid, it could get stuck inside the pipette tip.
13. At this point organoids can be stored in PBS for several days at 4 °C in the wet chamber.
14. Most antibodies are too concentrated for direct preparation of the small volumes of the staining solution required by this protocol. Preparation of diluted antibody solutions is recommended to obtain solutions amenable to pipetting. If an antibody is to be used at 1:100–1:1000 dilution in the staining solution, prepare a ten times diluted solution of the antibody by mixing 1 μL of the antibody with 9 μL of the blocking solution. If an antibody is to be used at dilution higher than 1:1000, prepare a 1:100 dilution by mixing 1 μL of the antibody with 99 μL of the blocking solution. The diluted antibodies can be used for several weeks if stored at 4 °C.
15. Similar to the primary antibody staining solution, the preparation of the secondary antibody staining solution might require preparation of diluted secondary antibody solutions and DAPI in blocking solution. Secondary antibodies are usually used at 1:500–1:1000 dilution, DAPI at 1:3000 dilution in the final

solution. Protect both the diluted solution and the secondary antibody staining solution from light.

16. From this point on, avoid direct exposure of organoids to light. Work in a room with lights off and decrease lamp intensity of the stereoscope.
17. Successful clearing results in decreased visibility (increased translucency) of organoids. Be careful when aspirating the solutions to not lose the organoid.
18. When clearing the organoids with 100% glycerol, keep the staining chamber in dark, but avoid incubation in a wet chamber. Moistened glycerol has decreased clearing abilities.
19. Adjust the number of layers of the double-sided tape based on the size of the organoids embedded. For most tapes, one layer of tape should be used per every 200 μm of the sample.
20. Make sure to take the samples out from the fridge well in advance to imaging to allow glycerol to reach RT for the best clearing results.

Acknowledgments

We acknowledge the core facility CELLIM of CEITEC supported by the Czech-BioImaging large RI project (LM2018129 funded by MEYS CR) for their support with obtaining scientific data presented in this paper. This work was supported by the Grant Agency of Masaryk University (projects no. MUNI/G/1446/2018 and MUNI/A/1382/2019), and by funds from the Faculty of Medicine MU to junior researcher Z.K. (ROZV/28/LF/2020).

References

1. Sumbal J, Budkova Z, Traustadóttir GA, Koledova Z (2020) Mammary organoids and 3D cell cultures: old dogs with new tricks. J Mammary Gland Biol Neoplasia 25(4):273–288
2. Ewald AJ, Brenot A, Duong M, Chan BS, Werb Z (2008) Collective epithelial migration and cell rearrangements drive mammary branching morphogenesis. Dev Cell 14: 570–581
3. Neumann NM, Perrone MC, Veldhuis JH, Huebner RJ, Zhan H, Devreotes PN et al (2018) Coordination of receptor tyrosine kinase signaling and interfacial tension dynamics drives radial intercalation and tube elongation. Dev Cell 45:67–82.e6
4. Zhang X, Martinez D, Koledova Z, Qiao G, Streuli CH, Lu P (2014) FGF ligands of the postnatal mammary stroma regulate distinct aspects of epithelial morphogenesis. Development 141:3352–3362
5. Simian M, Hirai Y, Navre M, Werb Z, Lochter A, Bissell MJ (2001) The interplay of matrix metalloproteinases, morphogens and growth factors is necessary for branching of mammary epithelial cells. Development 128: 3117–3131
6. Centonze A, Lin S, Tika E, Sifrim A, Fioramonti M, Malfait M et al (2020) Heterotypic cell–cell communication regulates glandular stem cell multipotency. Nature 584: 608–613
7. Sumbal J, Chiche A, Charifou E, Koledova Z, Li H (2020) Primary mammary organoid model of lactation and involution. Front Cell Dev Biol 8:68

8. Alladin A, Chaible L, Garcia del Valle L, Sabine R, Loeschinger M, Wachsmuth M et al (2020) Tracking cells in epithelial acini by light sheet microscopy reveals proximity effects in breast cancer initiation. elife 9:e54066
9. Sirka OK, Shamir ER, Ewald AJ (2018) Myoepithelial cells are a dynamic barrier to epithelial dissemination. J Cell Biol 217:3368–3381
10. Georgess D, Padmanaban V, Sirka OK, Coutinho K, Choi A, Frid G et al (2020) Twist1-induced epithelial dissemination requires Prkd1 signaling. Cancer Res 80: 204–218
11. Cheung KJ, Gabrielson E, Werb Z, Ewald AJ (2013) Collective invasion in breast cancer requires a conserved basal epithelial program. Cell 155:1639–1651
12. Havas KM, Milchevskaya V, Radic K, Alladin A, Kafkia E, Garcia M et al (2017) Metabolic shifts in residual breast cancer drive tumor recurrence. J Clin Invest 127(6):2091–2105
13. Duarte AA, Gogola E, Sachs N, Barazas M, Annunziato S, de Ruiter JR et al (2018) BRCA-deficient mouse mammary tumor organoids to study cancer-drug resistance. Nat Methods 15:134–140
14. Sumbal J, Koledova Z (2019) FGF signaling in mammary gland fibroblasts regulates multiple fibroblast functions and mammary epithelial morphogenesis. Development 146(23): dev185306
15. Koledova Z, Zhang X, Streuli C, Clarke RB, Klein OD, Werb Z et al (2016) SPRY1 regulates mammary epithelial morphogenesis by modulating EGFR-dependent stromal paracrine signaling and ECM remodeling. Proc Natl Acad Sci U S A 113(39):E5731–E5740
16. Sachs N, de Ligt J, Kopper O, Gogola E, Bounova G, Weeber F et al (2018) A living biobank of breast cancer organoids captures disease heterogeneity. Cell 172:373–386.e10
17. Padmanaban V, Grasset EM, Neumann NM, Fraser AK, Henriet E, Matsui W et al (2020) (2020) Organotypic culture assays for murine and human primary and metastatic-site tumors. Nat Protoc 15(8):2413–2442
18. Nguyen-Ngoc K-V, Shamir ER, Huebner RJ, Beck JN, Cheung KJ, Ewald AJ (2015) 3D culture assays of murine mammary branching morphogenesis and epithelial invasion. Methods Mol Biol 1189:135–162
19. Koledova Z, Lu P (2017) A 3D fibroblast-epithelium co-culture model for understanding microenvironmental role in branching morphogenesis of the mammary gland. Methods Mol Biol 1501:217–231
20. Miller DH, Sokol ES, Gupta PB (2017) 3D primary culture model to study human mammary development. Methods Mol Biol 1612: 139–147
21. Linnemann JR, Miura H, Meixner LK, Irmler M, Kloos UJ, Hirschi B et al (2015) Quantification of regenerative potential in primary human mammary epithelial cells. Development 142:3239–3251
22. Koledova Z (2017) 3D coculture of mammary organoids with fibrospheres: a model for studying epithelial-stromal interactions during mammary branching morphogenesis. Methods Mol Biol 1612:107–124
23. Dekkers JF, Alieva M, Wellens LM, Ariese HCR, Jamieson PR, Vonk AM et al (2019) High-resolution 3D imaging of fixed and cleared organoids. Nat Protoc 14:1756–1771

Chapter 15

Divide and Conquer: Isolating Cell Populations to Investigate How Breast Cancer Risk Factors Alter the Breast Microenvironment

Lisa M. Arendt

Abstract

Breast cancer is a multifactorial disease with risk factors that are fixed or modifiable. Understanding how these risk factors interact within breast tissue may provide insight into how to improve interventions or chemoprevention strategies to reduce breast cancer incidence. Here we describe methods to utilize breast tissue from patients with defined risk factors undergoing reduction mammoplasty or prophylactic mastectomy to isolate epithelial cells, stromal cells, adipocytes, and macrophages to investigate how risk factors impact distinct cell populations within breast tissue. Following enzymatic digestion of breast tissue, adipocyte-enriched, stromal cell, and epithelial organoid fractions can be isolated. Using antibody-conjugated beads, further cell populations, such as macrophages, can be isolated for molecular analysis. These methods can be adapted to sequentially isolate other cell populations based on specific cell surface markers and are useful for small-sized breast tissue specimens.

Key words Breast, Macrophages, Adipocytes, Stromal cells, Reduction mammoplasty, Mammary epithelial cells

1 Introduction

Breast cancer is the most common cancer in women worldwide and the second most common cancer overall [1]. Multiple risk factors have been identified that lead to increased breast cancer risk. Although gender and age remain the most critical risk factors [2], risk is also enhanced by reproductive traits, genetic factors, breast density, and lifestyle factors, such as obesity and excess alcohol consumption [2–6]. Much of our understanding of how these factors interact to enhance breast cancer risk has been uncovered through clinical observations and epidemiological studies, and direct impacts of these risk factors in human breast tissue are not well understood. Investigating how risk factors alter breast tissue

Maria dM. Vivanco (ed.), *Mammary Stem Cells: Methods and Protocols*,
Methods in Molecular Biology, vol. 2471, https://doi.org/10.1007/978-1-0716-2193-6_15,

may improve interventions or chemoprevention strategies to reduce breast cancer incidence.

Many environmental exposures and risk factors for cancer are associated with chronic inflammation [7]. Chronic macrophage-induced inflammation within the mammary gland is a component of multiple risk factors examined in preclinical models of breast cancer [8–11]. Within the mammary gland, macrophages have been shown to play a role in regulation of the mammary epithelial stem cell niche [12, 13], and inflammatory macrophages have been shown to alter the secretome and function of stromal cells in breast tissue, including adipocytes and adipose stem cells [14, 15]. These observations suggest that changes in macrophage-induced inflammation due to breast cancer risk factors may alter both the epithelial stem cell niche as well as surrounding stromal cells. However, relationships among these cell populations and how they are altered in the context of risk factors for breast cancer have not been clearly defined within human breast tissue.

We have developed simple methods to dissect changes in cell populations within breast tissue utilizing non-diseased breast tissue isolated from women with defined breast cancer risk factors. Breast tissue is collected from women undergoing either reduction mammoplasty surgeries or prophylactic mastectomies due to genetic or familial risk factors. These breast tissue specimens are subjected to mechanical dissociation using a scissors under sterile conditions followed by enzymatic digestion to isolate cell populations of mature adipocytes, stromal cells and breast epithelial cell organoids (Fig. 1). Following the initial digestion, these cell fractions can be cultured to generate adherent cells for use in cell-based assays (Fig. 2) or further fractionated to isolate CD68-positive macrophages. Similar to recent studies in murine mammary glands [16], we have observed CD68-positive macrophages within breast lobules both in the stroma as well as interdigitating between epithelial cells (Fig. 3a). Utilizing epithelial and stromal cell-enriched fractions, macrophages can be isolated with high viability for molecular applications to identify functional changes (Fig. 3b). Although we have focused on macrophages, these methods can be adapted to sequentially isolate other immune cell populations based on cell type-specific surface markers. Further, this protocol is useful for breast tissue samples of limited quantity, which may make traditional methods of isolating cells using fluorescence-activated cell sorting (FACS) challenging.

2 Materials

2.1 Equipment

1. Biosafety cabinet.
2. 37 °C hybridization oven with tube rotator.

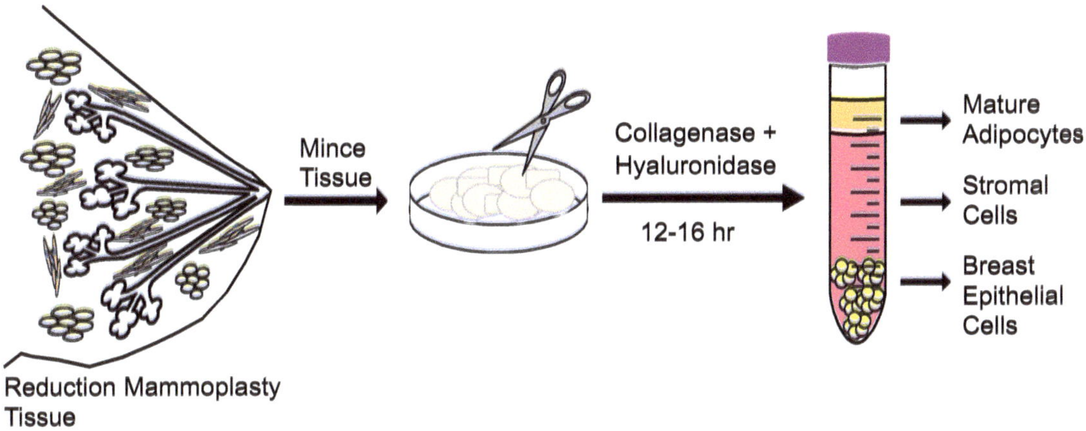

Fig. 1 Digestion of breast tissue to isolate distinct cell types. Breast tissue collected from reduction mammoplasty surgeries or prophylactic mastectomies can be minced into tissue fragments prior to enzymatic digestion with collagenase A and hyaluronidase. Following digestion, the tissue separates into 3 layers that can be utilized to isolate enriched fractions of mature adipocytes, stromal cells including adipose stem cells and fibroblasts, and breast epithelial cells. Within the stromal and breast epithelial cell fractions, further cell populations, including macrophages can be isolated

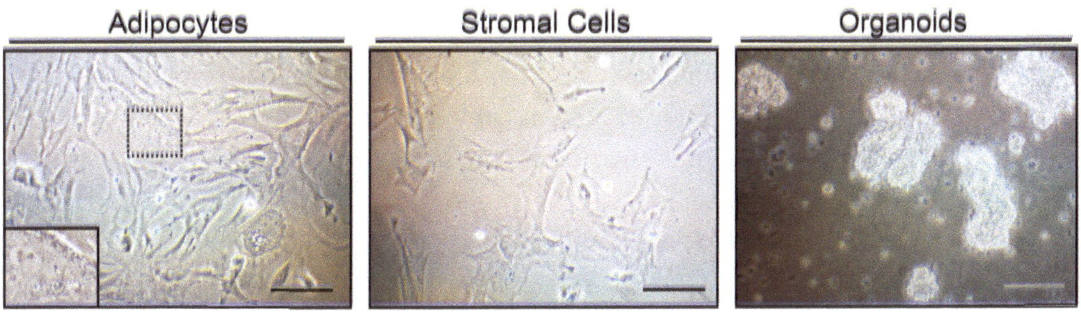

Fig. 2 Appearance of cells in each digestion fraction. Adipocytes isolated from the lipid layer of the digestion media and grown in culture lose their central lipid droplet within 1–5 days. Residual lipid can be observed in the adipocyte cytoplasm (inset). The stromal cell population contains a heterogenous population of mesenchymal cells, which is enriched for adipose stem cells and fibroblasts. The organoid fraction is enriched for clusters of breast epithelial cells, which can be isolated using further digestion with trypsin. Magnification bar = 50 μm

3. Tissue culture incubator set at 37 °C with 5% CO_2.
4. Water bath set to 37 °C.
5. Table-top centrifuge.
6. Refrigerated microcentrifuge.
7. DynaMag-5 magnet (Thermo Fisher Scientific, cat. #12303D).
8. Hemocytometer or automated cell counter.

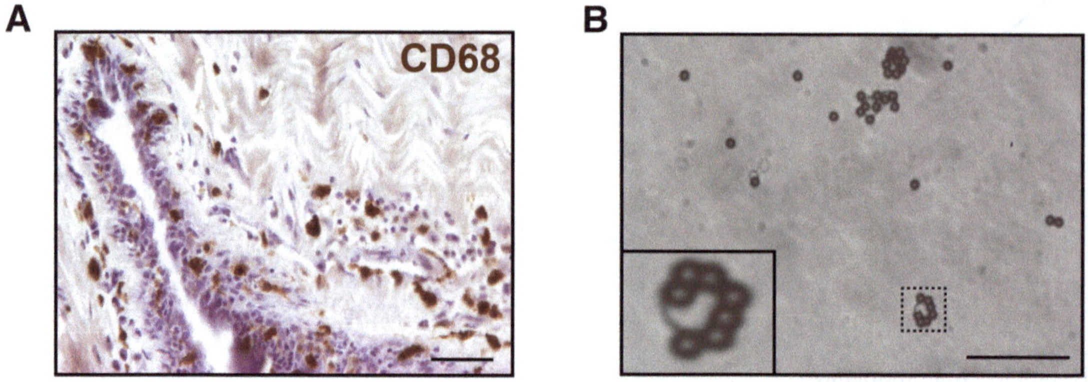

Fig. 3 Localization and isolation of macrophages from breast tissue. (**a**) Representative image of CD68-positive macrophages in breast tissue isolated from reduction mammoplasty surgery. CD68-positive macrophages are observed interdigitating among epithelial cells within the breast duct as well as in the surrounding extracellular matrix. (**b**) CD68-positive macrophages attached to beads following isolation from organoid and stromal cell fractions. Magnification bar = 50 μm

2.2 Reagents

1. Reduction mammoplasty tissue (*see* **Notes 1** and **2**).
2. Sterile Mayo scissors or razor blades.
3. 15 and 50 mL polypropylene conical tubes (*see* **Note 3**).
4. 12 mm × 75 mm 5 mL round bottom polystyrene tubes.
5. 100 mm × 20 mm tissue culture plates.
6. 1.5 mL microcentrifuge tubes.
7. Parafilm.
8. Plastic aspirating pipettes (*see* **Note 4**).
9. 0.05% trypsin 0.53 mM EDTA.
10. Bovine serum albumin (BSA).
11. Red blood cell lysis buffer: 0.15 M NH_4Cl, 0.01 M $KHCO_3$, 0.03 mM EDTA.
12. 40 μm basket cell strainer.
13. 10 mL syringes and 18 g needles.
14. DNAse I.
15. CD16/32 rat anti-mouse/human monoclonal antibody (clone 93).
16. CD68 mouse anti-human monoclonal antibody (clone Y1/82A) (*see* **Note 5**).
17. Dynabeads Pan Mouse IgG beads (Thermo Fisher Scientific, cat. #11041).

2.3 Media

1. Organoid Media: DMEM-Ham's F12 medium supplemented with 5% fetal calf serum, 10 μg/mL insulin, 10 ng/mL EGF,

0.5 μg/mL hydrocortisone, and 1% antibiotic-antimycotic (*see* **Note 6**).

2. Digestion Media: Organoid Media supplemented with 0.5 mg/mL collagenase A and 50 units/mL hyaluronidase.
3. Stromal Cell Media: DMEM supplemented with 10% fetal calf serum and 1% antibiotic-antimycotic.
4. Wash Buffer: 1 × Ca^{2+} and Mg^{2+} free phosphate buffered saline (PBS) supplemented with 5% fetal calf serum.
5. Isolation Buffer: PBS supplemented with 0.1% BSA and 2 mM EDTA, filter sterilized.

3 Methods

3.1 Digestion of Reduction Mammoplasty Tissue

1. Transfer reduction mammoplasty tissue into a biosafety cabinet and mince the sample with sterile razor blades or Mayo scissors until tissue pieces measure approximately 2–3 mm^3 (*see* **Notes** 7 and **8**).
2. Fill 15 mL conical tubes with 10 mL of Digestion Media. Add minced tissue to fill the tube to the 15 mL marker (approximately 2 g of minced tissue) leaving space at the top to facilitate movement of the liquid during incubation. Cover the tube and seal with Parafilm. Invert the tube to make sure that the tissue can mix with the media. If the tissue does not mix, remove tissue from the top of the tube to create more space (*see* **Note 8**).
3. Incubate the filled 15 mL tubes at 37 °C with rotation in a hybridization oven for approximately 12–16 h or until the tissue is homogenously digested (*see* **Note 9**).
4. Remove the tubes from the incubator, then centrifuge the tubes for 1 min at 9 × *g* to separate the dissociated tissue into an oil/fat layer, middle aqueous layer enriched for stromal cells, and bottom organoid pellet.

3.2 Isolation of Adipocytes

1. Remove the oil/fat layer with a 5 mL pipette and transfer to 15 mL conical tubes containing 10 mL of sterile PBS.
2. Cover the tubes, then gently invert 3–6 times to mix the contents of the tubes.
3. Centrifuge tubes for 1 min at 9 × *g* to separate into an oil/fat layer and aqueous layer (*see* **Note 10**).
4. Remove the oil/fat layer with a 5 mL pipette and plate on 100 mm × 20 mm tissue culture plates with 8 mL of Stromal Cell Media (*see* **Note 11**).

5. Aqueous layer and any organoids that pelleted at the bottom of the tubes can be combined with the aqueous and organoid layers from the initial digestion.

3.3 Separation of Stromal Cells and Epithelial Cells

1. Remove the aqueous stromal fractions from the organoid pellets. Combine stromal fractions from multiple tubes if they are from the same patient sample in a separate 50 mL conical tube with a 10 mL pipette.
2. Resuspend organoid pellets with 10 mL of Wash Buffer into a 50 mL tube.
3. Centrifuge stromal cells and organoids at 300 × *g* for 5 min and aspirate the supernatant.
4. Resuspend the stromal cell pellet and organoid pellet each in 5 mL red blood cell lysis buffer. Incubate at room temperature for 5 min.
5. Add 10 mL of Wash Buffer to each tube and centrifuge at 300 × *g* for 5 min. Aspirate the supernatant.
6. Repeat **step 5** two additional times.
7. Resuspend the organoids in 10–30 mL of Stromal Cell Media and plate on 1–3 100 mm × 20 mm tissue culture plates (*see* **Note 12**). Incubate for 1–2 h at 37 °C in a tissue culture incubator. Dissociated stromal cells in the epithelial cell organoid fraction will adhere to the tissue culture plates, while the organoids will remain floating in the culture media (*see* **Note 13**).
8. Resuspend the stromal cells in 10 mL of Isolation Buffer and filter through a 40 μm basket filter into a 50 mL conical tube.
9. Count the stromal cells using a hemocytometer or cell counter. Centrifuge the cells at 300 × *g* for 5 min, remove the supernatant and resuspend the cells at a concentration of 1×10^7 cells/mL in Isolation Buffer.
10. Hold stromal cells on ice until starting the macrophage isolation.

3.4 Dissociation of Breast Epithelial Cells

1. Collect the non-adherent organoids into a 50 mL conical tube with a pipette and wash the plate with an additional 10 mL of PBS to remove any remaining organoids (*see* **Note 14**).
2. Centrifuge cells at 300 × *g* for 5 min. Aspirate and discard the supernatant.
3. Add 10 mL of 0.1% BSA in PBS and mechanically dissociate the pellet by passing it through an 18 g needle attached to a 10 mL syringe 8–10 times. Centrifuge at 300 × *g* for 5 min. Aspirate and discard the supernatant.

4. Resuspend the pellet in 2 mL 0.5% trypsin-EDTA. Triturate the solution with a 1 mL pipette for 1 min to mechanically dissociate the organoids. Incubate the tube in a water bath warmed to 37 °C for 5 min. Triturate the solution again for 1 min, then return the tube to the water bath for an additional 5 min.
5. Add 10 mL of Stromal Cell Media to the tube to quench the trypsin and 100 μL of a 5 mg/mL stock of DNAse to remove aggregates of cells due to cell death. Mix thoroughly with a 10 mL pipette until the cell clusters disperse, then filter through a 40 μm basket filter into a 50 mL conical tube. Wash the filter with an additional 10 mL of Stromal Cell Media.
6. Centrifuge the cells at 300 × *g* for 5 min. Resuspend cells in Isolation Buffer and count the cells using a hemocytometer or cell counter. Centrifuge cells at 300 × *g* for 5 min and resuspend the cells at a concentration of 1×10^7 cells/mL.
7. Keep cells on ice prior to macrophage isolation.

3.5 Macrophage Isolation

1. To block Fc receptors on immune cells, add 0.5 μg of CD16/32 antibodies/1×10^6 cells to the tubes containing the stromal and epithelial cell fractions (*see* **Note 15**). Gently mix the tubes, then incubate cells with antibodies on ice for 30 min. Add 10 mL of Wash Buffer to the cells, then centrifuge at 300 × *g* for 5 min. Aspirate the supernatant, then resuspend each cell fraction at a concentration of 1×10^7 cells/mL in Isolation Buffer. Hold the cells on ice.
2. While the cells are incubating with the CD16/32 antibodies, prepare the CD68 antibody-coated beads. Resuspend the Dynabeads by vortexing the vial for >30 s and removing beads for sorting. The concentration of beads should be 25 μL/1×10^7 cells for sorting. Place beads in a 5 mL round bottom tube.
3. Add 1 mL of Isolation Buffer to the beads and vortex for 30 s. Place the tube in the magnet for 1 min and discard the supernatant.
4. Remove the tube from the magnet and resuspend the beads at the starting volume in Isolation Buffer. Transfer to a microcentrifuge tube.
5. Add 1 μg of CD68 antibody/25 μL of beads and incubate with rotation for 30 min at room temperature (*see* **Note 16** and **17**).
6. Transfer the antibody-coated beads to a 5 mL round bottom tube and place the tube in the magnet for 1 min.
7. Remove the supernatant, then remove the tube from the magnet. Resuspend the beads in 1 mL of Isolation Buffer. Place the tube in the magnet for 1 min.

8. Remove the supernatant, then remove the tube from the magnet. Resuspend the beads at the starting volume in Isolation Buffer.
9. Combine CD68 antibody-coated beads with the stromal cell and epithelial cell fractions in separate microcentrifuge tubes. Incubate the cells with the beads with rotation at 2–4 °C for 20 min (*see* **Note 18**).
10. Transfer cells and beads to 5 mL round bottom tubes, then add an additional 1 mL of Isolation Buffer to each tube. Place the tubes in the magnet for 1 min. Remove the supernatant from each tube into a separate 15 mL conical tube. The supernatants contain the stromal cell-enriched fraction and the epithelial cell-enriched fraction.
11. Remove the tubes from the magnet and resuspend the beads in 2 mL of Isolation Buffer with gentle pipetting. Place the tubes in the magnet for 1 min. Remove the supernatants and combine with the stromal and epithelial cell fractions.
12. Repeat **step 10** one additional time.
13. Remove the tubes from the magnet. The bead pellets from the two fractions can be combined. Resuspend the beads in 1 mL of Isolation Buffer. Transfer bead-conjugated cells to microcentrifuge tubes and centrifuge at 300 × *g* for 5 min in a refrigerated microcentrifuge. Remove the supernatant. The macrophage cell pellet can be snap frozen in liquid nitrogen and stored at −80 °C for molecular analysis (*see* **Notes 19** and **20**).
14. Centrifuge the stromal and epithelial cell fractions at 300 × *g* for 5 min. Aspirate the supernatant. The cell pellets can be snap frozen in liquid nitrogen for molecular analysis or plated and expanded in culture for further study (*see* **Note 21**).

4 Notes

1. Human breast tissue is obtained from patients undergoing elective breast tissue reduction mammoplasty surgeries or prophylactic mastectomies due to genetic diagnosis or familial breast cancer risk. All studies should be performed in accordance with guidelines for human studies research and institutional policies.
2. Working with cells from human breast tissue is a potential biohazard and requires the utilization of biosafety containment procedures. Human breast tissue may be a source of bloodborne pathogens. Risk assessments should be made in conjunction with institutional guidelines for biosafety prior to starting any procedures.

3. We recommend using polypropylene centrifuge tubes with a plug cap. This type of tube is less likely to leak or crack during rotation, reducing the potential for laboratory contamination from the human breast tissue.
4. Use of plastic aspirating pipettes reduces the potential for sharp injuries due to broken glass pipettes to limit potential for exposure to blood-borne pathogens.
5. We have optimized the concentration of the CD68 antibody (Thermo Fisher, cat # 14-0689-82) for these conditions. However, other anti-human mouse IgG antibodies can easily be substituted in this system. The primary antibody should be titrated to optimize the antibody concentration used. Some antibodies show reduced antigen-binding efficiency when coated onto beads, even when the antibody shows efficacy in other assays. Mouse IgG antibodies should not be used in this system on mouse-derived tissues due to cross-reaction with endogenous IgG antibodies.
6. We suggest a combination of 10,000 U/mL penicillin G, 10 mg/mL streptomycin sulfate, and 25 μg/mL amphotericin B, such as Corning Cellgro, cat. #30-004-CI.
7. The breast tissue should remain in a closed specimen cup on ice prior to mincing and collagenase digestion. Recovery of viable cells is maximized if the tissue is processed on the same day as the surgical excision.
8. Following collagenase digestion, if the tissue pieces are >1 mm^3, the tissue may not have been minced sufficiently in Subheading 3.1 **step 1**, or the tubes were filled with too much tissue which inhibited efficient movement of tissue within the tubes during the collagenase digestion (Subheading 3.1 **step 2**). Ensure that the Digestion Media in the tubes can move freely when rotated.
9. Tissue variability will determine the length of time for digestion. Tissue that is challenging to mince may need a longer digestion time due to high breast tissue collagen content. Tissue that has a higher fat content may need a shorter time. Over digestion of the breast tissue will reduce the size of the organoid clusters and increase the presence of epithelial cells in the aqueous stromal cell fraction.
10. The oil/fat layer frequently contains contaminating organoid clusters that were not removed during the initial centrifugation. Washing the oil/fat layer with PBS will allow the organoids to settle to the bottom of the tube. The organoids can be removed and combined with the rest of the organoid layer prior to isolation of single epithelial cells.

11. The oil/fat layer is enriched for mature adipocytes. The adipocytes will grow in culture and lose their lipid droplet during the first 1–5 days in culture (Fig. 2).
12. The organoids should be spread out on the plates, and the number of plates will vary depending on the size of the organoid pellet.
13. The organoid pellet is enriched for breast epithelial cells. However, there are still stromal cells suspended in the media surrounding the organoids. Plating cells in Stromal Cell Media for 2 h depletes stromal cells by approximately 50%, although this is dependent both on the individual patient samples and the efficiency of the collagenase digestion.
14. The stromal cells depleted from the organoid fraction can be maintained by feeding the cells that adhered to the plate with Stromal Cell Media. The complement of cells depleted from the organoid pellet is similar in composition to the stromal cells isolated in the aqueous layer.
15. Blocking the Fc receptor reduces nonspecific binding from other immune cells, such as dendritic cells and B cells, which improves the purity of the macrophage population.
16. When incubating the beads with the antibody or in later steps with the cells, the mixer should provide tilting and rotation of the tubes to ensure that the beads do not settle at the bottom of the tube.
17. Because the antibodies are conjugated directly to the beads, additional antibodies to detect cell surface proteins of other cell populations can be conjugated to beads for sequential sorting of different cell populations.
18. Because macrophages are phagocytic cells, the stromal cells should be maintained at 4 °C to reduce phagocytosis of the beads. The isolation buffer should be kept on ice for subsequent washing steps.
19. The bead-conjugated macrophages can be counted using a hemocytometer at this step. This will only give an estimate of the total number of macrophages isolated, because during the magnet steps, the macrophages will cluster due to close interactions with the beads.
20. The beads attached to the cells do not interfere with molecular analyses that include lysis of the cells in the first steps. We have observed that isolation of macrophages using the beads significantly improves cell viability compared to FACS and subsequent RNA quality. The macrophages should not be cultured following isolation due to phagocytosis of the beads in culture.

21. The stromal cell fraction contains a mixed population of cells, including adipose stem cells, fibroblasts, endothelial cells, and immune cells. Adipose stem cells can be isolated from this fraction through the use of FACS based on cell surface markers according to guidelines [17].

Acknowledgments

The author would like to thank Victoria Thompson for technical support. This work was supported by funds from the Department of Comparative Biosciences.

References

1. Bray F, Ferlay J, Soerjomataram I, Siegel RL, Torre LA, Jemal A (2018) Global cancer statistics 2018: GLOBOCAN estimates of incidence and mortality worldwide for 36 cancers in 185 countries. CA Cancer J Clin 68(6): 394–424. https://doi.org/10.3322/caac.21492
2. Thakur P, Seam RK, Gupta MK, Gupta M, Sharma M, Fotedar V (2017) Breast cancer risk factor evaluation in a Western Himalayan state: a case-control study and comparison with the Western World. South Asian J Cancer 6(3): 106–109. https://doi.org/10.4103/sajc.sajc_157_16
3. Miller ER, Wilson C, Chapman J, Flight I, Nguyen AM, Fletcher C, Ramsey I (2018) Connecting the dots between breast cancer, obesity and alcohol consumption in middle-aged women: ecological and case control studies. BMC Public Health 18(1):460. https://doi.org/10.1186/s12889-018-5357-1
4. Bissell MCS, Kerlikowske K, Sprague BL, Tice JA, Gard CC, Tossas KY, Rauscher GH, Trentham-Dietz A, Henderson LM, Onega T, Keegan THM, Miglioretti DL (2020) Breast cancer population attributable risk proportions associated with body mass index and breast density by race/ethnicity and menopausal status. Cancer Epidemiol Biomark Prev 29(10): 2048–2056. https://doi.org/10.1158/1055-9965.Epi-20-0358
5. Bravi F, Decarli A, Russo AG (2018) Risk factors for breast cancer in a cohort of mammographic screening program: a nested case-control study within the FRiCaM study. Cancer Med 7(5):2145–2152. https://doi.org/10.1002/cam4.1427
6. Godet I, Gilkes DM (2017) BRCA1 and BRCA2 mutations and treatment strategies for breast cancer. Integr Cancer Sci Ther 4(1): 10. https://doi.org/10.15761/icst.1000228
7. Aggarwal BB, Vijayalekshmi RV, Sung B (2009) Targeting inflammatory pathways for prevention and therapy of cancer: short-term friend, long-term foe. Clin Cancer Res 15(2): 425–430. https://doi.org/10.1158/1078-0432.Ccr-08-0149
8. Arendt LM, McCready J, Keller PJ, Baker DD, Naber SP, Seewaldt V, Kuperwasser C (2013) Obesity promotes breast cancer by CCL2-mediated macrophage recruitment and angiogenesis. Cancer Res 73(19):6080–6093. https://doi.org/10.1158/0008-5472.can-13-0926
9. Kiecolt-Glaser JK, Preacher KJ, MacCallum RC, Atkinson C, Malarkey WB, Glaser R (2003) Chronic stress and age-related increases in the proinflammatory cytokine IL-6. Proc Natl Acad Sci U S A 100(15):9090–9095. https://doi.org/10.1073/pnas.1531903100
10. Lyons TR, O'Brien J, Borges VF, Conklin MW, Keely PJ, Eliceiri KW, Marusyk A, Tan AC, Schedin P (2011) Postpartum mammary gland involution drives progression of ductal carcinoma in situ through collagen and COX-2. Nat Med 17(9):1109–1115. https://doi.org/10.1038/nm.2416
11. Provenzano PP, Inman DR, Eliceiri KW, Knittel JG, Yan L, Rueden CT, White JG, Keely PJ (2008) Collagen density promotes mammary tumor initiation and progression. BMC Med 6: 11. https://doi.org/10.1186/1741-7015-6-11
12. Gyorki DE, Asselin-Labat ML, van Rooijen N, Lindeman GJ, Visvader JE (2009) Resident macrophages influence stem cell activity in the mammary gland. Breast Cancer Res 11(4): R62. https://doi.org/10.1186/bcr2353

13. Chakrabarti R, Celia-Terrassa T, Kumar S, Hang X, Wei Y, Choudhury A, Hwang J, Peng J, Nixon B, Grady JJ, DeCoste C, Gao J, van Es JH, Li MO, Aifantis I, Clevers H, Kang Y (2018) Notch ligand Dll1 mediates cross-talk between mammary stem cells and the macrophageal niche. Science 360(6396):eaan4153. https://doi.org/10.1126/science.aan4153
14. Permana PA, Menge C, Reaven PD (2006) Macrophage-secreted factors induce adipocyte inflammation and insulin resistance. Biochem Biophys Res Commun 341(2):507–514. https://doi.org/10.1016/j.bbrc.2006.01.012
15. Ye J, Gimble JM (2011) Regulation of stem cell differentiation in adipose tissue by chronic inflammation. Clin Exp Pharmacol Physiol 38(12):872–878. https://doi.org/10.1111/j.1440-1681.2011.05596.x
16. Dawson CA, Pal B, Vaillant F, Gandolfo LC, Liu Z, Bleriot C, Ginhoux F, Smyth GK, Lindeman GJ, Mueller SN, Rios AC, Visvader JE (2020) Tissue-resident ductal macrophages survey the mammary epithelium and facilitate tissue remodelling. Nat Cell Biol 22(5):546–558. https://doi.org/10.1038/s41556-020-0505-0
17. Bourin P, Bunnell BA, Casteilla L, Dominici M, Katz AJ, March KL, Redl H, Rubin JP, Yoshimura K, Gimble JM (2013) Stromal cells from the adipose tissue-derived stromal vascular fraction and culture expanded adipose tissue-derived stromal/stem cells: a joint statement of the International Federation for Adipose Therapeutics and Science (IFATS) and the International Society for Cellular Therapy (ISCT). Cytotherapy 15(6):641–648. https://doi.org/10.1016/j.jcyt.2013.02.006

Chapter 16

An Organotypic Assay to Study Epithelial-Fibroblast Interactions in Human Breast

Jakub Sumbal, Thorarinn Gudjonsson, Gunnhildur Asta Traustadottir, and Zuzana Koledova

Abstract

Epithelial–stromal interactions play an essential role in regulation of mammary gland development, homeostasis, and tumorigenesis. Fibroblasts constitute a substantial proportion of mammary gland stromal cells in human breast and have been recognized for their paracrine signaling and extracellular matrix production and remodeling roles during normal breast development as well as in breast cancer. However, our current knowledge on human breast fibroblast functions is incomplete. Here we provide a detailed protocol for an organotypic human breast assay to facilitate research in the roles of human breast fibroblasts in mammary epithelial morphogenesis and early tumorigenesis.

Key words 3D culture, Breast, Epithelial-stromal interactions, Fibroblast, Mammary gland, Spheroid

1 Introduction

The mammary gland is a peculiar, highly specialized organ that achieves its full form and function only in females, postnatally and upon pregnancy. It is composed of two main compartments: the epithelium and stroma. The epithelium is built of two layers of epithelial cells (luminal and myoepithelial), forms branched ductal structures, and fulfills milk producing function during lactation. Stroma is composed of multiple stromal cell types, including fibroblasts, adipocytes, endothelial cells, and immune cells, and extracellular matrix (ECM). In human breast, fibroblasts surround mammary epithelium organized into terminal ductal lobular units (TDLUs) and form a substantial part of the intralobular stroma. Moreover, fibroblasts are present also in the interlobular stroma, which is highly collagenous and very compact, and separates individual TDLUs from each other as well as from the adipose tissue [1].

Maria dM. Vivanco (ed.), *Mammary Stem Cells: Methods and Protocols*,
Methods in Molecular Biology, vol. 2471, https://doi.org/10.1007/978-1-0716-2193-6_16,

Stroma plays an essential role in regulation of mammary gland development and homeostasis through continuous bidirectional interactions, which include paracrine signaling, ECM remodeling, and mechanosignaling [2]. Studies using in vivo mouse models and in vitro co-culture assays, including advanced organoid and organotypic cocultures, have revealed that mammary fibroblasts are one of the key regulators of mammary gland development [3]. They control epithelial morphogenesis through secretion of paracrine signals [4, 5] and production and organization of ECM proteins, such as collagen [4–7]. Importantly, these functions become hijacked in breast tumors to support tumor growth and progression to metastasis by cancer-associated fibroblasts (CAFs) [8]. Therefore, it is of major clinical relevance to investigate normal fibroblast functions, their regulation, and the differences between normal fibroblasts and CAFs to provide rationale for development of precise CAF-targeting breast cancer therapeutic strategies with low toxicities [3].

Here, we provide a detailed protocol for an organotypic assay that enables investigation of human breast fibroblast interactions with human breast epithelium. This assay utilizes the D492 cell line and primary human breast fibroblasts. The D492 cell line originates from $MUC1^{-}$/$EpCAM^{+}$ cells of suprabasal epithelial population from human breast, which was immortalized by E6 and E7 oncogenes from human papilloma virus type 16 [9]. The D492 cell line is estrogen receptor negative but expresses both luminal and myoepithelial markers (KRT19 and KRT14, respectively) and yields both luminal and myoepithelial cells in culture. In 3D Matrigel cultures, D492 cells can form branching structures akin to TDLUs of the human breast as well as rounded spheroids [9]. The D492 cell model has been used to study branching morphogenesis, cellular plasticity, cancer progression, and epithelial–endothelial crosstalk in the breast [10–12]. Primary human breast fibroblasts are isolated from human breast tissue obtained by reduction mammoplasty using a procedure that is a modification of our previously published protocols for mouse mammary tissue [13, 14]. The organotypic coculture assay presented here (Fig. 1) is useful for studies on the roles of fibroblasts in normal mammary epithelial development and as well as in early breast tumorigenesis.

2 Materials

All tools, materials, and solutions for human breast cell isolation, and for cell culture have to be sterile. Sterilize surgical instruments with dry heat and solutions with autoclave, filter-sterilize media and solutions for cell culture (using 0.22 μm filters), and use gamma-irradiated cell culture plastic.

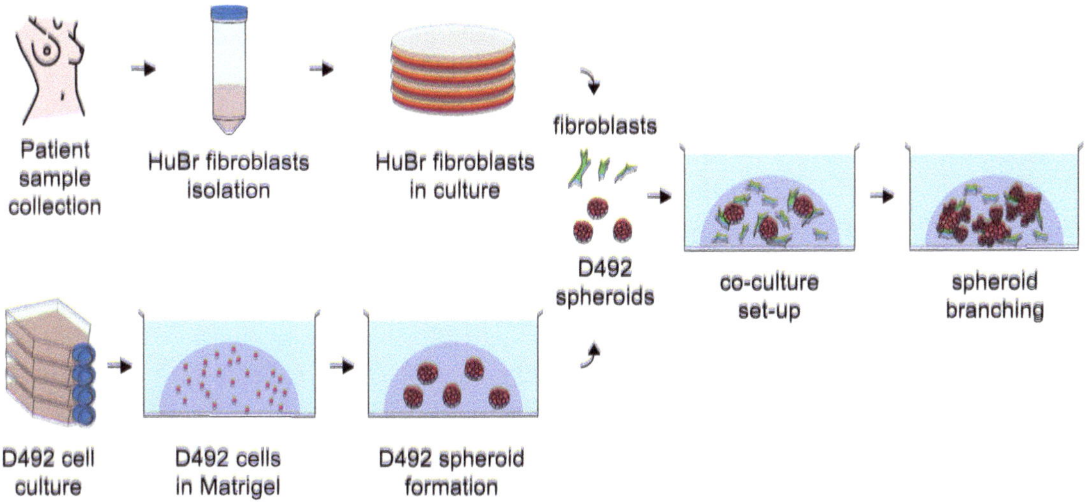

Fig. 1 Scheme of the protocol. Fibroblasts are isolated from human breast (HuBr) tissue and cultured in 2D for expansion. D492 cells are routinely cultured in 2D. For spheroid formation, D492 cells are cultured in 3D Matrigel. D492 spheroids are released from Matrigel and seeded in fresh Matrigel in coculture with fibroblasts. Over time, the effect of fibroblasts on epithelial morphogenesis is evaluated

2.1 Isolation of Fibroblasts from Human Breast

1. Human breast tissue from reduction mammoplasty (*see* **Notes 1** and **2**).
2. Dulbecco's Modified Eagle Medium: Nutrient Mixture F-12 (DMEM/F12).
3. Collagenase digestion solution: 2 mg/mL collagenase, 2 mg/mL trypsin, 5% (v/v) fetal bovine serum (FBS), 5 μg/mL insulin, 50 μg/mL gentamicin in DMEM/F12. For each 0.3 g of human breast tissue depleted of visible adipose tissue, prepare 10 mL of digestion solution by combining 8.7 mL of DMEM/F12, 0.5 mL of FBS, 10 μL of 50 mg/mL gentamicin, 400 μL of 50 mg/mL collagenase (*see* **Note 3**), 400 μL of 50 mg/mL trypsin (*see* **Note 4**), 50 μL of 1 mg/mL insulin (*see* **Note 5**). Filter-sterilize and warm up to 37 °C in a water bath before use. Prepare freshly for each primary cell isolation (*see* **Note 6**).
4. Coating solution: 2.5% (v/v) bovine serum albumin in PBS.
5. Deoxyribonuclease I (DNase I) solution: 20 U/mL DNase I in DMEM/F12. Add 40 μL of 2 U/μL DNase I to 4 mL of DMEM/F12 (*see* **Note 7**). Prepare fresh for each experiment.
6. Fibroblast culture medium: DMEM, 10% FBS, 1× Insulin-Transferin-Selenium (ITS), 100 U/mL of penicillin and 100 μg/mL of streptomycin. Store at 4 °C and use within a month.
7. Phosphate buffered saline (PBS), without calcium and magnesium.

8. Polystyrene Petri dish, 100 mm.
9. Cell culture dishes, 100 mm.
10. Forceps.
11. Disposable scalpel blades, no. 20 or 21.
12. Scalpel blade holders.
13. Syringes, 20 mL.
14. Syringe filters, 0.22 μm pore size.
15. Heated orbital shaker.

2.2 Primary Fibroblast Culture and Freezing

1. PBS.
2. 0.05% trypsin solution: 0.05% (v/v) trypsin and 0.2% (w/v) EDTA in PBS.
3. Trypsin neutralizing solution: 10% (v/v) FBS in PBS.
4. Fibroblast culture medium.
5. Fibroblast freezing medium: 10% (v/v) DMSO in heat-inactivated FBS (*see* **Note 8**). Store at −20 °C.
6. Cell culture dishes, 100 mm.

2.3 Culture and Freezing of D492 Cells

1. D492 cells.
2. Pure-Col Type I collagen solution, 3 mg/mL (Advanced BioMatrix).
3. PBS.
4. 0.25% trypsin solution: 0.25% (w/v) trypsin, 1 mM EDTA in PBS.
5. Soybean trypsin inhibitor (STI) solution, 10 mg/mL.
6. H14 medium: DMEM/F12, 2.5 mM L-Glutamine, 15 mM HEPES, 50 U/mL penicillin, 50 μg/mL streptomycin, 250 ng/mL insulin, 10 ng/mL EGF, 10 μg/mL transferrin, 2.6 ng/mL sodium selenite, 0.1 nM estradiol, 1.4 μM hydrocortisone and 0.15 U (5 μg/mL) prolactin.
7. D492 freezing medium: DMEM/F12, 10% (v/v) DMSO, 40% (v/v) heat-inactivated FBS (*see* **Note 8**). Store at −20 °C.
8. Cell culture flasks (T-25, T-75), or dishes (60 mm, 100 mm).

2.4 D492 Spheroid Formation and 3D Coculture Assay

1. Matrigel, growth factor reduced (*see* **Note 9**).
2. Dispase, 50 U/mL in 10 mM sodium acetate (pH 7.5), 5 mM calcium acetate. Store at −20 °C.
3. PBS.
4. 24-well cell culture dishes.
5. Hemocytometer, or another cell counting device.

2.5 General Materials and Equipment

1. Centrifuge tubes, 15 and 50 mL.
2. Pipettes (5 mL, 10 mL, 50 mL) and micropipettes.
3. Cryovials, 1.8 mL.
4. Freezing container (e.g., CoolCell or Mr. Frosty).
5. Centrifuge.
6. Cell culture incubator.
7. Ice bucket, shallow (suitable for use in cell culture hood).
8. Thermal block.
9. Bright-field microscope.

3 Methods

3.1 Isolation of Fibroblasts from Human Breast

1. Using sterile forceps, put the human breast tissue into a non-treated Petri dish. Using a scalpel, cut off any excess fat (yellow).
2. Weigh the tissue on scales aseptically (*see* **Note 10**). Calculate the amount of digestion solution required and prepare it.
3. Put the tissue back on the non-treated Petri dish. While holding the tissue with forceps, cut the tissue using a scalpel into small pieces of approximately 1 mm^3.
4. With the scalpel used for cutting, carefully transfer the minced tissue to a 50 mL tube containing digestion solution.
5. Incubate the suspension in an incubated shaker at 37 °C for 45 min while shaking at 100 rpm.
6. After digestion, centrifuge the digestion suspension at 450 × *g* for 10 min at RT (*see* **Note 11**).
7. For all subsequent steps, use pipettes, tips, and tubes coated with coating solution to avoid cell sticking to the plastic, thereby preventing loss of cells during the procedure (*see* **Note 12**).
8. Centrifugation yields 3 layers of the digestion suspension: a top white fatty layer, a middle aqueous phase (with the largest volume) and cell pellet at the bottom of the tube. Using a coated pipette, transfer the fatty layer into a 15 mL tube. Add 5 mL of DMEM/F12, mix well with the pipette and centrifuge at 450 × *g* for 10 min (*see* **Note 13**).
9. In the meantime, carefully aspirate the remaining supernatant from the 50 mL tube (*see* **Note 14**) and resuspend the pellet in 5 mL of DMEM/F12.
10. After centrifugation, aspirate the supernatant (both the top fatty and middle aqueous layers) from the 15 mL tube and to the pellet add the cells suspension from the 50 mL tube.

11. Using 5 mL of DMEM/F12, wash out any remaining cells from the 50 mL tube, combine the resulting suspension with the suspension in the 15 mL tube and mix well.
12. Centrifuge the 15 mL tube at 450 × *g* for 10 min.
13. Discard the supernatant, resuspend the pellet in 4 mL of DNase I solution and shake by hand or on an orbital shaker (50 rpm) for 3–5 min.
14. Add 6 mL of DMEM/F12, mix well and centrifuge at 450 × *g* for 10 min.
15. Discard the supernatant and perform differential centrifugation (*see* **Note 15**):
    (a) Resuspend the pellet in 8 mL of DMEM/F12.
    (b) Centrifuge the tube at 450 × *g* for 10 s at RT (*see* **Note 16**).
    (c) Transfer the supernatant into a 50 mL tube.
16. Repeat previous step four more times (5 times in total), while collecting the supernatants in one 50 mL tube.
17. After performing 5 rounds of differential centrifugation, the stromal fraction is separated (*see* **Note 17**). To collect the cells, centrifuge the 50 mL tube with the pooled supernatant at 600 × *g* for 3 min.
18. Aspirate the supernatant, resuspend the cell pellet in 10 mL of warm fibroblast culture medium and transfer the suspension onto a 100 mm cell culture dish (*see* **Note 18**).
19. Incubate the dish in cell culture incubator at 37 °C for 30 min (*see* **Note 19**).
20. After the incubation, use a microscope to check that the fibroblasts have attached to the dish. Most other stromal cells as well as epithelial organoids should remain in suspension (selective adhesion). Remove the unattached cells by aspirating the medium from the dish, wash the cell culture dish three times with PBS (5 mL per dish) to get rid of any unattached cells and add 10 mL of fresh fibroblast culture medium.
21. Incubate the plate with fibroblasts in a cell culture incubator (37 °C, 5% CO_2) for 24 h. This is passage 1 (p1).
22. Maintain the fibroblast culture in a cell culture incubator until the cells reach about 80% confluence while changing the medium every 2–3 days.

3.2 Culture of Primary Human Breast Fibroblasts

When the cells reach about 80% confluence, they need to be subcultured. The first subculture also serves to remove any epithelial contamination from the culture by a second round of selective adhesion procedure.

3.2.1 Passaging of Human Breast Fibroblasts

1. Aspirate the medium from the cell culture dish and wash the dish twice with 5 mL of PBS.
2. Add 1.5 mL of 0.05% trypsin solution to the dish, spread it all over the cell culture area.
3. Incubate the dish in cell culture incubator for 8 min.
4. Check the cell dissociation process under a microscope. Gently tap on side of the dish to help releasing cells from attachment. If the cells are not detaching, incubate the dish in cell culture incubator for a few minutes longer (*see* **Note 20**).
5. When the cells detach, add 5 mL of trypsin neutralizing solution to the dish and pipette the suspension several times to detach remaining cells from the bottom and to resuspend the cells.
6. Transfer the cell suspension to a 15 mL tube.
7. Wash the plate once with 5 mL of PBS and transfer the solution to the 15 mL tube to the cell suspension (*see* **Note 21**).
8. Centrifuge the 15 mL tube at 600 × *g* for 2 min.
9. Aspirate the supernatant and resuspend the pellet in fibroblast culture medium and split (1:3 to 1:5 according to the area) into new dishes with fresh fibroblast culture medium. Mark the culture dishes with passage number increased by 1.
10. To perform the selective adhesion procedure, place the dishes in the cell culture incubator and incubate for 20 min (*see* **Note 22**).
11. After 20 min, use a microscope to check that the fibroblasts have attached to the dish.
12. Aspirate the medium from the dishes, wash the dishes twice with PBS (5 mL per each dish per each wash) and add fresh fibroblast culture medium.
13. Maintain the cells in a cell culture incubator. Change the medium every 2–3 days. When the cells reach about 80% confluence, subculture them into new dishes, use for an experiment, or freeze for later use.

3.2.2 Freezing of Primary Human Breast Fibroblasts

1. To prepare frozen stocks of fibroblasts, use fibroblasts in passage two or higher, after the selective adhesion has been applied. Ensure that the fibroblasts for freezing are in top condition—healthy looking and not overgrown (not more that about 80% confluent). Preferentially, change the fibroblast culture medium 2 h before the planned processing for freezing.
2. To freeze fibroblasts, follow the **steps 2–8** in Subheading 3.2.1.

3. Aspirate the supernatant and resuspend the pellet in fibroblast freezing medium. Use 1 mL of fibroblast freezing medium for cells from one 100 mm dish.
4. Transfer the cell suspension into cryovials (*see* **Note 23**), 1 mL per vial.
5. Move the vials into a freezing container and incubate them for 10 min at RT.
6. Move the freezing container to a −80 °C freezer.
7. For long-term storage, move the vials from the −80 °C to liquid nitrogen (*see* **Note 24**).

3.3 Culture of D492 Cells

D492 cells are routinely maintained in collagen I-coated flasks in H14 medium at 37 °C, 5% CO_2 (*see* **Notes 25** and **26**). The medium is changed every 3 days.

3.3.1 Coating of Culture Vessels with Collagen I

1. Dilute 1 mL of cold Pure-Col Type I collagen solution in 44 mL cold PBS and mix.
2. Add 3 mL of diluted collagen I solution per each T-25 flask (*see* **Notes 27–29**).
3. Incubate for 30 min at 37 °C or for 1 h at RT.
4. Remove collagen I solution (*see* **Note 30**) and rinse the culture vessel with PBS.
5. Remove PBS and seed the cells immediately (*see* **Note 31**).

3.3.2 Culture and Passaging of D492 Cells

1. Seed the cells in collagen I-coated vessels (such as T-25 flasks) in pre-warmed H14 medium and incubate at 37 °C, 5% CO_2.
2. Change medium every 3 days.
3. Passage cells when they reach 90% confluency (*see* Fig. 2a, left).
4. Aspirate culture medium and rinse flask twice with PBS (3 mL per each T-25 flask).
5. Add 0.25% trypsin solution, 1 mL to each T-25 flask.
6. Incubate for 3–5 min at 37 °C until cells detach. Check cell detachment using a microscope (*see* **Note 32**).
7. Stop trypsinization by adding STI solution to each flask, 20 μL per 1 mL of trypsin.
8. Add 4 mL of PBS and gently pipet up and down to ensure all cells are detached.
9. Transfer the cell suspension to a 15 mL tube. Rinse the flask with PBS (3 mL per T-25 flask) and add the solution to the cell suspension in the 15 mL tube.

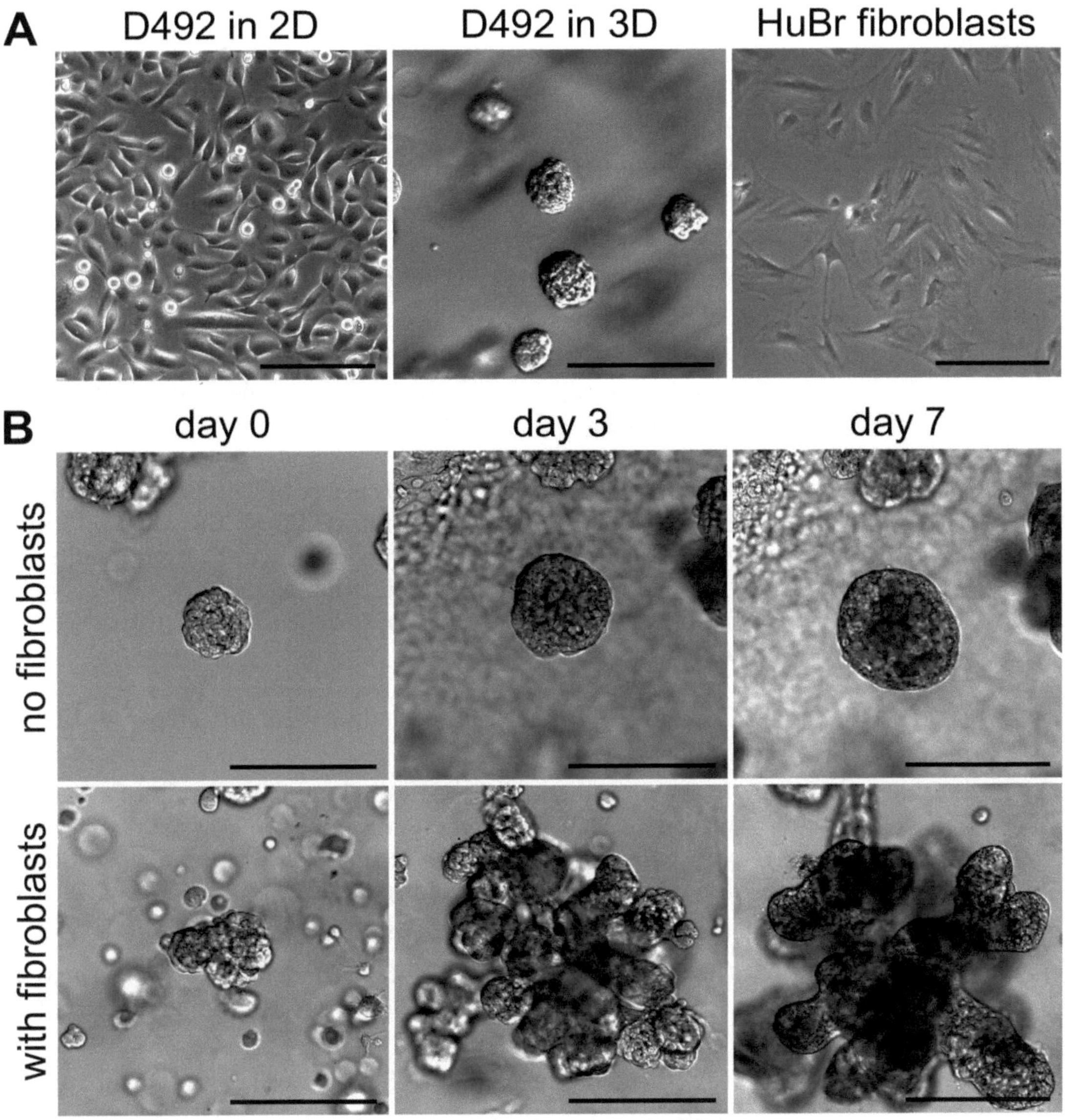

Fig. 2 Representative photographs of D492 cell culture and coculture with human breast fibroblasts. (**a**) Photographs of D492 cells cultured in 2D on collagen I-coated dishes (left), in 3D Matrigel (middle) and Primary Human Breast (HuBr) Fibroblasts cultured in 2D on plastic. Scale bars: 200 μm (2D) and 100 μm (3D). (**b**) The photographs from time-lapse imaging of D492 spheroids in 3D culture alone (top) or in coculture with human breast fibroblasts (bottom). Scale bars: 100 μm

10. Centrifuge at 200 × *g* for 3 min.
11. Aspirate supernatant and resuspend the cell pellet in pre-warmed H14 medium.
12. Subculture the cells at the ratio 1:10 to 1:20 ratio (*see* **Note 33**).

3.3.3 Freezing of D492 Cells

1. Allow the cells to reach 70–80% confluency.
2. Follow the **steps 4–10** in Subheading 3.3.2.
3. Aspirate the supernatant and resuspend the pellet in D492 freezing medium (e.g., 3 mL of D492 freezing medium for cells from one T-25 flask; *see* **Note 34**).
4. Transfer the cell suspension into marked cryovials, 1 mL per vial (*see* **Note 23**).
5. Move the vials into a freezing container and incubate them for 10 min at RT.
6. Move the freezing container to a −80 °C freezer.
7. For long-term storage, move the vials from the −80 °C to liquid nitrogen.

3.4 3D Coculture of D492 Spheroids with Human Breast Fibroblasts

D492 spheroids are formed by 3D culture of D492 cells in Matrigel for 1 week. Then, spheroids are released from the Matrigel using enzymatic digestion of the ECM and counted. Desired number of D492 spheroids are then cultured with human breast fibroblasts in fresh 3D Matrigel for several days, depending on the assay (*see* **Notes 35** and **36**).

3.4.1 D492 Spheroid Formation

1. Trypsinize D492 cells as described in **steps 4–11** in Subheading 3.3.2.
2. In the meantime, coat the desired number of wells of a 24-well plate with Matrigel patches, depending on how many spheroids you wish to produce. Using this protocol, you can produce hundreds to 3000 spheroids per well (*see* **Note 37**).
   - (a) Place the 24-well plate on ice.
   - (b) Carefully pipet 20 μL of Matrigel in the center of a well, spreading it into a circle of about 5–8 mm in diameter (*see* **Note 38**).
   - (c) Incubate the plate for 15 min in cell culture incubator.
3. Perform cell counting, e.g., using a hemocytometer.
4. Calculate the volume of D492 cell suspension needed for the spheroid formation, when 3000 cells are used for one well.
5. Transfer the required volume of D492 suspension into a 1.5 mL tube.
6. Centrifuge at 200 × *g* for 3 min.
7. Carefully aspirate the supernatant, leaving only a small drop of medium in the tube, and place the tube on ice.
8. Resuspend the cells in Matrigel, 50 μL per each well.
9. Plate the Matrigel in domes onto the Matrigel patches, 50 μL per each well.

10. Incubate the plate for 45 min in cell culture incubator (37 °C, 5% CO_2) to solidify the Matrigel.
11. Add pre-warmed H14 medium to the wells with plated cells, 1 mL per each well of a 24-well plate.
12. Culture the cells for 1 week. Change medium every 3 days. Check cell growth and spheroid formation regularly using a microscope.

3.4.2 3D Coculture Assay

1. When D492 spheroids are formed (*see* Fig. 2a, bottom), aspirate medium from the wells.
2. Using a 1000 μL micropipette with a coated tip (*see* **Note 12**), add 1 mL of pre-warmed DMEM/F12 to each well and, detach the Matrigel drop from the bottom, resuspend it a few times, and transfer the suspension into a 15 mL tube. Collect the suspension from all wells in one tube.
3. Wash out the wells used for D492 spheroid culture with approximately 100 μL of DMEM/F12 to collect any remaining spheroids in Matrigel and transfer the suspension to the 15 mL tube.
4. Add dispase to the suspension, 5 μL per each well of culture.
5. Incubate at 37 °C for 5 min while shaking (*see* **Notes 39**).
6. Check by eye or under a microscope the status of Matrigel dissolution. If pieces of Matrigel with entrapped spheroids are still visible, continue incubation for a few minutes.
7. When Matrigel is dissolved, add 5 mL of DMEM/F12 to the suspension and pipette repetitively three times.
8. Centrifuge at 450 × g for 3 min.
9. Aspirate the supernatant and resuspend the spheroids in 1 mL of H14 medium (*see* **Note 40**).
10. Count the spheroids:
    (a) Carefully resuspend the D492 suspension.
    (b) Immediately take 10 μL of the spheroid suspension and plate it in a drop on a cell culture dish (*see* **Note 41**). Place the stock D492 spheroid suspension in ice.
    (c) Count the number of spheroids (n) in the drop.
    (d) Calculate total spheroid number = n(spheroid in the drop)/V(drop) × V(total suspension).
    (e) Calculate the volume of D492 spheroid suspension required to obtain desired number of cocultures with desired number of spheroids. For instance, to obtain 4 cocultures with 300 spheroids in each well, calculate the volume of suspension that contains 1200 spheroids.
11. Prepare fibroblast suspension:

(a) Trypsinize fibroblast culture as described in **steps 2–8** in Subheading 3.2.1.

(b) Aspirate the supernatant and resuspend the cells in H14 medium, 1 mL per one 100 mm plate of fibroblast culture. Keep fibroblasts on ice until further use.

(c) Count fibroblast number and calculate the fibroblast concentration in the suspension and total number of fibroblasts in the suspension.

(d) Calculate the volume of fibroblast suspension required to obtain desired number of cocultures with desired number of fibroblasts. For instance, to obtain 4 cocultures with 30,000 fibroblasts in each well, calculate the volume of suspension that contains 120,000 fibroblasts.

12. In the meantime, coat the desired number of wells of a 24-well plate with Matrigel patches as described in **step 2** in Subheading 3.4.1. For instance, when 4 cocultures are planned, coat 4 wells with Matrigel.

13. Mix the desired number of spheroids with the desired number of fibroblasts in a 1.5 mL tube.

14. Centrifuge at 450 × *g* for 3 min.

15. Carefully aspirate the supernatant, leaving about 10 μL of medium in the tube.

16. Using a 10 μL micropipette, resuspend the spheroids with fibroblasts in the medium remaining in the 1.5 mL tube (*see* **Note 42**) and place the tube on ice.

17. Add Matrigel, 50 μL for each coculture (i.e., 200 μL for 4 co-cultures) and carefully pipette repetitively to homogenously distribute the cells and spheroids through the whole volume of Matrigel (*see* **Note 43**). Keep the tube on ice.

18. Plate the cocultures:

(a) Place the Matrigel-coated 24-well plate on thermal block pre-heated to 37 °C.

(b) Pipette the cell-Matrigel suspension onto Matrigel patches, 50 μL per well.

19. Incubate the plate with cocultures in cell culture incubator for 45 min to solidify Matrigel.

20. Add pre-warmed H14 medium and return the cell culture plate to the incubator.

21. Culture the cocultures for desired time according to the experimental requirements. Change the medium every 3 days and regularly check the co-cultures using a microscope. Alternatively, the coculture development can be imaged using a time-lapse microscope (*see* Fig. 2b).

4 Notes

1. Use of human patient tissue requires ethical committee approval and patient's informed consent.
2. Aseptically excised human breast tissue should be immediately transferred into sterile DMEM/F12 medium in a 50 mL tube and transferred into the laboratory on ice.
3. To prepare 50 mg/mL collagenase, dissolve 500 mg of collagenase in 10 mL of DMEM/F12 and make 400 μL and 800 μL aliquots. Store at −20 °C.
4. To prepare 50 mg/mL trypsin, dissolve 500 mg of trypsin in 10 mL DMEM/F12 and make 400 μL and 800 μL aliquots. Store at −20 °C.
5. To prepare 1 mg/mL insulin, dissolve 50 mg of insulin in 50 mL of 0.1 M HCl, pH 2–3. Filter-sterilize and make aliquots of 100–250 μL. Store at −20 °C.
6. The basis of collagenase digestion solution without the enzymes and insulin can be prepared in advance and stored at 4 °C for up to 1 month. Add the enzymes and insulin to the solution on the day of tissue digestion.
7. To prepare 2 U/μL DNase I, dissolve 2000 U of DNase I in 1 mL of sterile 0.15 M NaCl. Make 40 μL aliquots and store at −20 °C.
8. To heat inactivate FBS, incubate an aliquot (e.g., 50 mL) of FBS (thawed in advance at 4 °C) in a tightly sealed centrifuge tube in a water bath at 56 °C for 30 min. Then rapidly cool the heat-inactivated FBS in ice.
9. Thaw the 10 mL vial of Matrigel in ice in refrigerator overnight. Avoid exposing Matrigel to temperatures over 10 °C because it would lead to its premature and irreversible gelation. Using chilled pipette tips, aliquot Matrigel into pre-chilled 1.5 mL tubes on ice and store the aliquots at −20 °C. Avoid repeated freeze-thaw cycles.
10. To weigh the tissue while practicing an aseptic technique, you can use a scale placed in a cell culture hood. Alternatively, to use a scale on laboratory bench, first weigh a closed sterile tube, then take the tube to the cell culture hood and put the tissue in the weighted tube, close the tube and weigh the tube together with the tissue. Then calculate the weight of the tissue: m (tissue) = m(tube with tissue) − m(tube).
11. All centrifuge times are indicated as time of full-speed rotation. If your centrifuge does not support the function of counting time of full-speed rotation, add 1 min to all centrifuge times

except for the differential centrifugation described in **step 15a–c** in Subheading 3.1.

12. To coat a pipette/tip, aspirate the coating solution into the pipette/tip until almost all of the pipette's/tip's working surface is covered, then return the used coating solution back into the stock tube/bottle containing the rest of the coating solution. To coat a tube, carefully pipette the coating solution on the inner walls of the tube or fill the tube with the coating solution, then aspirate the coating solution with a pipette and return it back into the stock tube/bottle. The coating solution can be reused many times as long as it is kept sterile. Preferentially, filter-sterilize the solution after each experiment.
13. This step serves to obtain maximal cell yields from the digested tissue because during the procedure some cells become trapped in the fat. These cells can be released by resuspending the fatty layer in fresh DMEM/F12 and collected by centrifugation. If you wish, this step can be skipped and the fatty layer can be discarded.
14. To aspirate the supernatant, use a pipette attached to a pipettor rather than a vacuum aspiration system. The pipettor offers more control. Moreover, if the pellet is aspirated into the pipette by accident, it can be returned back in the tube – which is not possible if vacuum aspiration is used.
15. Differential centrifugation separates epithelial organoids from the stromal cells based on differences in their density. Epithelial organoids are larger and heavier and sink to the bottom of the tube faster than single-celled stromal cells. Stromal cells will remain in the supernatant. During the 5 rounds of differential centrifugation, the pellet becomes successively devoid of single cells. However, this technique is not absolute and some organoids (smaller fraction) will not sink to the bottom and will remain in the supernatant with stromal cells.
16. The time applies to the actual time of centrifugation at exactly $450 \times g$. If your centrifuge counts the total time of centrifugation instead of the time of centrifugation at the set speed, set the time of centrifugation to 1 min and watch carefully when the rotor reaches $450 \times g$. From this moment on, measure 10 s, and then stop the centrifuge manually by pressing the stop button.
17. The pellet in 15 mL tube contains epithelial organoids. It can be used for organoid culture, including co-culture with fibroblasts, or frozen in cell freezing medium for later use.
18. The suggested volume of fibroblast medium and one 100 mm cell culture dish are suitable for cell yield from 0.3 to 0.6 g of human breast tissue. If a higher amount of tissue is processed, a

higher cell yield is expected. Adjust the number of dishes for culture and the volume of fibroblast medium accordingly.

19. If you have more than one dish, place all dishes in a direct contact with the warm shelves of incubator. During the incubation, avoid moving the dishes or exposing them to vibrations, such as from harsh opening and closing of cell culture incubator.
20. Fibroblasts tend to increase their adherence with the time spent in culture. In such case, increased time of trypsin treatment is needed. Avoid keeping the cells on one dish for more than 2 weeks because then the cells become very difficult to detach and the required prolonged trypsin treatment would be harmful for the cells.
21. This step is optional. It serves to collect all cells from the dish.
22. Similarly to the first round of selective adhesion, place all dishes directly in contact with the warm shelves of the incubator, and avoid any disturbance to the plates.
23. Mark the cryovials with cell type, passage number, number of cells or cell culture dish size, and date of freezing.
24. Do not keep the cells at −80 °C for prolonged time period because their viability dramatically decreases after months spent at −80 °C.
25. Alternatively, the culture vessels can be coated also with Matrigel (1% Matrigel in DMEM/F12; keep at 4 °C) or gelatin (0.1% in PBS, autoclave and keep at RT).
26. D492 cells are very sensitive to CO_2 levels and humidity in the incubator. At sub-optimal growth conditions, their growth will halt.
27. For other types of cell culture vessels, adjust the amount of diluted collagen I solution to the culture area.
28. Store the unused diluted collagen I solution at 4 °C.
29. Alternatively, the vessels with diluted collagen I solution can be stored for up to 2 weeks at 4 °C. To use the stored dishes, first incubate them at 37 °C or RT and then rinse them as described in **steps 3–5** in Subheading 3.3.1).
30. The collagen I solution can be reused two times. Store the used solution at 4 °C with indicated number of use. The reused solution will become slightly cloudy due to solidified collagen.
31. The procedure for coating with Matrigel or gelatin is similar to coating with collagen I.
32. If D492 cells are overconfluent or starved, detaching the cells can be difficult. In such case, trypsinization can be facilitated by either rinsing the cells once with the 0.25% trypsin solution

before the 5 min trypsinization, or by trypsinizing the cells twice and collecting both suspensions in the same tube.

33. At this ratio, the cells will need to be split once a week.
34. Cells from one T-25 flask can be divided into three cryovials.
35. Similarly to human fibroblasts, other stromal cells can be co-cultured with D492 spheroids.
36. As required by the experiment, inhibitors of selected pathways or growth factors of interests can be added to the culture medium to test their effects on fibroblast-induced morphogenesis of D492 spheroids.
37. In theory, each D492 cell can produce a spheroid.
38. To work with Matrigel, always use pre-chilled tips. The tips can be chilled in a freezer. Alternatively, use tips kept at RT but cool each tip in 4 °C sterile PBS (by pipetting PBS in and out of the tip) immediately before use.
39. Use an incubated shaker (set at 37 °C, 100 rpm) or shake the tube submerged in water bath (37 °C) by hand.
40. Increase the amount of medium by 1 mL for each well of the original D492 spheroid culture.
41. To save on plastic use in the laboratory, we often keep lids from the used cell culture plates for this purpose.
42. This helps to homogenize the suspension before adding more viscous Matrigel, using which it is more difficult to homogenously distribute the spheroids and fibroblasts.
43. Pipette the suspension repetitively taking care not to introduce any air bubbles.

Acknowledgments

We thank present and past members of our laboratories for their technical support. This work was supported by the Icelandic Cancer Society Science fund (G.A.T.), the Grant Agency of Masaryk University (projects no. MUNI/G/1446/2018 to Z.K.), and funds from the Faculty of Medicine MU to junior researcher Z.K. (ROZV/28/LF/2020).

References

1. McNally S, Stein T (2017) Overview of mammary gland development: a comparison of mouse and human, methods Mol. Biol 1501: 1–17. https://doi.org/10.1007/978-1-4939-6475-8_1
2. Gjorevski N, Nelson CM (2011) Integrated morphodynamic signalling of the mammary gland. Nat Rev Mol Cell Biol 12:581–593. https://doi.org/10.1038/nrm3168
3. Sumbal J, Belisova D, Koledova Z (2020) Fibroblasts: the grey eminence of mammary

gland development. Semin Cell Dev Biol 111: 134–142. https://doi.org/10.1016/j.semcdb.2020.10.012

4. Koledova Z, Zhang X, Streuli C, Clarke RB, Klein OD, Werb Z, Lu P (2016) SPRY1 regulates mammary epithelial morphogenesis by modulating EGFR-dependent stromal paracrine signaling and ECM remodeling. Proc Natl Acad Sci U S A 113:E5731–E5740. https://doi.org/10.1073/pnas.1611532113
5. Sumbal J, Koledova Z (2019) FGF signaling in mammary gland fibroblasts regulates multiple fibroblast functions and mammary epithelial morphogenesis. Development 146(23): dev185306. https://doi.org/10.1242/dev.185306
6. Peuhu E, Kaukonen R, Lerche M, Saari M, Guzmán C, Rantakari P, De Franceschi N, Wärri A, Georgiadou M, Jacquemet G, Mattila E, Virtakoivu R, Liu Y, Attieh Y, Silva KA, Betz T, Sundberg JP, Salmi M, Deugnier M-A, Eliceiri KW, Ivaska J (2017) SHARPIN regulates collagen architecture and ductal outgrowth in the developing mouse mammary gland. EMBO J 36:165–182. https://doi.org/10.15252/embj.201694387
7. Hammer AM, Sizemore GM, Shukla VC, Avendano A, Sizemore ST, Chang JJ, Kladney RD, Cuitiño MC, Thies KA, Verfurth Q, Chakravarti A, Yee LD, Leone G, Song JW, Ghadiali SN, Ostrowski MC (2017) Stromal PDGFR-α activation enhances matrix stiffness, impedes mammary ductal development, and accelerates tumor growth. Neoplasia 19: 496–508. https://doi.org/10.1016/j.neo.2017.04.004
8. Houthuijzen JM, Jonkers J (2018) Cancer-associated fibroblasts as key regulators of the breast cancer tumor microenvironment. Cancer Metastasis Rev 37:577–597. https://doi.org/10.1007/s10555-018-9768-3
9. Gudjonsson T, Villadsen R, Nielsen HL, Rønnov-Jessen L, Bissell MJ, Petersen OW (2002) Isolation, immortalization, and characterization of a human breast epithelial cell line with stem cell properties. Genes Dev 16:693–706. https://doi.org/10.1101/gad.952602
10. Briem E, Ingthorsson S, Traustadottir GA, Hilmarsdottir B, Gudjonsson T (2019) Application of the D492 cell lines to explore breast morphogenesis, EMT and cancer progression in 3D culture. J Mammary Gland Biol Neoplasia 24:139–147. https://doi.org/10.1007/s10911-018-09424-w
11. Morera E, Steinhäuser SS, Budkova Z, Ingthorsson S, Kricker J, Krueger A, Traustadottir GA, Gudjonsson T (2019) YKL-40/CHI3L1 facilitates migration and invasion in HER2 overexpressing breast epithelial progenitor cells and generates a niche for capillary-like network formation. In Vitro Cell Dev Biol Anim 55:838–853. https://doi.org/10.1007/s11626-019-00403-x
12. Steinhaeuser SS, Morera E, Budkova Z, Schepsky A, Wang Q, Rolfsson O, Riedel A, Krueger A, Hilmarsdottir B, Maelandsmo GM, Valdimarsdottir B, Sigurdardottir AK, Agnarsson BA, Jonasson JG, Ingthorsson S, Traustadottir GA, Oskarsson T, Gudjonsson T (2020) ECM1 secreted by HER2-overexpressing breast cancer cells promotes formation of a vascular niche accelerating cancer cell migration and invasion. Lab Investig 100:928–944. https://doi.org/10.1038/s41374-020-0415-6
13. Koledova Z (2017) 3D coculture of mammary organoids with Fibrospheres: a model for studying epithelial-stromal interactions during mammary branching morphogenesis. Methods Mol Biol 1612:107–124. https://doi.org/10.1007/978-1-4939-7021-6_8
14. Koledova Z, Lu P (2017) A 3D fibroblast-epithelium co-culture model for understanding microenvironmental role in branching morphogenesis of the mammary gland. Methods Mol Biol 1501:217–231. https://doi.org/10.1007/978-1-4939-6475-8_10

Chapter 17

Patient-Derived Explant Cultures of Normal and Tumor Human Breast Tissue

Isabel Gris-Cárdenas, Miriam Rábano, and Maria dM. Vivanco

Abstract

Tissue culture has evolved considerably over the last few years, including cell culture in three dimensions, organoids, cocultures of different cell types and the use of diverse types of matrices in an attempt to mimic conditions that more closely resemble those found in the original tissue or organ. In this chapter, we describe how patient-derived breast tissue can be cultured on sponges for several days, maintaining their original architecture and with the capacity to respond to treatments. This protocol facilitates the study of the tissue responses without the need for extensive tissue manipulation, cell digestion or use of a biomaterial as scaffold, while maintaining the stroma and extracellular matrix organization. This method has the potential to improve preclinical testing by contributing to provide more accurate data reflecting cell–cell and cell–matrix interactions, tumor microenvironment, drug effects or stem cell function in normal- and pathophysiology of the breast.

Key words Breast tissue, 3D cultures, Ex vivo culture, Sponge

1 Introduction

Over the last decades, much has been learnt by studying cancer cells grown on monolayers on plastic surfaces. In an attempt to improve conditions over the limitations inherent to this method, and to develop model systems that truly recapitulate this heterogeneous disease, researchers have been using a variety of approaches. These include cocultures of at least two different cell types on monolayer, the use of three-dimensional cultures where cells are embedded in different types of matrices [1] or organoid cultures [2], among others.

An alternative approach is the use of organotypic cultures, which has been used for culturing freshly resected tissue pieces, including from brain [3], pancreas [4], prostrate [5] or breast [6, 7] cancer tissues. These types of cultures have the advantage that

Maria dM. Vivanco (ed.), *Mammary Stem Cells: Methods and Protocols*,
Methods in Molecular Biology, vol. 2471, https://doi.org/10.1007/978-1-0716-2193-6_17,

provides the contribution of the tumor-associated microenvironment, which is increasingly recognized as a strong modulator of disease progression and responses to therapy [8]. Recapitulation of the molecular basis of the disease should consider alterations in tumor vasculature, ECM density, stromal composition and other factors that may affect tumor development as well as the efficacy of anticancer treatments, in order to advance our understanding of cancer biology and improve predictability of novel therapies [9].

In this chapter, we present a method for *ex vivo* culture of breast tissue specimens, as patient-derived explants, that retains tissue architecture, thus providing a valuable approach to study the alterations of the complex interactions that occur during tumor development and in response to therapies. The platform we use is a gelatin sponge, an absorbable hemostatic agent that is characterized by its powerful absorption capacity and porosity [10]. Although originally this type of sponge was used for hemorrhage control in dental surgery [11], its use as scaffold has been extended to the biomedicine research area, from tissue engineering [12], to studies of angiogenesis [13], to preclinical studies of human tissue samples from different organs, such as the ovary [14], or the breast [15, 16]. Here, we present a protocol for culture of breast tissue specimens, both normal and tumor biopsies, but it is also applicable and adaptable to other type of tissue samples. After culturing, tissue explants can be harvested and used for a variety of assays, including immunohistochemistry, RNA or protein extraction.

2 Materials

1. Phosphate-buffered saline (PBS) with 10% penicillin and streptomycin.
2. Sponges: Ethicon Spongostan (i.e., Cat#MS0005, Johnson and Johnson).
3. 10 cm tissue culture dishes.
4. Sterile forceps.
5. Disposable scalpel.
6. Sterile surgical scissors.
7. 24-well tissue culture plate.
8. Cell incubator.
9. Laminar flow hood.
10. Culture media: DMEM:F12 medium, 10% fetal bovine serum, 2–5 mM L-Glutamine, 10 μg/ml hydrocortisone, 10 μg/ml human recombinant insulin, 1% of penicillin and streptomycin, 250 ng/ml amphotericin B.

3 Methods

Follow national biosafety level 2 guidelines when working with human tissue samples. *See* **Note 1** for additional information on tissue collection.

1. Prepare the culture medium (*see* **Note 2**).
2. Prepare the medium-soaked gelatin sponges by placing one sponge per well in a sterile 24-well tissue culture plate with 0.5 ml culture media (*see* **Note 3**).
3. Warm the sponges in a humidified incubator during 30 min at 37 °C and 5% CO_2.
4. Sterilize non-sterile material with 70% of ethanol, place it on the laminar flow hood and use 15 min UV sterile cycle.
5. Receive samples from the hospital and deliver to the lab on ice (*see* **Note 4**).
6. Wash the tissue with PBS containing antibiotics on a 10-cm tissue culture plate in a laminar flow hood (*see* **Note 5**).
7. Remove the surrounding adipose tissue (*see* **Note 6**) using clean forceps and a disposable sterile scalpel.
8. Dissect the tissue into many small ~1–2 mm^3 pieces in preparation for culture (*see* **Note 7**).
9. Place tissue chunks on sponges (Fig. 1), up to four pieces of tissue are placed on each sponge tissue (*see* **Note 8**).
10. Add culture media to a final volume of 1 ml.
11. Culture tissue explants at 37 °C and 5% CO_2 for the desired time point(s) (*see* **Note 9**).
12. Harvest the explants for the required downstream assays accordingly (i.e., RNA extraction, protein extraction or immunohistochemistry) (*see* **Note 10**). Figure 2.

4 Notes

1. Please ensure that you follow local regulations regarding human tissue collection and manipulation. All research using tissue samples should conform to the standards set by the Declaration of Helsinki. The use of freshly resected human tissue samples should be approved by the required local Ethics Committees and the material collected with written informed consent from patients and data should be de-identified to ensure Data Protection.
2. The medium presented here is the one routinely used in our laboratory for both normal and tumor breast tissue samples,

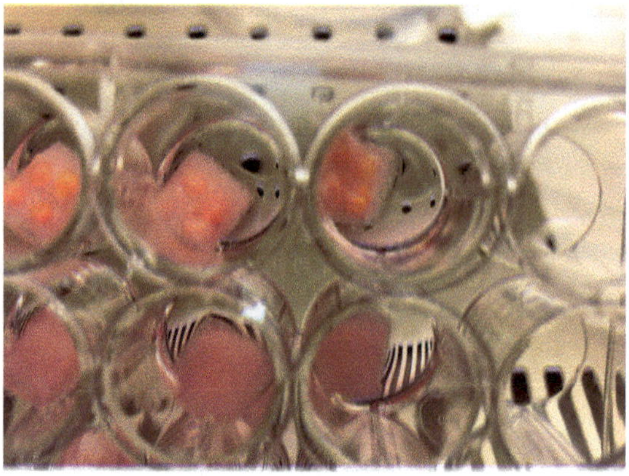

Fig. 1 Breast tissue fragments on top of medium-soaked sponges cultured in 24-well plates

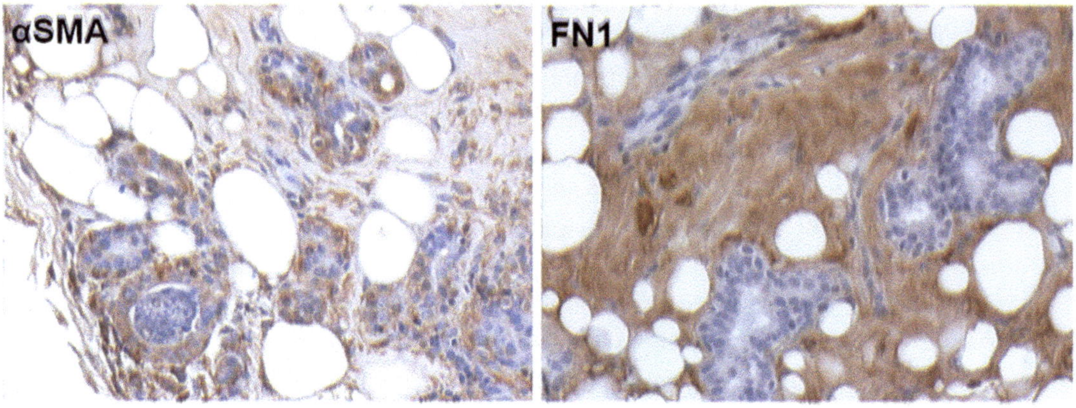

Fig. 2 Patient-derived explant culture maintains tissue morphology and cell viability. Example of immunostaining for alfa smooth muscle actin (αSMA), as marker of stromal and myoepithelial cells, and fibronectin-1 (FN1) as marker of stromal cells, in breast patient-derived explants after 72 h cultured on gelatin sponges

however, any culture medium can be selected here, depending on the choice of the researcher. Indeed, mammosphere medium or organoid medium, for example, can also be used in this type of assays. The composition of the culture medium is also going to depend on the type of experimental assay that is going to be performed afterwards. For example, if experiments are going to be testing the effects of estrogen, then the DMEM medium should be phenol red-free to avoid the estrogenic activity present in phenol red. In addition, the FBS should be charcoal treated to remove endogenous hormones. The presence of antimycotic and antibiotic solutions is recommended to reduce the possibility of contamination, although it can be avoided if sterile conditions have been constantly maintained after the surgical procedure.

3. The final volume in each well should not exceed 1 ml. Sponges tend not to remain flat after hydration with culture medium, so it is easier to try and position them in the well so that the tissue pieces stay on a flat surface when the volume is only 0.5 ml. In addition, if the tissue samples are going to be treated with some sort of growth factor, inhibitor, etc., then the appropriate vehicle and treatments can already be added at this step to each well at the desired concentration, to allow comparison of the effects on the tissue explants.
4. After the surgical procedure, the clinical pathologist resects a sample of the desired region (non-malignant or malignant) and the specimen is transported to the research laboratory in a sterile container in cold PBS on ice as soon as possible (ideally no more than 1–2 h after surgery). The samples that are dissected and placed into culture quickly have less degradation and better tissue morphology than those that have been sitting on ice for a while.
5. Depending on the hospital, the clinical pathologist may not always resect the tissue sample inside a laminar flow hood. Therefore, to minimize the risk of contamination, the tissue samples are washed in PBS with antibiotic and antimycotic before proceeding to tissue dissection.
6. Depending on the type of experimental assay to be performed afterwards, it may be important to remove fat tissue (easily identifiable yellow part) to select the glandular tissue. For example, certain proteomics procedures may be affected by excess fat in the sample.
7. Discard the scalpel when blade blunts to avoid damage to the tissue structure.
8. A maximum of four pieces are placed on each sponge to increase the representation of the heterogeneity of the breast tissue. However, tissue may be scarce, and not every little piece is going to contain the same percentage of epithelial cells and the rest of the cell components. Therefore, consider how many sponges and pieces of tissue per sponge you need to get an optimal representation of tissue heterogeneity, response to the potential treatments used and enough sample size for the experimental assays. At this point, it may also be relevant to use one or various of the tissue pieces as an example of the situation at time 0 for comparison with later point(s). Furthermore, take into account number of sponges needed depending on treatments to be analyzed. The fragments can be equally distributed between two or more wells, allowing direct comparison of treatment effects in matched breast tissue.
9. Most of the analysis that we have performed in the laboratory have been after culture periods of up to 4 days. However, other

reports have shown different time courses and, for example, Kokkinos and colleagues, using a similar methodology, have shown that human patient derived pancreatic explants cultured for 12 days maintain a majority of cells with BrdU staining and that p53 protein status is also preserved [4].

10. After the selected time, the explants can be snap-frozen, processed for RNA (or maintained in RNA later) or protein extraction or formalin-fixed and paraffin-embedded, as required, depending on the type of analysis. As examples of some of the potential protocols to be used: when explants are collected for RNA extraction, samples are homogenized using a Tissue Homogenizer with a Soft Tissue Omni Tip™ Plastic Homogenizing Probe. The tissue is disrupted in the tube with 0.5 ml of lysis buffer with 1-thioglycerol. RNA is extracted from the homogenate using the ReliaPrep™ RNA Miniprep kit for fibrous tissue from Promega (up to 20 mg). When samples are collected for immunohistochemistry, they are fixed in 10% neutral buffered formalin overnight at 4 °C. Then they are cleaned three times with PBS, placed in a histology cassette and then be processed and embedded in paraffin for histological staining. When samples are collected for protein extraction, the solubilization of the protein content can be done in Eppendorf tubes by addition of SDS buffer (1.5% SDS). Add ten times the weight of the tissue and subsequent incubation at 90 °C and shaking at 1400 rpm (Eppendorf, Thermomixer Compact, 5350) for 10 min. Any remaining solid residue was sedimented at 14,000 × *g* for 10 min 4 °C.

Acknowledgments

We are grateful for the generosity of many women and patients that collaborated in this project. This work was supported by grants from the Ministry of Science and Innovation (SAF2017-84934-R) and the Department of Industry and Innovation Technology of the Government of the Autonomous Community of the Basque Country.

References

1. Weigelt B, Ghajar CM, Bissell MJ (2014) The need for complex 3D culture models to unravel novel pathways and identify accurate biomarkers in breast cancer. Adv Drug Deliv Rev 69-70:42–51
2. Rosenbluth JM, Schackmann RCJ, Gray GK, Selfors LM, Li CM, Boedicker M, Kuiken HJ, Richardson A, Brock J, Garber J, Dillon D, Sachs N, Clevers H, Brugge JS (2020) Organoid cultures from normal and cancer-prone human breast tissues preserve complex epithelial lineages. Nat Commun 11(1):1711
3. Zhu L, Valiente M (2021) Organotypic brain cultures for metastasis research, Neuromethods book, Brain tumors. Springer Protocols, New York
4. Kokkinos J, Sharbeen G, Haghighi KS, Ignacio RMC, Kopecky C, Gonzales-Aloy E,

Youkhana J, Timpson P, Pereira BA, Ritchie S, Pandzic E, Boyer C, Davis TP, Butler LM, Goldstein D, McCarroll JA, Phillips PA (2021) Ex vivo culture of intact human patient derived pancreatic tumour tissue. Sci Rep 11(1):1944

5. Centenera MM, Raj GV, Knudsen KE, Tilley WD, Butler LM (2013) Ex vivo culture of human prostate tissue and drug development. Nat Rev Urol 10:483–487
6. Centenera MM, Hickey TE, Jindal S, Ryan NK, Ravindranathan P, Mohammed H, Robinson JL, Schiewer MJ, Ma S, Kapur P, Sutherland PD, Hoffmann CE, Roehrborn CG, Gomella LG, Carroll JS, Birrell SN, Knudsen KE, Raj GV, Butler LM, Tilley WD (2018) A patient-derived explant (PDE) model of hormone-dependent cancer. Mol Oncol 12(9):1608–1622
7. Gustafsson A, Garre E, Leiva MC, Salerno S, Ståhlberg A, Landberg G (2021) Patient-derived scaffolds as a drug-testing platform for endocrine therapies in breast cancer. Sci Rep 11(1):13334
8. Hirata E, Sahai E (2017) Tumor microenvironment and differential responses to therapy. Cold Spring Harb Perspect Med 7(7):a026781
9. Boix-Montesinos P, Soriano-Teruel PM, Armiñán A, Orzáez M, Vicent MJ (2021) The past, present, and future of breast cancer models for nanomedicine development. Adv Drug Deliv Rev 173:306–330
10. Petersen JK, Krogsgaard J, Nielsen KM, Nørgaard EB (1984) A comparison between 2 absorbable hemostatic agents: gelatin sponge (Spongostan) and oxidized regenerated cellulose (Surgicel). Int J Oral Surg 13(5):406–410
11. Jensen SS, Yazdi PM, Hjørting-Hansen E, Bosshardt DD, von Arx T (2010) Haemostatic effect and tissue reactions of methods and agents used for haemorrhage control in apical surgery. Int Endod J 43(1):57–63
12. Klangjorhor J, Nimkingratana P, Settakorn J, Pruksakorn D, Leerapun T, Arpornchayanon O, Rojanasthien S, Kongtawelert P, Pothacharoen P (2012) Hyaluronan production and chondrogenic properties of primary human chondrocyte on gelatin based hematostatic spongostan scaffold. J Orthop Surg Res 7:40
13. Ribatti D, Nico B, Vacca A, Presta M (2006) The gelatin sponge-chorioallantoic membrane assay. Nat Protoc 1(1):85–91
14. Ricciardelli C, Lokman NA, Sabit I, Gunasegaran K, Bonner WM, Pyragius CE, Macpherson AM, Oehler MK (2018) Novel ex vivo ovarian cancer tissue explant assay for prediction of chemosensitivity and response to novel therapeutics. Cancer Lett 421:51–58
15. Leiva MC, Garre E, Gustafsson A, Svanström A, Bogestål Y, Håkansson J, Ståhlberg A, Landberg G (2021) Breast cancer patient-derived scaffolds as a tool to monitor chemotherapy responses in human tumor microenvironments. J Cell Physiol 236(6): 4709–4724
16. Kähkönen TE, Toriseva M, Petruk N, Virta AR, Maher A, Eigėlienė N, Kaivola J, Boström P, Koskivuo I, Nees M, Tuomela JM, Ivaska KK, Härkönen PL (2021) Effects of FGFR inhibitors TKI258, BGJ398 and AZD4547 on breast cancer cells in 2D, 3D and tissue explant cultures. Cell Oncol (Dordr) 44(1):205–218

Chapter 18

Negative-Selection Enrichment of Circulating Tumor Cells from Peripheral Blood Using the Microfluidic CTC-iChip

Risa Burr, Jon F. Edd, Brian Chirn, Avanish Mishra, Daniel A. Haber, Mehmet Toner, and Shyamala Maheswaran

Abstract

The ability to isolate and analyze rare circulating tumor cells (CTCs) holds the potential to increase our understanding of cancer evolution and allows monitoring of disease and therapeutic responses through a relatively non-invasive blood-based biopsy. While many methods have been described to isolate CTCs from the blood, the vast majority rely on size-based sorting or positive selection of CTCs based on surface markers, which introduces bias into the downstream product by making assumptions about these heterogenous cells. Here we describe a negative-selection protocol for enrichment of CTCs through removal of blood components including red blood cells, platelets, and white blood cells. This procedure results in a product that is amenable to downstream single-cell analytics including RNA-Seq, ATAC-Seq and DNA methylation, droplet digital PCR (ddPCR) for tumor specific transcripts, staining and extensive image analysis, and ex vivo culture of patient-derived CTCs.

Key words Circulating tumor cell, Metastasis, Microfluidics, Negative selection, iChip, Blood-based biopsy

1 Introduction

Circulating tumor cells (CTCs) are tumor-derived cells released into the bloodstream from both primary and metastatic tumors [1, 2]. CTCs are the putative precursors of metastasis, with circulation in the bloodstream resulting in dissemination of metastasis-competent cells from the primary tumor to secondary tissues [1, 3]. Therefore, the study of CTCs provides a window into important aspects of metastasis, particularly survival in the circulation and extravasation into the secondary site. Additionally, the presence of CTCs in the circulation of cancer patients means that they can be relatively easily collected and interrogated through a simple peripheral blood draw ("blood-based biopsy"), a much less invasive procedure than a primary tumor biopsy. This increased ease

Maria dM. Vivanco (ed.), *Mammary Stem Cells: Methods and Protocols*,
Methods in Molecular Biology, vol. 2471, https://doi.org/10.1007/978-1-0716-2193-6_18,

of collection is crucial for settings requiring repeated tumor sampling including longitudinal analysis of disease evolution and treatment responses as well as early detection of disease in high-risk populations [4]. Analysis of CTCs further has the advantage of sampling all sites of disease and providing a more holistic picture of a patient's cancer. A single liquid biopsy reports on multiple tumors at once, tumors at sites that are not amenable to traditional biopsy, and the heterogeneity of a single tumor, all of which are difficult to query with current biopsy technologies [5].

CTCs isolated from the blood can be used for downstream single-cell omics interrogation and to generate long-term *ex vivo* cultures of CTC cell lines which can be used for mechanistic studies [6–8]. Single cell RNA-Sequencing of patient-derived CTCs has identified mechanisms of prostate cancer antiandrogen resistance [9], and shown high expression of stromal-derived extracellular matrix genes in pancreatic ductal adenocarcinoma [10]. Additional studies detected acquired T790M EGFR mutations in patients with Non-Small Cell Lung Cancer [11] and described prostate and breast tumor heterogeneity [12–16]. *Ex vivo* cultures of CTCs enriched from breast cancer patients retain their heterogeneity and have helped uncover the interconversion of HER2+/HER2- populations of CTCs in HER2- breast cancer patients [17–19]. More recently, isolation of CTCs from the peripheral blood of cancer patients has been used to generate biomarker panels that can be used to predict drug response and early dissemination in prostate cancer [20], to predict therapeutic response in localized and metastatic breast cancer [21], to diagnose hepatocellular carcinoma [22], and to monitor response to immune checkpoint therapy in melanoma [23].

Numerous methods to isolate CTCs from whole blood have been described in the literature. These technologies make use of microfluidics, centrifugation, flow cytometry, and immunomagnetic bead-based separation depending on the application. These methods basically take advantage of three broad characteristics of CTCs – physical characteristics, such as size [24–28], density [29, 30], deformability and charge [31], positive selection for surface antigens or gene expression on CTCs [2, 32–44], and negative selection for surface antigens of blood and/or epithelial cells [45–47]. CellSearch® (Veridex, Menarini Silicon Biosystems), is the only CTC isolation methodology FDA-approved for in vitro diagnostic use and relies on positive selection against EpCAM. However, because CTCs are highly heterogeneous, positive selection methods that make assumptions about the nature of CTCs risk selecting for a subset of the population in circulation and thereby generating biased results. Additionally, many of the above methods are not compatible with downstream culturing of patient-derived CTC cell lines, either because they capture the CTCs on a surface where it is difficult to remove them, or because the process of enriching the CTCs makes them nonviable [48]. For this reason, our CTC-iChip technology combines size-based selection microfluidics (to separate all nucleated cells from blood) with negative selection against white blood cell markers to enrich viable CTCs in the product without assumption as to the surface antigens of the CTC population.

Enrichment of CTCs using the CTC-iChip begins with immunomagnetic bead labeling of white blood cells in whole blood using biotinylated antibodies against CD45, CD66b, and CD16 and streptavidin magnetic beads. This labeled whole blood then enters the CTC-iChip where size-based separation of red blood cells, platelets, and free magnetic beads is performed using deterministic lateral displacement [49]. The remaining cells in buffer then go through inertial focusing so that they form a single-file line in a micro-channel for separation by microfluidic magnetophoresis and deflection of the magnetic-bead labeled white blood cells. The CTC-iChip runs at a rate of 13 mL blood/hour and achieves an average 4.0 base-10 log depletion of white blood cells, 3.3 log depletion of platelets, and 5.1 log depletion of red blood cells (median for EDTA blood). The yield of spiked CTCs averages >95%.

2 Materials

2.1 Blood Collection and Incubation

1. EDTA blood collection tubes.
2. 50 mL conical Falcon tubes.
3. Serological pipettes.
4. Orbital rotating tube mixer.
5. Mesh cloth.
6. CBC machine.

2.2 Reagents and Buffers

1. IFD running buffer: 1% Kolliphor P188 (Pluronic-F68) in PBS (*see* **Note 1**).
2. Antibody Cocktail (4× AB Cocktail) CD45, CD66b, and CD16 (10:1:1 ratio) (*see* **Note 2**).
3. Magnetic Beads (*see* **Note 3**).
4. Tween20.
5. PBS.
6. BSA.
7. 50% Ethanol: 125 mL diH20, 125 mL 200 Proof (100% Ethanol) (*see* **Note 4**).

2.3 IFD Tower (See Note 5)

1. IFD Chip (Monolithic Chip).
2. Buffer Cartridge (Nordson EFD Cartridge PN: 7012416).
3. Blood holder and stoppers (Nordson EFD Barrel PN:7012155, Stopper: PN7014598).
4. 2 waste collection bags.
5. 15 mL conical tube.
6. Parafilm.
7. Ice bucket.
8. Male to male adapter.
9. Teflon tape.
10. Luer adapter (Male Luer Lock to ¼ -18 NPT F-14MTLL-1).
11. Priming kit.
12. Priming syringe.
13. 50% ethanol.
14. Check valve (Nordsonefd Medical ref.: VP55401068N).
15. Magnifying glass.

3 Methods

Carry out all procedures at room temperature unless otherwise specified.

3.1 Preparing the Blood Sample

1. Collect 10–20 mLs of peripheral venous blood from a patient in purple EDTA tubes and store at room temperature on a hula mixer rotating at 12 RPM speed (*see* **Note 6**).
2. Pool blood into a 50 mL conical tube and measure total blood volume (*see* **Note** 7).
3. Take a Complete Blood Count (CBC; Sysmex KX-21N) of the blood (*see* **Note 8**).
4. Spike blood with 18ul AB Cocktail per mL of blood (*see* **Note 9**).
5. Incubate blood and AB cocktail for 30 min, rotating on hula mixer at 12 RPM (*see* **Note 10**).
6. Spike blood with 120 μL of magnetic beads per ml of blood (*see* **Note 11**).

3.2 Setting Up CTC-iChip Tower

1. Power on the IFD tower via the power button on the machine.
2. Open the door and insert a new CTC-iChip into the IFD tower and lock all of the lines into place before closing the door (Fig. 1a, part 1, b) (*see* **Note 12**).

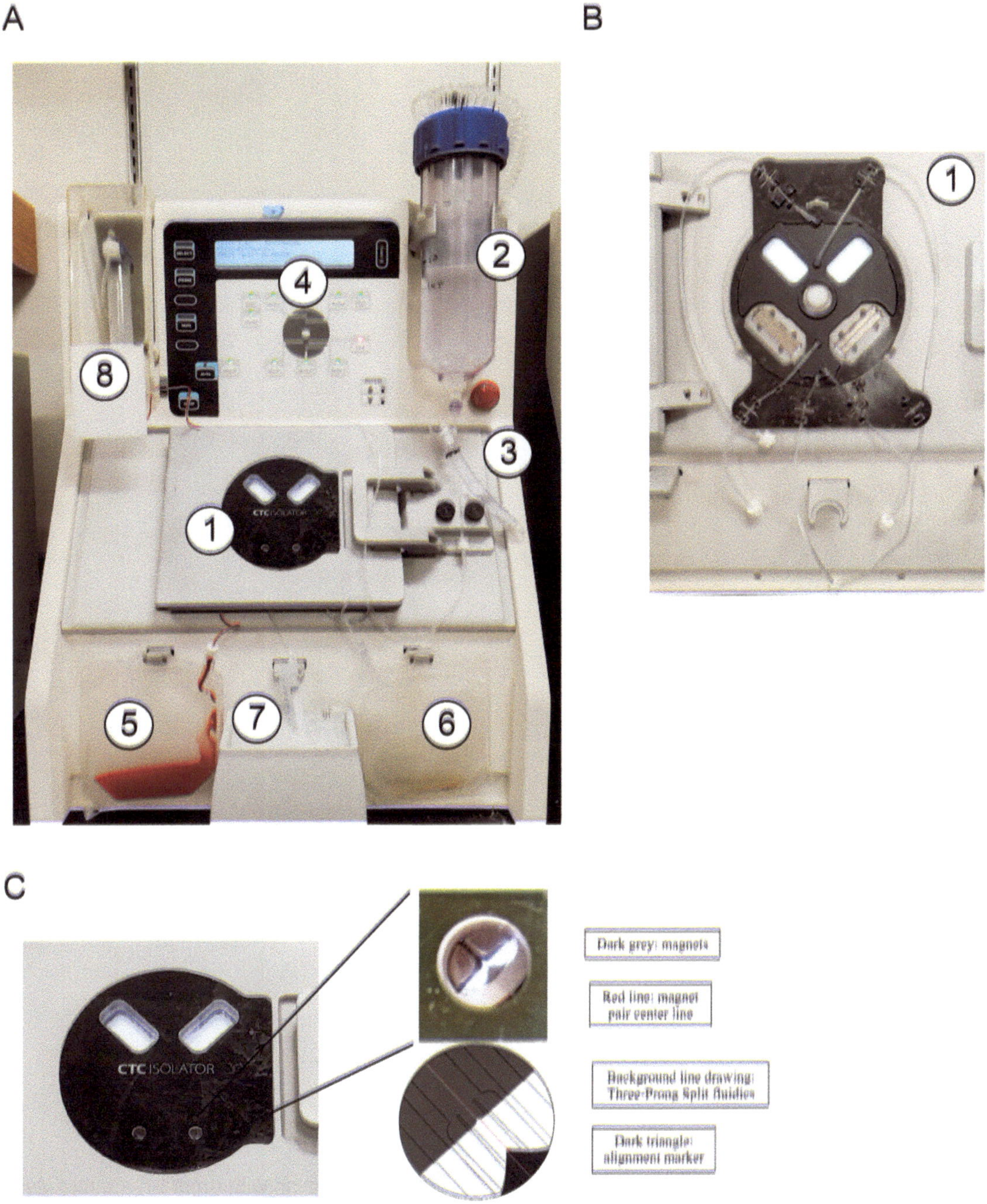

Fig. 1 Setting up the tower. (**a**) Picture of CTC-iChip tower in operation corresponding to Subheading 3.2. Part numbers shown in white circles correspond to those referenced in the methods section. Setup of iChip with input and output lines is shown in (**b**). Blood and buffer are pumped into the iChip and RBC waste, WBC waste, and enriched CTC product flow out of the iChip through the indicated lines. Proper alignment of the iChip over the magnets is crucial for success. Alignment can be evaluated by centering the notch on the magnet over the middle marking printed on the chip while looking through the window on the door of the tower with a magnifying glass, as shown in (**c**)

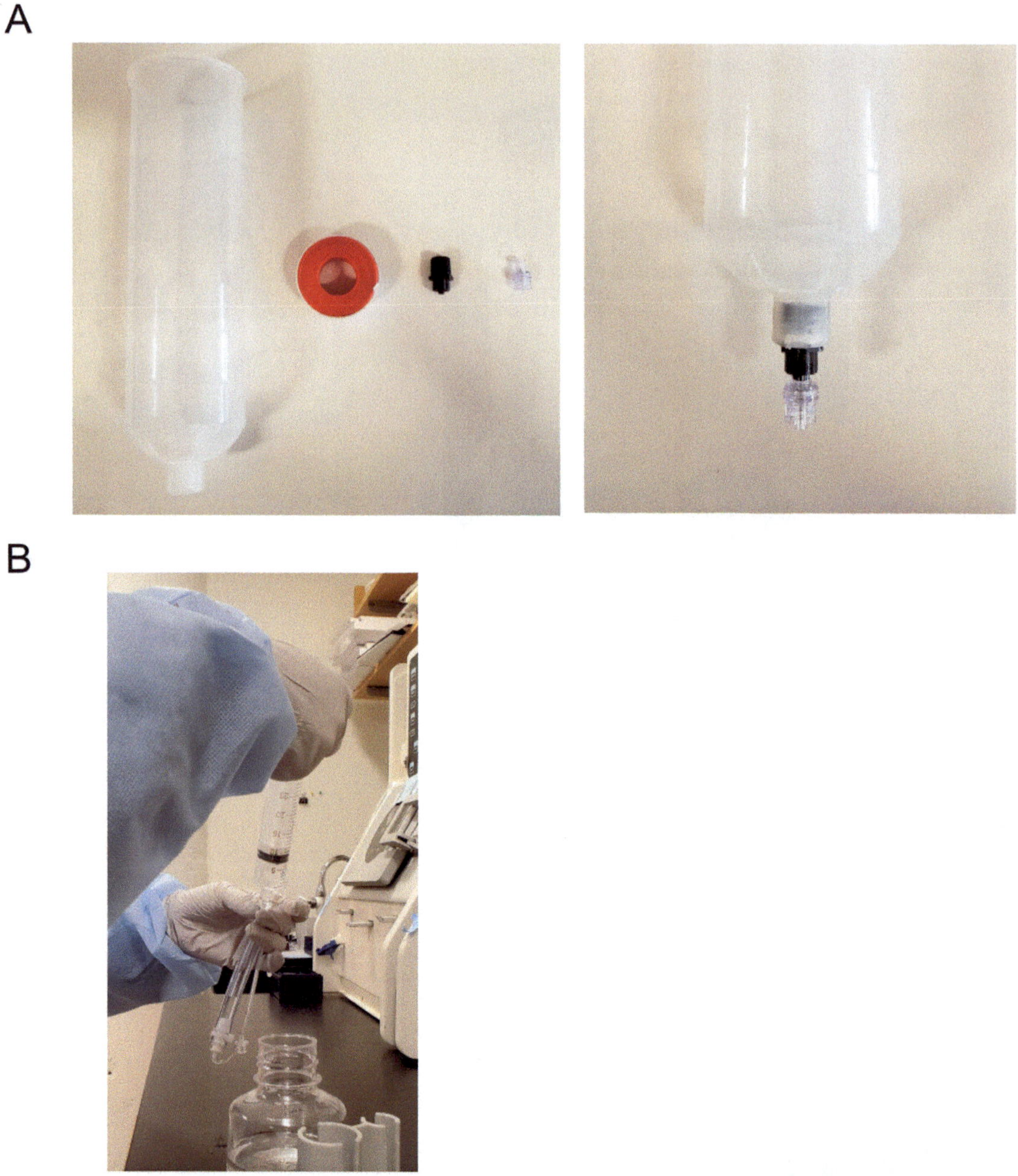

Fig. 2 Setting up the buffer cartridge. (**a**, **b**) The buffer cartridge is assembled with (from l-r) a buffer stein, Teflon tape, a Luer adapter, and a check valve, using a wrench to tighten, as described in Subheading 3.2, **step 4**. (**c**) The priming kit is filled three-fourth of the way with 50% ethanol by attaching the syringe to the bottom lead and holding the kit upright, as described in Subheading 3.2, **step 6**

3. Check the alignment of the chip by looking through the door window with a magnifying glass (Fig. 1b, c) (*see* **Note 13**).
4. Assemble the buffer cartridge with Luer adapter/Teflon tape and check valve (Fig. 2a):

   (a) Wrap 5 in. of Teflon tape around the Luer adapter (*see* **Note 14**).
   (b) Wrench the Luer adapter into the buffer cartridge until tight. Attach the check value at the end of the Luer adapter.
5. Insert the assembled buffer cartridge into the buffer cartridge holder in the top right of the IFD machine (Fig. 1a, part 2).
6. Use a 60 mL syringe to fill the priming kit three-fourth the way with 50% ethanol from the bottom (Fig. 2b). Connect this side to the check valve (Fig. 1a, part 3) (*see* **Note 15**).
7. Fill buffer cartridge with 350 mL of running buffer. Place lid on and lock the cap.
8. Prime the chip (Fig. 1a, part 4).
   (a) The menu will read "CTC BETA INSTRUMENT. PRESS SELECT TO CHOOSE CHIP." Press Select.
   (b) The menu will read "CTC-iCHIP SELECTED. PRESS 'SELECT' AGAIN FOR OTHER OR LOAD INSTRU MENT . PRESS 'PRIME' FOR NEXT STEP." Press Prime.
   (c) Machine should pressurize and ethanol should run through each line individually (*see* **Note 16**).
9. After priming (takes around 10 min) the menu will read "PRIMING COMPLETE. LOAD SAMPLE. CLOSE DOORS THEN PRESS 'RUN' WHEN READY."
10. Connect pre-weighed waste bags to the left-most and right-most lines (Fig. 1a, part 5,6).
11. Secure a pre-weighed 15 mL conical product tube to the product tube holder, insert the middle output line and seal the top with parafilm (Fig. 1a, part 7).
12. Connect the blood sample holder to the chip via the top right line and place it in the upper left rocking holder (Fig. 1a, part 8; Fig. 2c).

3.3 Running the iChip

1. Dilute the blood: Subtract the bead and antibody cocktail volume from the original blood volume. Add this amount of buffer to the blood.
2. Pour diluted blood carefully into the blood sample holder (Fig. 1a, part 8) (*see* **Note17**).
3. Cap sample holder with blue plastic cap and then twist cap until it clicks shut.
4. Press *Run* on the tower (Fig. 1a, part 4) (*see* **Note 18**).

3.4 Cleanup

1. Drain the remaining buffer (*see* **Note 19**).
2. Weigh product tube and waste bags and record final mass.
3. Take a CBC of the RBC waste after mixing it in the bag.
4. Cap waste bags and then dispose in biohazard waste.
5. Store the product on ice for downstream processing.

4 Notes

1. To make IFD running buffer, add 5.00 g of Kolliphor P188 (Pluronic-F68) into 500 mL of sterile 1× PBS (-MgCl–CaCl) for a final concentration of 1% F68. Dissolve Kolliphor on rocker (~10 min) at low speed. Filter buffer through 500 ml Nalgene sterilization filter system, 0.2 μm pore size and store at room temperature. Typical to make 2 L at a time. Unused buffer can be stored for 1 week. Best practice is to make fresh buffer the day of the experiment.
2. Prepare 4× Antibody Cocktail (AB Cocktail) with CD45, CD66b, and CD16 (10:1:1 ratio). CD45 Biotin–Clone H130 (1.5 μg/mL WB), CD16–Clone 80H3 (0.15 μg/mL WB), CD66b–3G8 (0.15 μg/mL WB). Dilute 4× concentration to 1× working stock with filtered PBS and aliquot into 1.5 mL Eppendorf tubes. Can be stored at 4 °C for 1 week.
3. To prepare the magnetic beads, place a 10 mL aliquot (Invitrogen Dynabeads Myone Streptavidin T1 Ref: 35602) on magnet holders and gently invert up and down until solution is clear. We usually make 10 mL aliquots and store them in the freezer. Then, the supernatant should be aspirated. Be careful not to aspirate beads, be sure to push the tip of aspirator against the falcon tube wall opposite the magnet. Wash beads 3× with filtered 0.01% Tween20 in PBS and resuspend in 0.1% BSA and aliquot into 1.5 mL Eppendorf tubes. The resuspended beads can be stored at 4 °C for 2–3 weeks.
4. Filter 50% ethanol using a 0.2uM pore size Nalgene Nylon filtration system. Do not use cellulose, a cellulose membrane filtration system. Ethanol will dissolve the cellulose membrane.
5. Tellbio Inc. is preparing to make these commercially available.
6. Blood drawn in EDTA tubes should be run on iChip within 6 h of draw time. Wear proper PPE when handling blood, this includes a lab coat, protective eye equipment or face shield, and disposable gloves.
7. Blood clots will clog the chip. If there are visible blood clots pass the blood through a mesh cloth or 30-μm cell strainer.

8. This step is not necessary to run the protocol and serves for recording keeping.
9. For example 10 mL of blood is spiked with 180 μL of AB Cocktail. Do not vortex AB cocktail.
10. Antibody incubation can be left longer, if necessary, up to 2 h, without compromising the experiment. Set up iChip tower while blood is incubating with AB cocktail.
11. Vortex the beads well before pipetting. For example, 10 mL of blood is spiked with 120 μL of beads.
12. When closing the door, make sure not to pinch any of the lines.
13. 5 lanes should line up to the arrow on the machine.
14. In order to have a proper seal, apply Teflon tape clockwise.
15. It is important to fill from the bottom of the priming kit in order to preserve sterility. The 50% ethanol must be nylon filtered and syringe should not be reused.
16. Tape the lines close together and collect the drip into a catch container.
17. Double check that the sample holder is connected to the chip.
18. Some of the common problems and solutions are: *Chip Clogging*. If the chip clogs, the tower will be pressurized and the run continued, however there will be a stop in flow in the tubing. To attempt to unclog the chip, we recommend back flushing. Back flushing is clamping all lines except for the buffer and blood supply lines. This allows the buffer to flow backward through the chip into the blood supply and unclogging the chip. Clogging is typically caused by coagulated blood or bead aggregation. This can be avoided by running the samples as fresh as possible and not over-incubating with beads.
    (a) Switch the machine into manual mode by pressing manual on the tower.
    (b) Clamp down the Waste 1, Product, and Waste 2 lines by pressing the "Waste 1", "Product", and "Waste 2" buttons. At this point, only the blood supply line should be open.
    (c) Physically clip the Waste 1, Product, and Waste 2 lines closed (optional).
    (d) Depressurize the Sample by pressing the "Sample supply" button and watch the buffer move through the chip and flow into the blood supply.

    Tower fails to catch end of run. When the tower fails to catch the end of the run it will continue to flow sample (air) and buffer through the chip even though there is no more blood. If this happens, it can be stopped by simply pressing the red "Reset" button on the tower or turning off the tower. Either of these actions will end the run and depressurize the tower.

Loose tubing causing leakage. If any tubing is loose or poorly connected there is the potential for leakage. It may be necessary to use a new chip and start a new run. If the leakage is minimum (typically through the buffer supply) it is possible to continue with the run.

Red blood cell carry-over. If the product has an unusual amount of RBC carry-over, rerunning the product through the chip as a cleanup step can reverse this problem. This is not recommended for most runs as a significant amount of product will be lost.

19. Removing the tubing management device and bringing it and the buffer cartridge over to the sink will prevent any unnecessary spillage.

References

1. Steeg PS (2006) Tumor metastasis: mechanistic insights and clinical challenges. Nat Med 12(8):895–904. https://doi.org/10.1038/nm1469
2. Nagrath S, Sequist LV, Maheswaran S, Bell DW, Irimia D, Ulkus L, Smith MR, Kwak EL, Digumarthy S, Muzikansky A, Ryan P, Balis UJ, Tompkins RG, Haber DA, Toner M (2007) Isolation of rare circulating tumour cells in cancer patients by microchip technology. Nature 450(7173):1235–1239. https://doi.org/10.1038/nature06385
3. Lambert AW, Pattabiraman DR, Weinberg RA (2017) Emerging biological principles of metastasis. Cell 168(4):670–691. https://doi.org/10.1016/j.cell.2016.11.037
4. Maheswaran S, Haber DA (2010) Circulating tumor cells: a window into cancer biology and metastasis. Curr Opin Genet Dev 20(1): 96–99. https://doi.org/10.1016/j.gde.2009.12.002
5. Sundaresan TK, Haber DA (2015) Fantastic voyage: the future of cancer diagnostics. Lancet Oncol 16(16):1596–1598. https://doi.org/10.1016/S1470-2045(15)00469-6
6. Zheng Y, Miyamoto DT, Wittner BS, Sullivan JP, Aceto N, Jordan NV, Yu M, Karabacak NM, Comaills V, Morris R, Desai R, Desai N, Emmons E, Milner JD, Lee RJ, Wu CL, Sequist LV, Haas W, Ting DT, Toner M, Ramaswamy S, Maheswaran S, Haber DA (2017) Expression of β-globin by cancer cells promotes cell survival during blood-borne dissemination. Nat Commun 8:1–12. https://doi.org/10.1038/ncomms14344
7. Yu M, Bardia A, Aceto N, Bersani F, Madden MW, Donaldson MC et al (2014) Cancer therapy. *Ex vivo* culture of circulating breast tumor cells for individualized testing of drug susceptibility. Science 345(6193):216–220. https://doi.org/10.1126/science.1253533
8. Cayrefourcq L, Mazard T, Joosse S, Solassol J, Ramos J, Assenat E, Schumacher U, Costes V, Maudelonde T, Pantel K, Alix-Panabières C (2015) Establishment and characterization of a cell line from human circulating colon cancer cells. Cancer Res 75(5):892–901. https://doi.org/10.1158/0008-5472.CAN-14-2613
9. Miyamoto DT, Zheng Y, Wittner BS, Lee RJ, Zhu H, Broderick KT, Desai R, Fox DB, Brannigan BW, Trautwein J, Arora KS, Desai N, Dahl DM, Sequist LV, Smith MR, Kapur R, Wu CL, Shioda T, Ramaswamy S, Ting DT, Toner M, Maheswaran S, Haber DA (2015) RNA-Seq of single prostate CTCs implicates noncanonical Wnt signaling in antiandrogen resistance. Science 349(6254):1351–1356. https://doi.org/10.1126/science.aab0917
10. Ting DT, Wittner BS, Ligorio M, Vincent Jordan N, Shah AM, Miyamoto DT, Aceto N, Bersani F, Brannigan BW, Xega K, Ciciliano JC, Zhu H, MacKenzie OC, Trautwein J, Arora KS, Shahid M, Ellis HL, Qu N, Bardeesy N, Rivera MN, Deshpande V, Ferrone CR, Kapur R, Ramaswamy S, Shioda T, Toner M, Maheswaran S, Haber DA (2014) Single-cell RNA sequencing identifies extracellular matrix

gene expression by pancreatic circulating tumor cells. Cell Rep 8(6):1905–1918. https://doi.org/10.1016/j.celrep.2014.08.029

11. Sundaresan TK, Sequist LV, Heymach JV, Riely GJ, Jänne PA, Koch WH, Sullivan JP, Fox DB, Maher R, Muzikansky A, Webb A, Tran HT, Giri U, Fleisher M, Yu HA, Wei W, Johnson BE, Barber TA, Walsh JR, Engelman JA, Stott SL, Kapur R, Maheswaran S, Toner M, Haber DA (2016) Detection of T790M, the acquired resistance EGFR mutation, by tumor biopsy versus noninvasive blood-based analyses. Clin Cancer Res 22(5):1103–1110. https://doi.org/10.1158/1078-0432.CCR-15-1031
12. Miyamoto DT, Lee RJ, Stott SL, Ting DT, Wittner BS, Ulman M et al (2012) Androgen receptor signaling in circulating tumor cells as a marker of hormonally responsive prostate cancer. Cancer Discov 2(11):995–1003. https://doi.org/10.1158/2159-8290.CD-12-0222
13. Deng Y, Zhang Y, Sun S, Wang Z, Wang M, Yu B et al (2014) An integrated microfluidic chip system for single-cell secretion profiling of rare circulating tumor cells. Sci Rep 4:13–15. https://doi.org/10.1038/srep07499
14. De Luca F, Rotunno G, Salvianti F, Galardi F, Pestrin M, Gabellini S et al (2016) Mutational analysis of single circulating tumor cells by next generation sequencing in metastatic breast cancer. Oncotarget 7(18):26107–26119. https://doi.org/10.18632/oncotarget.8431
15. Chen JF, Ho H, Lichterman J, Lu YT, Zhang Y, Garcia MA et al (2015) Subclassification of prostate cancer circulating tumor cells by nuclear size reveals very small nuclear circulating tumor cells in patients with visceral metastases. Cancer 121(18):3240–3251. https://doi.org/10.1002/cncr.29455
16. Powell AA, Talasaz AH, Zhang H, Coram MA, Reddy A, Deng G et al (2012) Single cell profiling of circulating tumor cells: transcriptional heterogeneity and diversity from breast cancer cell lines. PLoS One 7(5):e33788. https://doi.org/10.1371/journal.pone.0033788
17. Jordan NV, Bardia A, Wittner BS, Benes C, Ligorio M, Zheng Y et al (2016) HER2 expression identifies dynamic functional states within circulating breast cancer cells. Nature 537(7618):102–106. https://doi.org/10.1038/nature19328
18. Fehm T, Muller V, Aktas B, Janni W, Schneeweiss A, Stickeler E et al (2010) HER2 status of circulating tumor cells in patients with metastatic breast cancer: a prospective, multicenter trial. Breast Cancer Res Treat 124(2):403–412. https://doi.org/10.1007/s10549-010-1163-x
19. Lindstrom LS, Karlsson E, Wilking UM, Johansson U, Hartman J, Lidbrink EK et al (2012) Clinically used breast cancer markers such as estrogen receptor, progesterone receptor, and human epidermal growth factor receptor 2 are unstable throughout tumor progression. J Clin Oncol 30(21):2601–2608. https://doi.org/10.1200/JCO.2011.37.2482
20. Miyamoto DT, Lee RJ, Kalinich M, LiCausi JA, Zheng Y, Chen T et al (2018) An RNA-based digital circulating tumor cell signature is predictive of drug response and early dissemination in prostate cancer. Cancer Discov 8(3):288–303. https://doi.org/10.1158/2159-8290.CD-16-1406
21. Kwan TT, Bardia A, Spring LM, Giobbie-Hurder A, Kalinich M, Dubash T et al (2018) A digital rna signature of circulating tumor cells predicting early therapeutic response in localized and metastatic breast cancer. Cancer Discov 8(10):1286. https://doi.org/10.1158/2159-8290.CD-18-0432
22. Kalinich M, Bhan I, Kwan TT, Miyamoto DT, Javaid S, LiCausi JA et al (2017) An RNA-based signature enables high specificity detection of circulating tumor cells in hepatocellular carcinoma. Proc Natl Acad Sci U S A 114(5):1123–1128. https://doi.org/10.1073/pnas.1617032114
23. Hong X, Sullivan RJ, Kalinich M, Kwan TT, Giobbie-Hurder A, Pan S et al (2018) Molecular signatures of circulating melanoma cells for monitoring early response to immune checkpoint therapy. Proc Natl Acad Sci U S A 115(10):2467–2472. https://doi.org/10.1073/pnas.1719264115
24. Hosokawa M, Kenmotsu H, Koh Y, Yoshino T, Yoshikawa T, Naito T et al (2013) Size-based isolation of circulating tumor cells in lung cancer patients using a microcavity array system. PLoS One 8(6):e67466. https://doi.org/10.1371/journal.pone.0067466
25. Kim TH, Lim M, Park J, Oh JM, Kim H, Jeong H et al (2017) FAST: size-selective, clog-free isolation of rare cancer cells from whole blood at a liquid-liquid interface. Anal Chem 89(2):1155–1162. https://doi.org/10.1021/acs.analchem.6b03534
26. Kim YJ, Kang YT, Cho YH (2016) Poly(ethylene glycol)-modified tapered-slit membrane filter for efficient release of captured viable circulating tumor cells. Anal Chem 88(16):7938–7945. https://doi.org/10.1021/acs.analchem.5b04927

27. Hayashi M, Zhu P, McCarty G, Meyer CF, Pratilas CA, Levin A et al (2017) Size-based detection of sarcoma circulating tumor cells and cell clusters. Oncotarget 8(45): 78965–78977. https://doi.org/10.18632/oncotarget.20697
28. Kim MS, Sim TS, Kim YJ, Kim SS, Jeong H, Park JM et al (2012) SSA-MOA: a novel CTC isolation platform using selective size amplification (SSA) and a multi-obstacle architecture (MOA) filter. Lab Chip 12(16):2874–2880. https://doi.org/10.1039/c2lc40065k
29. Rosenberg R, Gertler R, Friederichs J, Fuehrer K, Dahm M, Phelps R et al (2002) Comparison of two density gradient centrifugation systems for the enrichment of disseminated tumor cells in blood. Cytometry 49(4):150–158. https://doi.org/10.1002/cyto.10161
30. Wang ZP, Eisenberger MA, Carducci MA, Partin AW, Scher HI, Ts'o PO (2000) Identification and characterization of circulating prostate carcinoma cells. Cancer 88(12):2787–2795. https://doi.org/10.1002/1097-0142(20000615)88:12<2787::AID-CNCR18>3.0.CO;2-2
31. Friedlander TW, Premasekharan G, Paris PL (2014) Looking back, to the future of circulating tumor cells. Pharmacol Ther 142(3): 271–280. https://doi.org/10.1016/j.pharmthera.2013.12.011
32. Gogoi P, Sepehri S, Zhou Y, Gorin MA, Paolillo C, Capoluongo E et al (2016) Development of an automated and sensitive microfluidic device for capturing and characterizing circulating tumor cells (CTCs) from clinical blood samples. PLoS One 11(1):e0147400. https://doi.org/10.1371/journal.pone.0147400
33. Kim J, Cho H, Han SI, Han KH (2016) Single-cell isolation of circulating tumor cells from whole blood by lateral magnetophoretic microseparation and microfluidic dispensing. Anal Chem 88(9):4857–4863. https://doi.org/10.1021/acs.analchem.6b00570
34. Satelli A, Brownlee Z, Mitra A, Meng QH, Li S (2015) Circulating tumor cell enumeration with a combination of epithelial cell adhesion molecule-and cell-surface vimentin-based methods for monitoring breast cancer therapeutic response. Clin Chem 61(1):259–266. https://doi.org/10.1373/clinchem.2014.228122
35. Chen L, Peng M, Li N, Song Q, Yao Y, Xu B et al (2018) Combined use of EpCAM and FRα enables the high-efficiency capture of circulating tumor cells in non-small cell lung cancer. Sci Rep 8(1):1–10. https://doi.org/10.1038/s41598-018-19391-1
36. Krivacic RT, Ladanyi A, Curry DN, Hsieh HB, Kuhn P, Bergsrud DE et al (2004) A rare-cell detector for cancer. Proc Natl Acad Sci U S A 101(29):10501–10504. https://doi.org/10.1073/pnas.0404036101
37. Stott SL, Hsu CH, Tsukrov DI, Yu M, Miyamoto DT, Waltman BA et al (2010) Isolation of circulating tumor cells using a microvortex-generating herringbone-chip. Proc Natl Acad Sci U S A 107(43):18392–18397. https://doi.org/10.1073/pnas.1012539107
38. Wu S, Liu S, Liu Z, Huang J, Pu X, Li J et al (2015) Classification of circulating tumor cells by epithelial-mesenchymal transition markers. PLoS One 10(4):1–14. https://doi.org/10.1371/journal.pone.0123976
39. Magbanua MJM, Carey LA, DeLuca A, Hwang J, Scott JH, Rimawi MF et al (2015) Circulating tumor cell analysis in metastatic triple-negative breast cancers. Clin Cancer Res 21(5):1098–1105. https://doi.org/10.1158/1078-0432.CCR-14-1948
40. Hou JM, Krebs M, Ward T, Sloane R, Priest L, Hughes A et al (2011) Circulating tumor cells as a window on metastasis biology in lung cancer. Am J Pathol 178(3):989–996. https://doi.org/10.1016/j.ajpath.2010.12.003
41. Ge F, Zhang H, Wang DD, Li L, Lin PP (2015) Enhanced detection and comprehensive in situ phenotypic characterization of circulating and disseminated heteroploid epithelial and glioma tumor cells. Oncotarget 6(29):27049–27064. https://doi.org/10.18632/oncotarget.4819
42. Cho EH, Wendel M, Luttgen M, Yoshioka C, Marrinucci D, Lazar D et al (2012) Characterization of circulating tumor cell aggregates identified in patients with epithelial tumors. Phys Biol 9(1):016001. https://doi.org/10.1088/1478-3975/9/1/016001
43. Murlidhar V, Zeinali M, Grabauskiene S, Ghannad-Rezaie M, Wicha MS, Simeone DM et al (2014) A radial flow microfluidic device for ultra-high-throughput affinity-based isolation of circulating tumor cells. Small 10(23): 4895–4904. https://doi.org/10.1002/smll.201400719
44. Zhao M, Wei B, Chiu DT (2013) Imaging multiple biomarkers in captured rare cells by sequential immunostaining and photobleaching. Methods 64(2):108–113. https://doi.org/10.1016/j.ymeth.2013.08.006
45. Karabacak NM, Spuhler PS, Fachin F, Lim EJ, Pai V, Ozkumur E et al (2014) Microfluidic, marker-free isolation of circulating tumor cells

from blood samples. Nat Protoc 9(3): 694–710. https://doi.org/10.1038/nprot.2014.044

46. Naume B, Borgen E, Tossvik S, Pavlak N, Oates D, Nesland JM (2004) Detection of isolated tumor cells in peripheral blood and in BM: evaluation of a new enrichment method. Cytotherapy 6(3):244–252. https://doi.org/10.1080/14653240410006086

47. Marrinucci D, Bethel K, Kolatkar A, Luttgen MS, Malchiodi M, Baehring F et al (2012) Fluid biopsy in patients with metastatic prostate, pancreatic and breast cancers. Phys Biol 9(1):016003. https://doi.org/10.1088/1478-3975/9/1/016003

48. Andree KC, van Dalum G, Terstappen LWMM (2016) Challenges in circulating tumor cell detection by the CellSearch system. Mol Oncol 10(3):395–407. https://doi.org/10.1016/j.molonc.2015.12.002

49. Ozkumur E, Shah AM, Ciciliano JC, Emmink BL, Miyamoto DT, Brachtel E et al (2013) Inertial focusing for tumor antigen-dependent and -independent sorting of rare circulating tumor cells. Sci Transl Med 5(179):179ra47. https://doi.org/10.1126/scitranslmed.3005616

Chapter 19

Measuring Mechanical Properties of Breast Cancer Cells with Atomic Force Microscopy

Barbara Zbiral, Andreas Weber, and José L. Toca-Herrera

Abstract

Using physical models to describe the response of cells to external stimuli has grown popular in the last three decades. The mechanical properties of cells are tightly linked to biochemical signaling pathways related to cell structure, proliferation, differentiation, motility, and cell fate. This chapter aims to describe how to perform mechanical experiments on MCF-7 breast cancer cells using the atomic force microscope. We present a stepwise procedure on sample preparation, force spectroscopy measurements and a guide on data evaluation using standard rheological models. We demonstrate how to derive all viscoelastic parameters of the cell by conducting stress relaxation and creep experiments. Additionally, the reader can find a sample dataset and the code required for data evaluation.

Key words Cell mechanics, Atomic force microscopy, Stress relaxation and creep, Young's Modulus, Relaxation time, Indentation, Viscosity

1 Introduction

Breast cancer is the most frequently diagnosed cancer and the leading cause of cancer death among women. Cancer progression happens gradually and is—aside from genetic and histological changes—accompanied by changes in the mechanical phenotype of the tissue and cells. Cells sense mechanical cues from their environment, such as the rigidity or elasticity of the extracellular matrix, and actively balance these forces by exerting contractile stresses. This reciprocal balance of forces influences a variety of cellular properties critical to tumorigenesis, such as structure, motility, proliferation, and differentiation. Biologically, mechanical signaling (called mechanotransduction) is governed by integrins, focal adhesions and cytoskeletal networks. This mechanical signaling between the extracellular matrix and the cells is tightly connected to biochemical signal transduction networks which control factors relevant to tumor transformation, invasion, and

Maria dM. Vivanco (ed.), *Mammary Stem Cells: Methods and Protocols*,
Methods in Molecular Biology, vol. 2471, https://doi.org/10.1007/978-1-0716-2193-6_19,

metastasis [1]. Interestingly, while malignant breast cancer tissue has been found stiffer than healthy tissue, on a single-cell level, malignant cells appear softer. This softening has been correlated with changes in cytoskeletal organization. The inner structure of malignant cells becomes increasingly disordered, rendering those cells more pliable than their healthy counterparts. Investigating the mechanical properties of cancer cells may therefore help to better understand the physical mechanisms responsible for cancer progression and malignancy [2, 3].

Atomic Force Microscopy (AFM) is one of the most frequently used methods for examining cell mechanics. A noteworthy advantage of this method is a flexible and multifaceted experimental setup with the possibility to derive several different cell mechanical parameters from an experiment. A general setup of an AFM can be seen in Fig. 1. The AFM itself consists of a voltage-sensitive piezoelectric element, a soft cantilever of micrometer length scale with an indenter (probe) that can vary in geometry and size, a laser, and a photodiode. The cantilever backside reflects the laser onto a photodiode, which reports the vertical and lateral deflection (arising from probe–sample interactions such as mechanical contact, repulsive or attractive forces) of the cantilever with nanometer resolution. The reaction of the cantilever to external force can be described plainly as the bending of an ideal linear elastic spring. Hence, as the probe extends toward or retracts from the sample, the cantilever deflection is monitored as a function of the piezo position and the vertical deflection is converted into units of force [4].

Force spectroscopy on living cells allows to record the force response of a cell as it is indented by the probe as force-distance (*F-d*) curve. Parameters such as maximum load, loading/unloading rates, cantilever stiffness, indenter geometry and indenter chemistry can be varied, thus enabling a wide variety of measurement modes. Indentations curves are used routinely to determine the sample elasticity (Young's modulus) via fitting mechanical models. Moreover, the force response or deformation of a cell can be recorded over time. Such experiments are equivalent to stress relaxation and creep compliance tests, respectively, and allow examining the rheological properties of a cell [5–9].

In this chapter, a stepwise instruction for force spectroscopy experiments on MCF-7 breast cancer cells is provided, alongside a guide for data evaluation.

2 Materials

2.1 Cell Culture

1. MCF-7 breast cancer cell line.
2. Culture medium: DMEM, 10% FBS, and 1% Penicillin/Streptomycin (*see* **Note 1**).

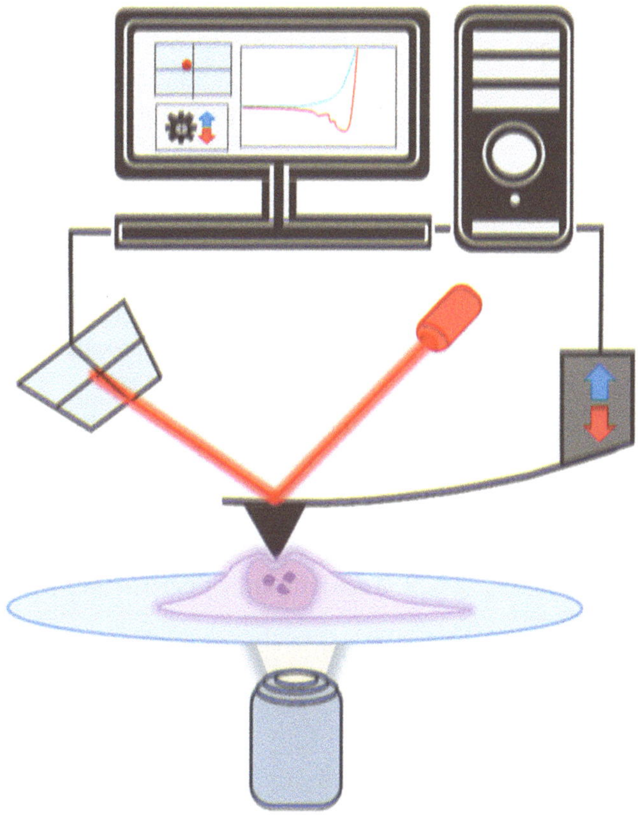

Fig. 1 Schematic drawing of an AFM. A cell is indented using a pyramidal probe fixed to a cantilever above the nucleus. Other important elements of the measurement system, such as the piezoelectric element, laser path and photodiode, inverted optical microscope, and computer controlling the AFM are shown

3. 1× Phosphate buffered saline pH 7.4.
4. Dissociation agent: TrypLE (*see* **Note 1**).
5. Serological cell culture flasks: T-25, for adherent cells, non-coated (*see* **Note 1**).
6. Cell Counter.

2.2 Sample Preparation for Atomic Force Microscopy

1. Glass cover slips for Atomic Force Microscopy (diameter of 24 mm, thickness 0.08–0.12 mm).
2. Cell culture plates: 6-well, uncoated, sterile.
3. Microscopy medium: CO_2-independent, serum-free medium (Leibovitz's) (*see* **Note 2**).
4. Rounded, flat (fine) stainless-steel tweezers.
5. Plasma cleaner.

2.3 Force Spectroscopy (See Note 3)

1. Atomic force microscope (JPK NanoWizard® III) mounted on inverted optical microscope Zeiss Axio Observer Z1 with at least 20× magnification (*see* **Notes 3** and **4**). AFM should be mounted in anti-vibrational way.
2. CellHesion® module.
3. Liquid sample chamber with temperature control.
4. Silicon nitride cantilevers, quadratic pyramidal indenter (Bruker, DNP-S B, nominal 23 kHz and 0.12 N/m) (*see* **Note 5**).
5. Antistatic stainless-steel tweezers: One pair with a sharp tip, one pair with an angled, flat tip (to handle cantilever chip), one pair with a rounded tip.
6. Cantilever holder (glass head) and metal clamp to fix cantilever chip (*see* **Note 5**).
7. Standard microscopy slide.
8. Dustless microfiber tissues.
9. Incident light microscope.

2.4 Chemicals

1. Ethanol (70%).
2. Acetone.
3. Milli-Q water.
4. N_2 drying gas.

2.5 Software

1. JPKSPM Data Processing in the current version.
2. RStudio with package R afmToolkit installed.

3 Methods

3.1 Cell Culture

1. Cultivate MCF-7 cells in culture medium in T-25 cell culture flasks at 37 °C, 95% humidity, and 5% CO_2.
2. Maintain routine culture at a maximum confluence of 80% (*see* **Note 6**).

3.2 Sample Preparation

The sample preparation comprises preparation of the AFM glass cover slips, seeding of the cells and incubation period.

3.2.1 Slide Preparation

1. Clean glass cover slips in 70% ethanol for 10 s and dry with N_2 drying gas.
2. Plasma cleaning of cover slips at 0.4 mbar and 15 mA for 1 min.
3. Add 2 mL sterile PBS to each cover slip in a laminar flow work bench. This step should be completed immediately after plasma-cleaning.

3.2.2 Cell Sample Preparation

1. Aspirate culture medium from the cell culture flask containing the cells to be measured. Wash cells once with PBS, detach with dissociation reagent for 3–5 min. Centrifuge cells and resuspend pellet in culture medium.
2. Count cells and prepare a dilution of 10^5 cells/mL in culture medium (*see* **Note 6**).
3. Aspirate PBS from cover slips and add 1 mL of the cell solution per well. Add 2 mL culture medium on top (*see* **Notes 6** and 7).
4. Incubate for 24 h at 37 °C and 5% CO_2 (*see* **Note 6**).
5. Shortly before AFM measurements, replace culture medium with 2 mL microscopy medium (Leibovitz's). Wash cells carefully 3× with PBS after aspirating the culture medium.

3.3 Force Spectroscopy

3.3.1 Assembling the System (See **Note 3**)

1. Clean the glass head (cantilever holder) with 70% ethanol and a microfiber tissue, dry with N_2. Check the cut glass surfaces for any visible impurities before mounting the cantilever (*see* **Note 8**).
2. With the antistatic flat, angled tweezers, carefully pick up the pre-cleaned cantilever chip, and place it onto the glass head (*see* **Note 8**).
3. With the sharp, fine tweezers, pick up the metal clamp and fix the cantilever to the glass head. While sliding the clamp on, hold the cantilever chip securely in place with the flat, angled tweezers. Check position of the cantilever in the incident light microscope (*see* **Note 9** for proper handling of cantilevers).
4. Mount the glass head with the cantilever onto the AFM and mount the AFM onto the sample stage (*see* **Note 10**).

3.3.2 Calibration of the System

The interaction of the sample with the cantilever is measured as the bending of the cantilever in response to attractive/repulsive forces. The cantilever is considered to be an ideal linear elastic spring of constant (k_c), therefore the bending is proportional to the applied force

$$F = k_c \times z \qquad (1)$$

The bending is recording as a change of the position of the deflected laser point on the photodiode (Fig. 1). Conversion of the measured voltage to a force is done by calibrating the system. First the sensitivity (how many nm per V) of the system needs to be determined. This is done by acquiring a force-distance curve on an "infinitely" stiff (compared to the cantilever) substrate thus measuring solely the cantilever deflection. Commonly, standard microscope glass slides are used for this purpose. An inverse linear fitting of the linear contact region yields the sensitivity in [nm/V] as can be seen in Figs. 2 and 3 for a curve acquired in air and in liquid,

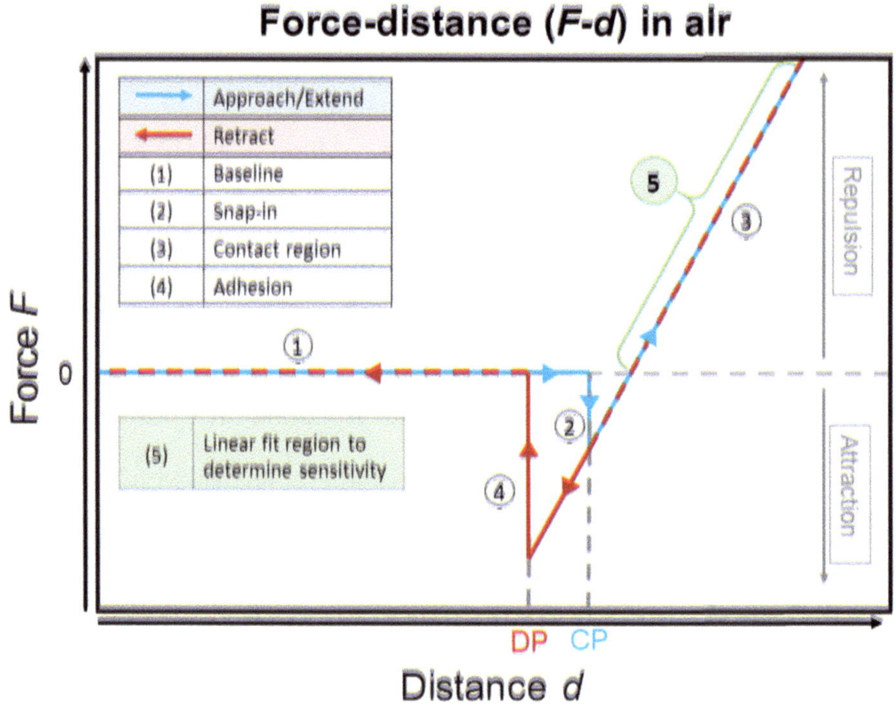

Fig. 2 Representative *F-d* curve on glass in air. The green brace indicates the part of the contact region on which to fit an inverse linear function to calculate the sensitivity

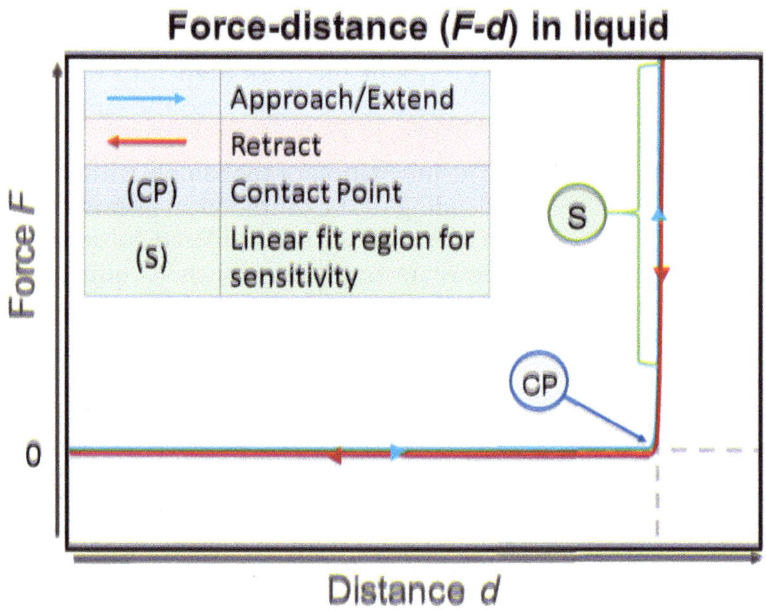

Fig. 3 Representative *F-d* curve on glass in liquid. The green brace indicates the part of the contact region on which to fit an inverse linear function to calculate the sensitivity

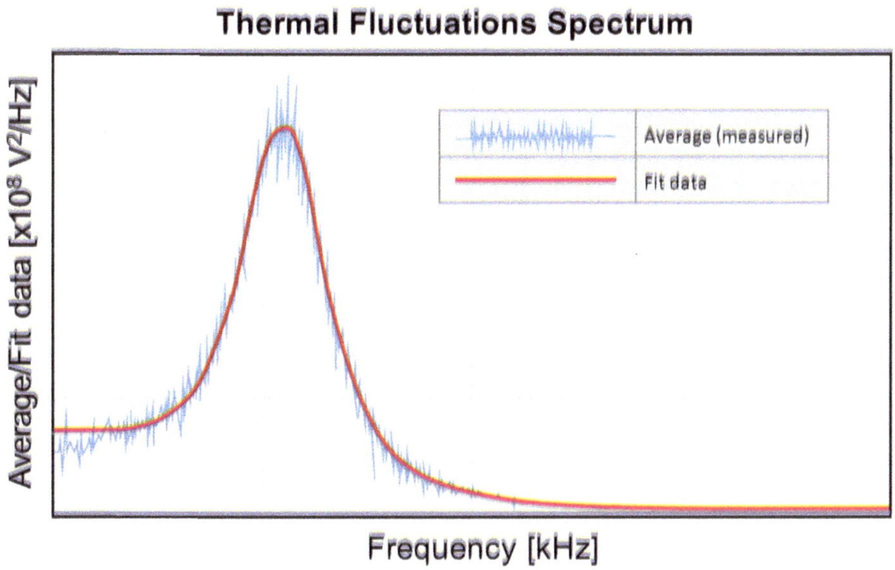

Fig. 4 Representative thermal noise spectrum (blue) fitted with a Lorentzian function (red)

respectively. The real spring constant of the cantilever is then determined using the equipartition theorem

$$\frac{1}{2}k_c\langle \Delta Z_c{}^2\rangle = \frac{1}{2}k_b T \quad (2)$$

by letting the cantilever vibrate freely under the thermal noise of the system away from the surface (note that here k_b is the constant of Boltzmann, and T the temperature of the system). An example of a fluctuation spectrum with a Lorentz fit to determine the cantilever properties is shown in Fig. 4 (*see* **Note 11**). Calibration can be performed in air and in liquid.

The values for sensitivity and spring constant determined in liquid are the ones used for the measurements [10]. To calibrate the system, following steps must be taken:

3.3.3 Calibration of System in Air

1. Focus the laser of the AFM toward the end of the cantilever (*see* **Note 12**).
2. Adjust the mirror to maximize the sum signal of the photodiode (*see* **Note 12**).
3. Adjust the laser position to zero vertical and zero lateral deflection on the photodiode to maximize the sum signal.
4. Approach the sample surface carefully (*see* **Note 10**).
5. Acquire a force–distance curve on the microscopy glass, using a set voltage of 0.3–1 V and a speed of 5 μm/s (*see* **Note 11**).

6. Calculate the sensitivity by fitting a linear function to the contact region of the approach curve and taking the inverse of the slope (Fig. 2).
7. To determine the spring constant of the cantilever, retract the cantilever by 200 μm or more from the surface, record a thermal noise spectrum and fit a Lorentz function to the first peak of the spectrum then calculate the spring constant using the equipartition theorem as can be seen in Fig. 4 (*see* **Note 11**).

3.3.4 Calibration of System in Liquid

1. Unmount the AFM and switch out the microscopy slide for the liquid sample chamber and mount the sample (*see* **Note 13**). Add 600–800 μL of CO_2-independent microscopy medium to the chamber and set the temperature in the chamber to 37 °C (*see* **Note 13**). Add a drop of the same medium to the cantilever to help break the surface tension and avoid air bubbles. Remount the AFM (*see* **Note 10**).
2. Repeat the calibration process as described previously in **steps 3–7** in liquid (*see* Fig. 3 for a representative *F-d* curve in on glass in liquid). Choose a cell-free and visually clean spot on the glass cover slip to record a *F-d* curve for determining the sensitivity. To determine the spring constant in liquid, retract the cantilever at least 50 μm from the surface before recording a thermal noise spectrum. Resonance frequency and stiffness should be roughly similar to the ones given by the manufacturers.
3. Use the calculated values from the liquid measurement in subsequent sample measurements.
4. Before starting measurements on cells, the sample needs to equilibrate as fluctuations arising from the temperature increase and liquid flow lead to a strong thermal drift in the system. After roughly 30 min, the signal should be stable enough for measurements. Retract the cantilever by 2–3 steps of 500 μm for safety.
5. If needed, switch to CellHesion mode (an extra module of the JPK-AFM), and manually enter the determined calibration values in liquid (*see* **Note 14**).

3.3.5 Force Spectroscopy Measurements

Once the system has been calibrated and reached equilibrium, force-spectroscopy measurements on cells can start. The following section covers three different experimental modes: force-distance measurements, constant height measurements, and constant force measurements. The first allows for determining mechanical properties such as the elasticity (Young's modulus) of the sample. The latter two additionally contain rheological information and consist of stress relaxation and creep compliance experiments.

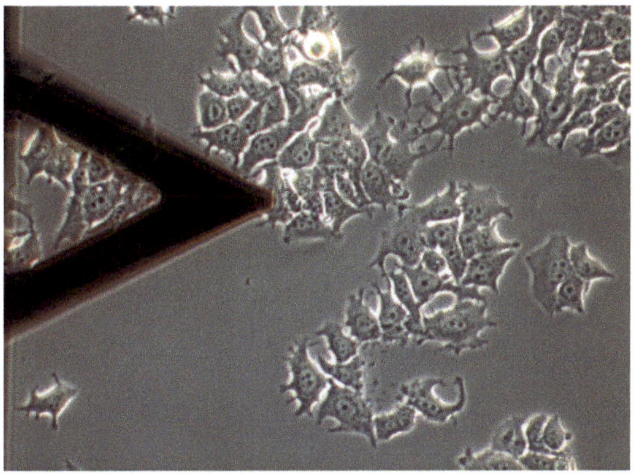

Fig. 5 MCF-7 cells (on borosilicate glass immersed in CO_2-independent microscopy medium at 37 °C) during an AFM experiment. Samples prepared in this way can be measured up to 4 h (*see* **Note** 5)

1. Using the inverted optical microscope, select a position of the sample with cells by moving the stage laterally, then approach the sample surface carefully with the cantilever (Fig. 5).
2. Move the cantilever tip over the nuclear region of a cell either by hand or using the x-y-piezo (*see* **Note 15**).
3. Adjust the laser position on the photodiode to zero vertical and zero lateral deflection.
4. Acquire a series of *F-d* measurements on the same cell with a set force of e.g., 1 nN, a speed of 5 μm/s and a curve length of 7.5 μm (*see* **Notes 16** and **17**).
5. Regularly acquire *F-d* curves on glass to check for impurities on the cantilever.
6. Select the next cell to measure and repeat **steps 1–4** until the desired number of measurements is made (*see* **Note 18**).
7. For measurements requiring longer z-ranges, switch to the CellHesion mode (e.g., for stress relaxation and creep measurements).
8. To acquire a force-time (*F-t*) curve in constant height mode (stress relaxation), a pause segment must be recorded in between the approach and retract segments. In this pause segment, the height of the piezoelectric element (the z-height of the cantilever) is held constant once the set force is reached and the change in recorded force over time is measured. For example, with a set force of 1 nN and a pause segment time of 10 s, the probe indents the segment until a force of 1 nN is reached, and records the force response (its relaxation) at this z-height for 10 s before retracting (*see* **Note 19**).
9. To acquire a *F-t* curve in constant force mode (creep), a pause segment between the approach and retract segments is defined

in which the force applied by the probe is held constant for the segment duration. For example, with a set force of 1 nN and a pause segment time of 10 s, the probe will indent the segment until a force of 1 nN is recorded, and then change the z-height to keep this force constant for 10 s before retracting, recording the deformation of the cell over time. For the constant force mode, application of the right feedback conditions is needed (*see* **Note 19**).

3.3.6 Disassembling the System and Cleaning

1. After completing the measurements, retract the cantilever far from the surface using the stepper motors.
2. Unmount the cantilever holder from the AFM and discard of the used cell sample. Clean the loose parts of the liquid sample chamber which were in contact with medium (i.e., the gasket and the metal fitting) with 70% ethanol.
3. Unmount the cantilever from the glass head and carefully clean it by suspending the chip in pure acetone for at least 10 s, moving it around slowly and carefully, using the flat-angled antistatic tweezers and dry carefully with a microfiber tissue (further cleaning *see* **Note 8**).
4. Clean the glass head well, using ethanol, then Milli-Q water and a microfiber tissue. Store glass head and metal clamp fully suspended in 70% ethanol (further cleaning *see* **Note 8**).

3.4 Data Evaluation

Force–distance curves can be divided into three parts: extension toward the sample, contact with the sample, and retraction from the sample (Fig. 6). The extend segment yields information about repulsive, attractive, or structural forces between the probe and the sample. The second part of the curve corresponds to the contact regime in which the cantilever is in physical contact with the sample. This segment provides information about the mechanical properties of the sample. Finally, the retracting curve contains information about mechanical properties due to unloading, tip-sample-adhesion or rupture forces and membrane tether formation. In this chapter the sole focus will be on the contact part, comprising of sample indentation, stress relaxation, and creep experiments.

To determine the contact region, two crucial steps must be performed: defining the contact point and defining the baseline. The easiest way to perform the former is by eye as the contact point is the first deviation of slope from the baseline. Since this is quite error-prone and depends highly on the experience of the user, various software solutions and homebuilt programs have been published for this purpose [4, 11–13]. Regardless the case, several processing steps are necessary to extract mechanical and rheological parameters. These are outlined in the following section.

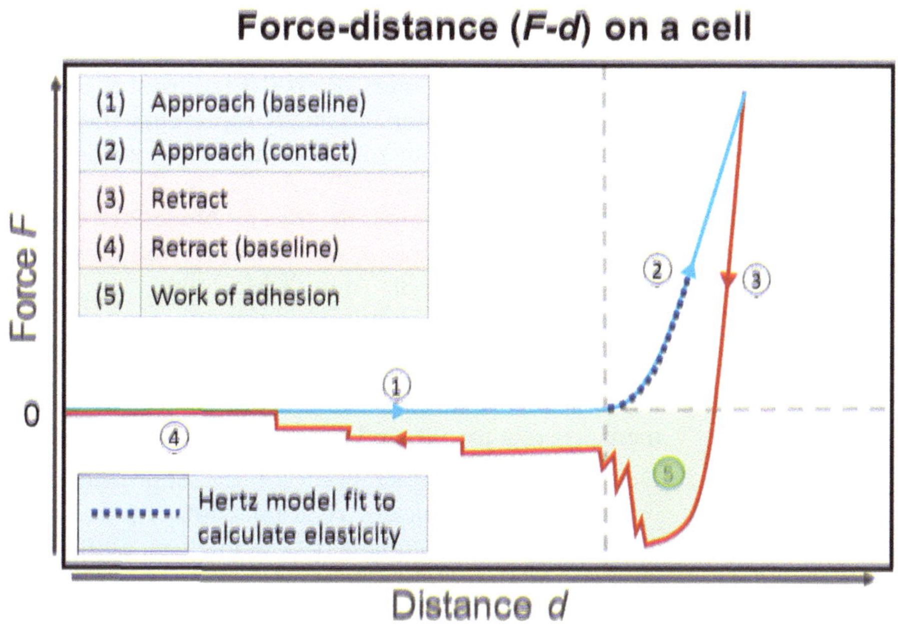

Fig. 6 Representative Force-distance (*F-d*) curve on a cell. The blue dotted line indicates the Hertz model fit to the contact region of the approach

3.4.1 Elasticity

The most common way to determine cell mechanics is calculation of the elastic modulus (Young's Modulus, E). The elastic modulus is calculated from the contact region of the approach segment. Normally, Hertz models with modifications for indenter geometries are used. For a four-sided pyramidal indenter, the Sneddon-modified Hertz model is used [14–16]:

$$F = \frac{E}{1-\nu^2}\frac{\tan(\alpha)}{\sqrt{2}}\delta^2 \tag{3}$$

Alternatively, the solution used for a conical indenter is:

$$F = \frac{E}{1-\nu^2}\frac{\tan(\alpha)}{\pi}\delta^2 \tag{4}$$

For a parabolic (or spherical at low indentations) indenter, the following relationship is used:

$$F = \frac{4\sqrt{R_C}}{3}\frac{E}{1-\nu^2}\delta^{3/2} \tag{5}$$

For a spherical indenter, the following model is used:

$$F = \frac{E}{1-\nu^2}\left(\frac{R^2+a^2}{2}\ln\left(\frac{R+a}{R-a}\right) - aR\right) \tag{6}$$

Here, E is the elastic (Young's) modulus, ν is the Poisson ratio of the cell (commonly assumed to be 0.5, corresponding to complete incompressibility, thus the volume of the material does not change), α is the half-opening angle of the pyramid and the cone, respectively, and δ is the deformation of the sample. In the modification for parabolic indenters, R_C is the radius of the tip curvature. For the spherical indenter model, R represents the radius of the sphere, and a is the contact radius. The Hertz model is only valid for small indentations. For a cell, commonly accepted indentations are up to 10% of the sample height (*see* **Note 20**). Since cells are viscoelastic bodies, a more accurate name for the calculated parameter would be the *apparent* Young's Modulus.

Determination of Young's Modulus using manufacturer software JPKSPM Data Processing

1. For baseline definition, define a portion of the approach segment, correct for tilt, and subtract the baseline from the curve.
2. Determine the contact point of the approach curve either visually or automatically. A good contact point definition is essential to all subsequent data evaluation steps. The contact point is the point where a significant deviation from the baseline is seen.
3. Correct the curve for cantilever bending to determine the indentation.
4. Calculate the elastic modulus according to Eqs. 3–6.

Determination of Young's Modulus using R afmToolkit (*see* **Note 21**)

1. Using JPKSPM Data Processing software, extract raw force spectroscopy curves with all headers.
2. Import curves in R according to the code description, determine the contact and detachment point, and perform the baseline correction.
3. Determine the slopes around the contact point and calculate the corrected indentation. Now, the apparent Young's Modulus can be calculated. Values can then be easily extracted or used in R for plotting.
4. A batch of curves acquired on cells can be found in the afm Toolkit together with sample code to demonstrate the analysis described above.

3.4.2 Viscoelastic Measurements: Stress Relaxation and Creep

Cells exhibit both viscous and elastic behavior. This means that under a given stress or strain, cells react in a time-dependent way. Classical experiments from material science to determine such behavior are either stress relaxation (keeping the indentation constant) or creep (applying the same force over time) experiments. An example of a stress relaxation experiment on cells can be seen in Fig. 7 as a force–time curve. Figures 8 and 9 show a creep experiment; first as force–time, then as indentation-time curve. In the

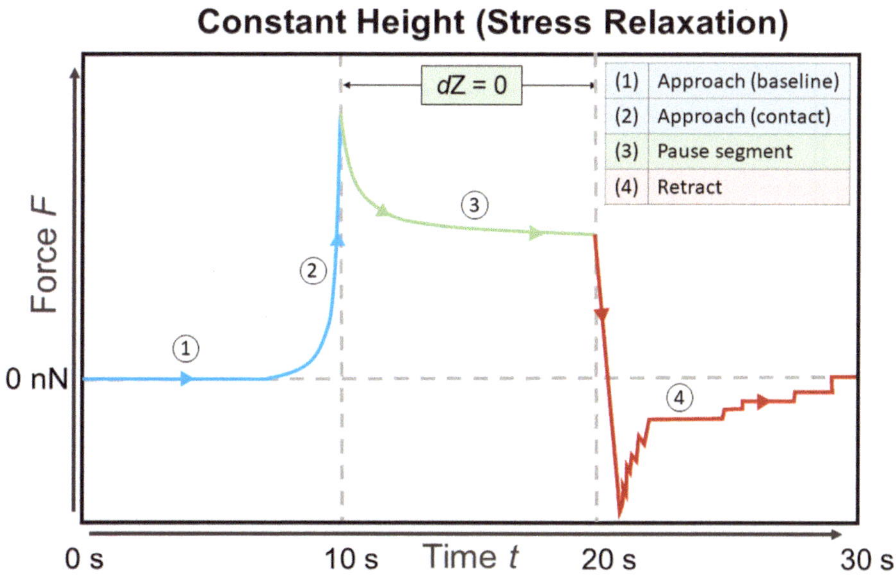

Fig. 7 Force-time (*F-t*) curve of a stress relaxation experiment with a constant height pause segment of 10 s

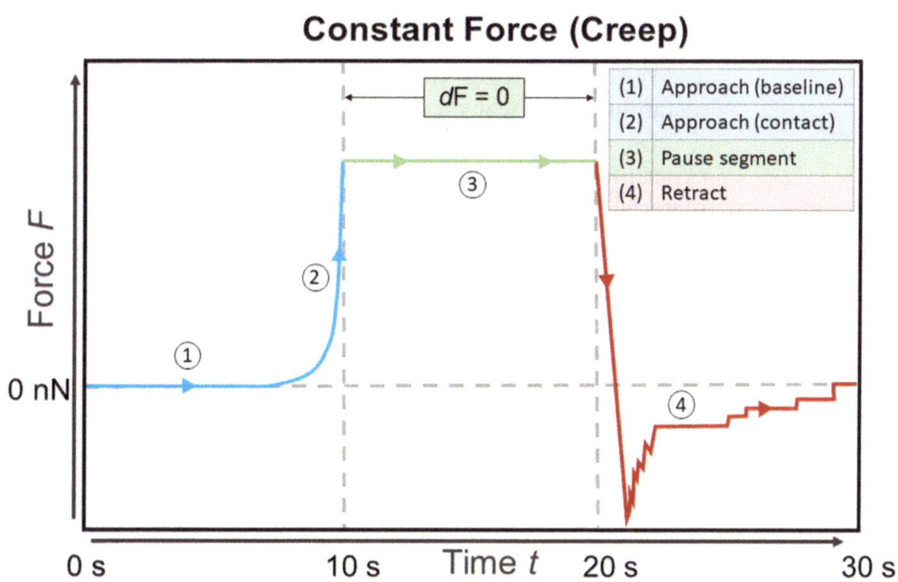

Fig. 8 Force-time (*F-t*) curve of a creep experiment with a constant force pause segment of 10 s

case of stress relaxation the force decays exponentially, while in the case of creep the indentation increases over time.

Various models exist to describe this behavior, ranging from continuum, to discrete (arrangement of elastic springs and viscous dashpots) to power law models. A general empirical description using discrete mechanical elements called Prony series is given for stress relaxation as relaxation shear modulus $G(t)$

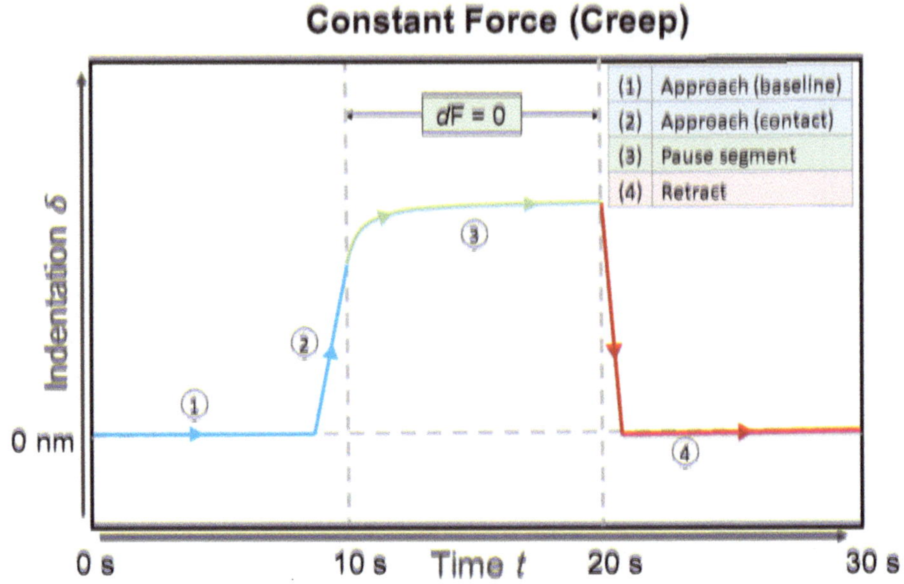

Fig. 9 Indentation-time (δ-t) curve of a creep experiment with a constant force pause segment of 10 s

$$G(t) = G_{\infty} + \sum G_i e^{-\frac{t}{\tau_i}} \tag{7}$$

With G_{∞} being the shear modulus at equilibrium and the instantaneous shear modulus.

$G_0 = G_{\infty} + \Sigma G_i$, and $E = 2\ G(1 + \nu)$. Note that t is the experimental time while τ refers to the relaxation time of the material. In similar fashion, for the creep compliance $J(t)$ the Prony series is given as

$$J(t) = C_0 + \sum C_i e^{-\frac{t}{\tau_i}} \tag{8}$$

Where the parameters are corresponding to the inverse of the shear modulus. A Prony series with $i = 1$ corresponds to a standard linear solid (spring in parallel to Maxwell element), while at $i = 2$ a spring is in parallel with two Maxwell elements. Both models fit well to stress relaxation and creep data on living cells and analytical solutions to determine mechanical properties exist (*see* **Note 22**). Figure 10 a and b represent a set of our own data from stress relaxation and creep experiments with maximum loads of 1, 2, and 4 nN on the nuclear region of MCF-7 cells. A four-sided pyramidal indenter was used. Figures 10 c and d present the parameters τ_1, τ_2, x_1, and x_2 derived from the biexponential decay fit (*see* Eqs. 7 and 8). Note that in the case of creep, the relaxation times denoted as τ_i in Eq. **8** are termed x_1 and x_2 in Fig. 10d.

Determination of cell viscoelastic properties using R afmToolkit (*see* **Note 23**)

1. Using JPKSPM Data Processing software, extract raw force spectroscopy curves with all headers.

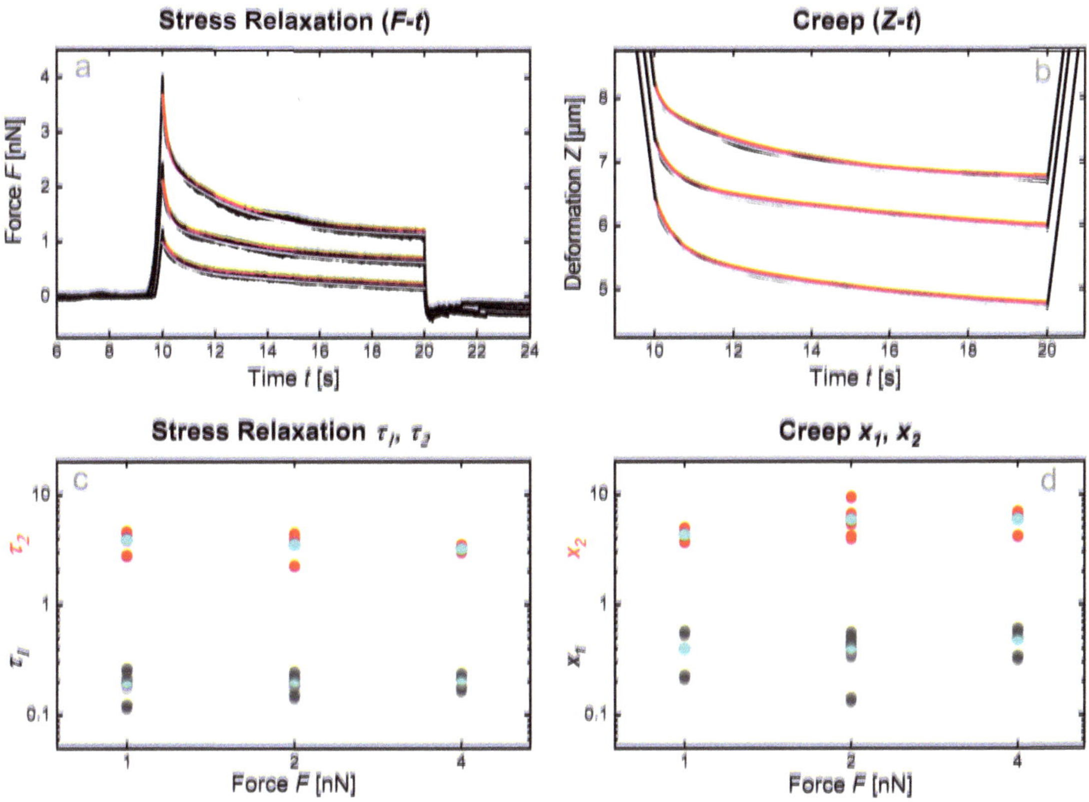

Fig. 10 Typical stress relaxation (**a**) and creep (**b**) curves acquired with a maximum applied force of 1, 2, and 4 nN on the nuclear region of MCF-7 cells. The black lines display experimental data and the red lines show a biexponential decay fit (*see* Eqs. 7 and 8) to the pause segments. The derived parameters τ_1 and τ_2 from the stress relaxation experiments and x_1 and x_2 from the creep experiments are shown in (**c**) and (**d**), respectively. The black and red dots refer to experimental data, with the blue dots indicating the respective mean values

2. Import curves into R afmToolkit, perform contact and detachment point definition, baseline correction, zero-point-slope calculation and indentation correction.
3. Fit a corresponding relaxation model (either exponential decay or power law decay) to the pause segment of the *F-t* curves. Model parameters such as relaxation times and decay amplitudes can then be extracted and used as inputs for calculation of mechanical properties (*see* **Note 22**).
4. A batch of curves acquired on cells can be found in the R afmToolkit together with sample code to demonstrate the analysis described above.

4 Notes

1. The Materials section lists reagents and media used in our lab. Depending on availability and needs, different cultivation media and trypsin in place of TrypLE can be used. Cells can be cultivated on different substrates and flasks as needed. All materials should be stored accordingly (4 °C for most components) and provided sterile. Cell mechanical properties also depend on confluence and substrate properties (stiffness, roughness, chemistry).
2. Some liquid sample chamber systems provide a CO_2-controlled atmosphere. In this case, CO_2-independent microscopy medium is not required.
3. In our laboratory, we use JPK AFMs for measuring cellular mechanics. Different AFM companies also provide AFMs suitable for such measurements. Instructions of the manufacturer should be followed. Some points to consider are z-range, availability of heated liquid chambers, instrument drift, instrument accuracy, compatibility with an optical/fluorescence microscope, open vs closed software, controller speed and more.
4. The AFM is very sensitive to noises and vibrations, which may become apparent in the measurements as artifacts or noisy baseline. While some amount of noise is inevitable, sources of noisiness must be minimized. An anti-vibration table decreases the effect of building vibrations. Setting up the AFM in a well-isolated room (e.g., cellar or suspended from bungee ropes) far from potential sources of disruptions (e.g., loud, slamming doors) is recommended. Acoustic boxes decrease external noise levels.
5. Proper choice of cantilever is crucial for performing good measurements. The stiffness of the cantilever should be roughly that of the sample. Normally, cantilevers with stiffness ranging from 0.005 to 0.15 N/m are used for cell mechanics. The resonance frequency should be as high as possible. Different cantilever geometries (rectangular and triangular) are available, both can be used. Tip geometry depends on the need of the user: Sharp tips (around 10 nm radius) are used to measure local mechanics while spherical, micrometer sized indenters measure the response of the whole cell. AFM manufacturers have different cantilever holder options. Normally the cantilever is fixed on the glass holder using either metal clamps or screws. In some cases, cantilevers must be glued to the holder. Before opening a cantilever box for the first time, briefly inspect each chip under an optical microscope for manufacturing errors.

6. Cell concentration and incubation time (e.g., for drug treatments) should be adapted according to the aim of the experiment. Cell mechanical properties also depend on confluence and substrate properties (stiffness, roughness, chemistry).
7. If possible, include an internal control in each plate by seeding the same number of cells in an empty well to assure homogeneity in morphology and density between cells grown on plastic versus glass substrate.
8. Do not touch the glass head on the cut glass surfaces. Stains, scrapes, and impurities on the cut glass surfaces disturb the laser path and lead to poor stability of the laser signal. If impurities are present, the glass head should be cleaned according to the manufacturer (ultrasound bath in detergent). Cantilevers should be cleaned prior to use with UV/Ozone or piranha solution.
9. When fixing the cantilever chip with a metal clamp, the cantilevers on one end of the chip are inevitably destroyed by the clamp. Double check the orientation of the chip (cantilever B is indicated for measuring in the Materials section) before fixing the chip with the clamp. Do not use the chip if it has fallen or flipped. Cantilever chips are sensitive to static. Even when working with antistatic tweezers, static from the operator can increase the risk of the cantilever chip flipping upside down—therefore breaking it—while handling it. Therefore, we recommend any person handling cantilever chips to ground themselves regularly to dissipate any static charge from the body.
10. When mounting the AFM, check if the distance between the cantilever and the glass slide is comfortably large (>1 mm). In doubt, retract the piezo by 500 μm before mounting. The AFM is suspended on stepper motors which are used to travel large distances. These motors are not controlled over a feedback loop and therefore care must be taken when approaching surfaces. For user with little experience, we suggest using automatic approach routines supplied by manufacturer software.
11. Cantilever calibration for rectangular cantilevers can also be done using the so-called contact-free Sader method. In addition, various correction factors to calibration values have been proposed for higher vibrational modes (*see* ref. 5 for further information). Due to hydrodynamic drag, the resonance frequency of cantilevers is reduced in liquid media. Real cantilever spring constant values can differ from the ones given by the manufacturer. When using a contact-based method, the force set point should be kept low to reduce tip/sample damage.
12. Positioning the laser farther out on the cantilever (i.e., closer to the tip) increases the sensitivity, but also the noise in the

system. Choosing a laser position a bit backed away from the tip is therefore preferable. When changing between media of different refractive index, the path of the laser through the media changes. Therefore, adjustment of the mirror is needed to counterbalance this.

13. Dry the bottom of the cover slip by gently dragging it over a microfiber towel. If the bottom of the slide is insufficiently dried, medium droplets or crystallized medium compounds may convolute the microscopy image and make detecting cells and their nuclear region more difficult. Nevertheless, once the cover slip is recovered from the well, a fast-paced workflow should be maintained to minimize the time the cells are out of medium. The area the glass head can move laterally is restricted in the liquid sample chamber. Before mounting the AFM, position the sample stage as centrally as possible. The microscope objective can be used as a guide for setting the sample stage position centrally.

14. CellHesion allows for a piezo z-range of up to 100 μm while at the same time employing a less precise (size and time response, creep) piezo element. Different manufacturers provide different solutions for such long-range AFM measurements.

15. Cells are heterogeneous, multilayered structures. When measuring cell mechanics, different properties are expected with respect to positioning of the indenter on the cell body. Normally, measurements are performed at the nuclear region to reduce substrate artifacts and measurement variability. Performing measurements in a two-dimensional grid (force-volume or force mapping) overcomes this problem but prolongs measurement time. Measurements using sharp tips are more prone to local differences in mechanics but enable visualizing such differences. Samples prepared in the way described can be measured for up to 4 h according to our experience.

16. Probably the most important parameters in AFM measurements are the loading/unloading rate of the measurement and the maximum load used. Cells are viscoelastic bodies; therefore, the measured mechanical properties depend on the frequency of the measurements. Further, since cells show a complex three-dimensional anisotropic structure, measurements at different indentations depths (as result of change in force) may probe other materials and thus results in altered values. We have shown in recent publications that loading rates ranging from 1 to 20 μm/s and forces ranging from 0.5 to 2.5 nN seem to be feasible for cell mechanical measurements. This also depends on the geometry of the indenter. Other parameters to consider are the length of curves (must be long

enough to remove any contact between cantilever and sample), data point frequency and pause in between curves.

17. How many cells should be measured and how often can one measure the same cell? In our opinion this is a question wide open for discussion in the field. Since cells are living systems, intercellular variability due to e.g., different cell cycle state or adherence to the substrate can be quite high (with cell populations normally showing a log-normal distribution). In addition, the measurement position influences the calculated parameters. Another source of error is the measurement error of the system itself, which leads to uncertainties when performing analysis. Therefore, we recommend measuring a minimum of 10 cells per population (optimally 25 to 100) and measuring each cell up to 5 times. Another possibility is acquiring force maps of cells to increase the number of data points per cell.

18. A small drift in the system is normal. This is due to thermal fluctuations, fluid flow and vibrations, and presents as the laser point drifting vertically or laterally on the photodiode over time. However, continuous, large drifts may indicate a contamination of the cantilever or the glass head. A contamination of the cantilever can sometimes be resolved by moving the cantilever through the medium (e.g., 20 μm up and down several times). If the drift remains, the cantilever must be unmounted and cleaned with ethanol and Milli-Q water. Cleaning the chip with ethanol will cause the cantilever to contract, therefore requiring the calibration to be redone. If this does not resolve the issue, UV/Ozone cleaning may be necessary. The glass head may also be dirty and disturb the laser path. In this case, unmount the cantilever and clean the glass head with Milli-Q water, 70% ethanol and—if possible—blow some drying gas on it. Repeat the calibration in liquid after remounting the system.

19. Prior to performing stress relaxation and creep experiments, the relaxation time of the cells needs to be determined. This is done by varying the duration of the pause segment to see when a linear plateau (which corresponds e.g., to a relaxed sample) is reached.

20. As mentioned above, cells are viscoelastic bodies. Hertz elastic theory assumes ideal linear elastic bodies, small indentations, the indenter being an isotropic infinite half-space and the indenter being infinitely harder than the sample. Clearly, cells violate many of these assumptions. Nevertheless, for indentations below 10% of cell height Hertz elastic theory fits well to the experimental data. For deeper indentations, thin-layer corrections need to be considered. Fitting ranges much larger will lead to an artificially high apparent Young's modulus, as the

substrate stiffness influences the shape of the curve. At much smaller indentations, fitting problems may occur due to poorly defined contact geometry. Cell height can be determined by comparing curves recorded on the cover slip to curves recorded on cells. The difference in contact point (vertical tip position, z-height) corresponds to the cell height [17].

21. The R afmToolkit [11] is a package written in the programming language R by our group to perform batch-analysis of force spectroscopy curves. Current functions are: import of large sets of AFM curves, determination of contact and detachment point, baseline correction, indentation correction, calculation of Young's Modulus, indentations, adhesion work, adhesion force, adhesion position, viscoelastic model fitting and extraction of all those parameters. We continually improve functions of the toolkit. The current version of the toolkit can be found on github (https://github.com/rbensua/afmToolkit).

22. A multitude of mathematical solutions for determination of viscoelastic properties exist in the literature. Regularly used are standard linear solid (SLS) [18], Zener model [9], and power law rheology models, or combinations of them. The articles *Nanobiomechanics of living cells: a review (2014)* [19] and *Measuring viscoelasticity of soft biological samples using atomic force microscopy (2019)* [20] can serve as instructive reviews.

23. Alternatively, pause segments can be extracted, and fitted by hand with whatever fitting software is available.

References

1. Kumar S, Weaver VM (2009) Mechanics, malignancy, and metastasis: the force journey of a tumor cell. Cancer Metastasis Rev 28:113–127
2. Li QS, Lee GYH, Ong CN et al (2008) AFM indentation study of breast cancer cells. Biochem Biophys Res Commun 374:609–613
3. Gal N, Weihs D (2012) Intracellular mechanics and activity of breast cancer cells correlate with metastatic potential. Cell Biochem Biophys 63: 199–209
4. Gavara N (2017) A beginner's guide to atomic force microscopy probing for cell mechanics. Microsc Res Tech 80:75–84
5. Butt HJ, Cappella B, Kappl M (2005) Force measurements with the atomic force microscope: technique, interpretation and applications. Surf Sci Rep 59:1–152
6. Benitez R, Toca-Herrera JL (2014) Looking at cell mechanics with atomic force microscopy: experiment and theory. Microsc Res Tech 77: 947–958
7. Alonso JL, Goldmann WH (2003) Feeling the forces: atomic force microscopy in cell biology. Life Sci 72:2553–2560
8. Roca-Cusachs P, Conte V, Trepat X (2017) Quantifying forces in cell biology. Nat Cell Biol 19:742–751
9. Moreno-Flores S, Benitez R, Vivanco MDM et al (2010) Stress relaxation and creep on living cells with the atomic force microscope: a means to calculate elastic moduli and viscosities of cell components. Nanotechnology 21: 445101
10. JPK Instruments (2016) A practical guide to AFM force spectroscopy and data analysis. JPK Instruments Tech Note 1–8
11. Benítez R, Moreno-Flores S, Bolós VJ et al (2013) A new automatic contact point

detection algorithm for AFM force curves. Microsc Res Tech 76:870–876

12. Benítez R, Bolós VJ, Toca-Herrera JL (2017) afmToolkit: an R package for automated AFM force-distance curves analysis. R J 9:291–308
13. Lin DC, Dimitriadis EK, Horkay F (2007) Robust strategies for automated AFM force curve analysis - I. non-adhesive indentation of soft, inhomogeneous materials. J Biomech Eng 129:430–440
14. Rico F, Roca-Cusachs P, Gavara N et al (2005) Probing mechanical properties of living cells by atomic force microscopy with blunted pyramidal cantilever tips. Phys Rev E Stat Nonlin Soft Matter Phys 72:021914
15. Hertz H (1881) Über die Berührung fester elastischer Körper. J für die reine und Angew Math 92:156–171
16. Sneddon IN (1965) The relation between load and penetration in the axisymmetric boussinesq problem for a punch of arbitrary profile. Int J Eng Sci 3:47–57
17. Weber A, Iturri J, Benitez R et al (2019) Measuring biomaterials mechanics with atomic force microscopy. 1. Influence of the loading rate and applied force (pyramidal tips). Microsc Res Tech 82:1392–1400
18. Darling EM, Zauscher S, Guilak F (2006) Viscoelastic properties of zonal articular chondrocytes measured by atomic force microscopy. Osteoarthr Cartil 14:571–579
19. Chen J (2014) Nanobiomechanics of living cells: a review. Interface Focus 4:20130055
20. Efremov YM, Okajima T, Raman A (2019) Measuring viscoelasticity of soft biological samples using atomic force microscopy. Soft Matter 16:64–81

Chapter 20

Breast Cancer Stem Cells: A Clinician's View

Jose Schneider and Francisco J. Vizoso

Abstract

Much has been revealed about breast cancer stem cells, their properties, and behavior and yet, a gap remains between the ever expanding knowledge and its effective clinical applications. The previous chapters have illustrated the extraordinary wealth of methodologies presently used to continue improving our understanding of breast cancer stem cells. The hope is that soon all this knowledge will translate into applicable improvements in the clinic.

Key words Breast cancer stem cells, Breast cancer, Therapy

Looking back at the previous edition of this book, it is depressingly obvious that the wishes expressed in its closing remarks by one of the present authors of this chapter (JS) have not been fulfilled. Those wishes were naïvely based upon relative ignorance concerning the huge problem posed by the central role of breast cancer stem cells (BCSCs) in breast cancer origin, survival, metastasis generation, and resistance to therapy. The only partial excuse for this admitted ignorance is that the bulk of knowledge related to all of these areas has been generated during the past decade, and with it has come a sobering humility cure. Gone are the days when a clinician treating breast cancer broadly understood the functioning of the medical weapons he had in his armamentarium against the disease: radiotherapy, chemotherapy, and much later the first targeted biological therapy, trastuzumab, followed by a legion of similar ones. Surgery, the first historical approach of all, is such a rude therapeutical tool, that it can be understood by even the most intellectually limited of clinicians. Nowadays, the situation has changed completely. To be able to apply even the most basic of biological therapies, the ones interfering with the tyrosine-kinase pathway, a clinician must (or should) have at least a working knowledge of this fundamental pathway and its ramifications, and this is only the beginning. With every passing day, new drugs interfering

Maria dM. Vivanco (ed.), *Mammary Stem Cells: Methods and Protocols*,
Methods in Molecular Biology, vol. 2471, https://doi.org/10.1007/978-1-0716-2193-6_20,

with elements of the tyrosine-kinase cascade are entering clinical trials, and to keep abreast of developments in just this particular area alone is a titanic enterprise. Add to it that the whole basic concept of breast cancer pathology and prognosis has been upturned during the last two to three decades due to the advancements in molecular biology, which have changed the concept of what a breast cancer is, and how it behaves, completely. Those of us of a certain age have had to reset our paradigms radically, in order to accept that what we thought was a single disease with differentiating pathological nuances, is now a family composed, following the most gross of molecular classifications, of at least five entirely different members, for which the therapeutical approach is equally different: Luminal-A, Luminal-B, HER2-enriched, claudin-low and basal-like tumors. Standard treatments for all, based on tumor size and axillary involvement, with the possible addition of hormonal treatment based upon the presence or not of hormone receptors are something that, although relatively recent, seems to belong to the Dark Ages. On the other hand, however, individualized treatment, which is the ultimate goal, hasn't arrived either, or at least not in an entirely satisfactory form. In most instances, we know the potential targets of biological therapy, but we still lack the sufficiently precise weapons to modulate them. Moreover, the biological pathways are so intricately intertwined, that interrupting a key one does not attain the desired effect in the long run, because the cancer cell finds alternative parallel routes to fulfil its metabolic, highly selfish purposes. In a certain sense, the more we know about the molecular biology of the cancer cell, the more helpless we feel, and this is highly frustrating. This applies in an even more extreme way to BCSCs, which show a metabolical flexibility, inherent to their "stemness," which makes them virtually invulnerable. In fact, one of their prominent features is that they show increased resistance to radio- and chemotherapy, if compared to their mature counterparts [1]. And from a biological point of view, this is necessarily so, because they are the ultimate source for cell regeneration, and if they should disappear completely, the whole organism would succumb with them. The perverse nature of cancer lies in that cancer cells are not "monsters" coming from outer space, but highly cunning, at first glance perfectly normal individuals immersed in the community, which at a certain point take advantage of all available tools this community uses to thrive, monopolizing them for their own purposes (uncontrolled proliferation and colonization of distant territories), ultimately destroying the whole ecosystem. In the case of breast cancer, furthermore, it seems now rather clear that it arises from the most sophisticated class of cells, in terms of their biological potential, namely BCSCs [2, 3]. One of them, at a certain point, seems, most probably by accumulation of spontaneous mutations during cell division, to cross the threshold into the "dark side," but conserving all inherent biological "skills" peculiar to stem cells, which gives them an enormous advantage over more differentiated cell lineages. All this, finally, without taking into consideration the interplay between tumor cells and the surrounding stroma, which in later times has gained prominence as playing a central role in cancer [4], as well as the phenomenon of epithelial-to-stromal transition (EST), which seems to facilitate that most identifying feature of malignant tumors, namely their ability to migrate from their original source to distant territories of the organism, and originate metastases in them [5, 6].

This is a highly simplified scheme of things, just for the practical use of simple minds like those of us clinicians, who have to deal in daily life primarily with other aspects concerning the management of breast cancer patients, like surgery, in the first place, which is comparatively straightforward, and all the complications arising from the other treatments applied, which range from the loss of hair, through cardiac and neurologic toxicity, to directly life-threatening conditions like medullary aplasia, neutropenic fever and thrombopenia, just to name a few. All of which certainly does not exempt us from the obligation of having at least a glimpse of the biological mechanisms underlying the disease we are dealing with, be it in an over-simple form like the one just outlined above. In this sense, and concerning BCSCs specifically, a clinician treating breast cancer in present days should have a basic knowledge concerning the three fundamental pathways involved in their function, as outlined in a really outstanding revision by Dittmer [7], if only to know where future treatment lines might look for their ideal targets:

1. The Wnt pathway, extremely complex, but essential for maintaining the subpopulation of ALDEFLOUR-positive BCSCs, which plays a pivotal role in the just described phenomenon of EST, the "disguise," or one of the "disguises," by which tumor cells elude recognition from the immune system, being thus able to pass through its barriers, migrate and metastasize [8]. One of the most ironical twists concerning this pathway is that, apparently, some of the chemotherapeutical compounds most widely used in breast cancer treatment (doxorubicin, paclitaxel) stimulate the predominance of ALDEFLOUR-positive BCSCs within the tumor, thus effectively promoting its growth and dissemination [7].
2. The Notch pathway, involved in tumor initiation, BCSC-mediated tumor progression, and resistance not only to both chemotherapy and radiotherapy, but also to hormone therapy (tamoxifen) [9–11].
3. The hedgehog pathway, in which the Gli proteins play a central role, being involved in the maintenance of the ALDEFLOUR-positive subpopulation, as well as indirectly in the resistance of BCSCs to therapy. This is, once again, an excellent example of

the interconnection of all molecular mechanisms related to cancer initiation, progression and resistance to therapy, and to the difficulties involved in targeting them decisively by simple means (i.e., targeting a single, or few particular genes).

A very recent, excellent review by Yang and colleagues [12] summarizes the "state of the art" concerning BCSC-targeted therapies, which still are, sorry for the spoiler, all at the experimental stage.

As a final remark, it might be that part of the solution to all the extremely complex, above-mentioned problems might lie in the properties of stem cells themselves, if properly identified and used to our own advantage: stem cells, and particularly stromal stem cells derived from healthy organs, show an array of tumor-modulating and immunomodulating features that might counterbalance some negative features of tumor stem cells, and in this particular case, BCSCs [13]. As an especially pertinent example, it has already been shown "in vitro" and "in vivo," in experimental models, that the secretome of stromal stem cells obtained from the human uterine cervix (hUCESCs) is able to inhibit breast cancer cell tumor growth [14, 15].

So, the bottomline to all the above-mentioned may be summarized, from a clinician's point of view, as follows:

1. Do stem cells matter in breast cancer?

   Yes, and a lot, indeed.

2. What is their role?

   They seem to be at the very origin of breast cancer. Furthermore, they are involved in tumor progression and the metastatic process, i.e., everything that defines a cancer as such.

3. Is there anything that can be done against them specifically?

   Not much, at the present moment, no. In fact, they have the ability to develop resistance against all known modes of conventional therapy (radiotherapy, chemotherapy, hormone therapy) at a level far above the one reached by mature tumor cells. From this therapy-resistant niche, the tumor is then regenerated, after having been treated by conventional means.

4. Are there any prospects for innovative treatments in the near future, targeting breast cancer stem cells?

   There are, but still only at the experimental level. Until large clinical trials show any efficacy, we are still at a complete loss.

   The only hope is that a future edition of this book may offer more optimistic answers.

References

1. Pavlopoulou Y, Oktay K, Vougas M, Louka CE, Vorgias AGG (2016) Determinants of resistance to chemotherapy and ionizing radiation in breast cancer stem cells. Cancer Lett 380:485–493
2. Pattabiraman DR, Weinberg RA (2014) Tackling the cancer stem cells - what challenges do they pose? Nat Rev Drug Discov 13:497–512
3. Peitzsch A, Tyutyunnykova K, Pantel A (2017) Dubrovska, cancer stem cells: the root of tumor recurrence and metastases. Semin Cancer Biol 44:10–24
4. Eiro N, Gonzalez LO, Fraile M, Cid S, Schneider J, Vizoso FJ (2019) Breast cancer tumor stroma: cellular components, phenotypic heterogeneity, intercellular communication, prognostic implications and therapeutic opportunities. Cancers (Basel) 11(5):664I
5. Baccelli A, Schneeweiss S, Riethdorf A, Stenzinger A, Schillert VV et al (2013) Identification of a population of blood circulating tumor cells from breast cancer patients that initiates metastasis in a xenograft assay. Nat Biotechnol 31:539–544
6. Scheel C, Weinberg RA (2012) Cancer stem cells and epithelial-mesenchymal transition: concepts and molecular links. Semin Cancer Biol 22:396–403
7. Dittmer J (2018) Breast cancer stem cells: features, key drivers and treatment options. Semin Cancer Biol 53:59–74
8. Liu S, Cong Y, Wang D, Sun Y, Deng L, Liu Y et al (2014) Breast cancer stem cells transition between epithelial and mesenchymal states reflective of their normal counterparts. Stem Cell Reports 2:78–91
9. D'Angelo RC, Ouzounova M, Davis A, Choi D, Tchuenkam SM, Kim G et al (2015) Notch reporter activity in breast cancer cell lines identifies a subset of cells with stem cell activity. Mol Cancer Ther 14:779
10. Lagadec C, Vlashi E, Della Donna L, Dekmezian C, Pajonk F (2012) Radiation-induced reprogramming of breast cancer cells. Stem Cells 30:833–844
11. Bui QT, Im JH, Jeong SB, Kim YM, Lim SC, Kim B et al (2017) Essential role of Notch4/STAT3 signaling in epithelial-mesenchymal transition of tamoxifen-resistant human breast cancer. Cancer Lett 390:115–125
12. Yang L, Shi P, Zhao G, Xu J, Peng W, Zhang J, Zhang G, Wang X, Dong Z, Chen F, Cui H (2020) Targeting stem cell pathways for cancer therapy. Signal Transduct Target Ther 5:8
13. Costa LA, Eiro N, Fraile M, Gonzalez LO, Saá J, Garcia-Portabella P, Vega B, Schneider J, Vizoso FJ (2020) Functional heterogeneity of mesenchymal stem cells from natural niches to culture conditions: implications for further clinical uses. Cell Mol Life Sci 22:1–21
14. Eiró N, Sendon-Lago J, Seoane S, Bermúdez MA, Lamelas ML, Garcia-Caballero T, Schneider J, Perez-Fernandez R, Vizoso FJ (2014) Potential therapeutic effect of the secretome from human uterine cervical stem cells against both cancer and stromal cells compared with adipose tissue stem cells. Oncotarget 5:10692–10708
15. Schneider J, Eiró N, Pérez-Fernández R, Martínez-Ordóñez A, Vizoso F (2016) Human uterine cervical stromal stem cells (hUCESCs): why and how they exert their antitumor activity. Cancer Genomics Proteomics 13:331–337

Index

Maria dM. Vivanco (ed.), *Mammary Stem Cells: Methods and Protocols*, Methods in Molecular Biology, vol. 2471, https://doi.org/10.1007/978-1-0716-2193-6,

N

O

P

R

S

MIX
Papier aus verantwortungsvollen Quellen
Paper from responsible sources
FSC® C105338

If you have any concerns about our products,
you can contact us on
ProductSafety@springernature.com

In case Publisher is established outside the EU,
the EU authorized representative is:
Springer Nature Customer Service Center GmbH
Europaplatz 3, 69115 Heidelberg, Germany

Printed by Libri Plureos GmbH
in Hamburg, Germany